Dragon fruit

Botany, Production and Uses

Dragon fruit

Botany, Production and Uses

Edited by

Sisir Mitra

*Former Dean, Faculty of Horticulture, and Dean, Post-Graduate Studies,
Bidhan Chandra Krishi Viswavidyalaya (State Agricultural University),
Mohanpur, West Bengal, India*

CABI

CABI is a trading name of CAB International

CABI
Nosworthy Way
Wallingford
Oxfordshire OX10 8DE
UK

Tel: +44 (0)1491 832111
E-mail: info@cabi.org
Website: www.cabi.org

CABI
200 Portland Street
Boston
MA 02114
USA

Tel: +1 (617)682-9015
E-mail: cabi-nao@cabi.org

A catalogue record for this book is available from the British Library, London, UK.

Library of Congress Cataloging-in-Publication Data

Names: Mitra, S. K. (Horticulturist), editor.
Title: Dragon fruit : botany, production and uses / edited by Sisir Mitra.
Other titles: Botany, production and uses
Description: Boston, MA : CAB International, [2024] | Series: Botany, production and uses | Includes bibliographical references and index. | Summary: "Dragon fruit (pitaya) is gaining consumer interest as a new, healthy exotic fruit. It has high nutritional value, health benefits, and undemanding cultivation requirements. This book reviews the current state of knowledge concerning the growing and cultivation, production technology, postharvest management and processing of dragon fruit"-- Provided by publisher.
Identifiers: LCCN 2023048318 (print) | LCCN 2023048319 (ebook) | ISBN 9781800623132 (hardback) | ISBN 9781800623149 (ebook) | ISBN 9781800623156 (epub)
Subjects: LCSH: Pitahayas.
Classification: LCC SB379.P53 D73 2024 (print) | LCC SB379.P53 (ebook) | DDC 634/.6--dc23/eng/20240209
LC record available at https://lccn.loc.gov/2023048318
LC ebook record available at https://lccn.loc.gov/2023048319

ISBN-13: 9781800623132 (hardback)
9781800623149 (ePDF)
9781800623156 (ePub)

DOI: 10.1079/9781800623156.0000

Commissioning Editor: Rebecca Stubbs
Editorial Assistant: Emma McCann
Production Editor: Shankari Wilford

Typeset by SPi, Pondicherry, India
Printed and bound in the UK by CPI Group (UK) Ltd, Croydon, CR0 4YY

Contents

Contributors

Lokman Altınkaya, Horticulture Program, Department of Plant and Animal Production, Kumluca Vocational School, Akdeniz University, Antalya, Türkiye. Email: altinkaya@akdeniz.edu.tr

Recep Balkıc, Department of Plant and Animal Production, Elmalı Vocational School, Akdeniz University, Antalya, Türkiye. Email: recepbalkic@akdeniz.edu.tr

Taner Bozkurt, Tekfen Agricultural Research Production and Marketing Inc., Adana, Türkiye. Email: taner.bozkurt@tekfen.com.tr

Phebe Ding, Department of Crop Science, Faculty of Agriculture, Universiti Putra Malaysia, 43400 UPM Serdang, Selangor, Malaysia. Email: phebe@upm.edu.my

Hamide Gübbük, Department of Horticulture, Faculty of Agriculture, Akdeniz University, Antalya, Türkiye. Email: gubbuk@akdeniz.edu.tr

Kundan Kishore, Central Horticultural Experiment Station (ICAR-Indian Institute of Horticultural Research), Bhubaneswar 751019, India. Email: kkhort12@gmail.com

Huadong Li, Sinochem (Hainan) Agricultural Ecology Limited Company, Haikou, China. Email: lihuadong3613@163.com

Haibo Long, Environment and Plant Protection Institute, Chinese Academy of Tropical Agricultural Sciences, Haikou 571101, China. Email: hb_long@yeah.net

Sisir Mitra, Former Professor, Faculty of Horticulture, Bidhan Chandra Krishi Viswavidyalaya (State Agricultural University), Mohanpur, West Bengal, India. Email: sisirm55@gmail.com

Renato Paiva, Department of Biology, Federal University of Lavras (UFLA), Brazil. Email: renpaiva @ufla.br

Pradyot K. Pathak, KVK, Ashoknagar, West Bengal University of Animal and Fishery Science, Kolkata, West Bengal, India. Email: pathakpradyot@gmail.com

Dinh Thi Yen Phuong, Applied Biotechnology Institute, 12, Tay Thanh, Tan Phu, Ho Chi Minh City, Vietnam. Email: phuong.dinh@wsu.edu

Leila Aparecida Salles Pio, Department of Agriculture, Federal University of Lavras (UFLA), Brazil. Email: leila.pio@ufla.br

Ankita Sahu, ICAR-Central Institute for Women in Agriculture, Bhubaneswar 751003, India. Email: sahuankita29@gmail.com

Özhan Şimşek, Horticulture Department, Agriculture Faculty, Erciyes University, Kayseri, Türkiye. Email: ozhansimsek@erciyes.edu.tr

Liangde Tang, National Key Laboratory of Green Pesticide, Guizhou University, Guiyang 550025, China. Email: ldtang@catas.cn

Nigel P. Taylor, Depto de Biologia, Universidade Federal São Carlos, Sorocaba, São Paulo 18052-780, Brazil. Email: nigelcactus.taylor@gmail.com

Mai Van Tri, Former Deputy Director, Southern Horticultural Research Institute (SOFRI), Tien Giang, Vietnam; Director, Southeast Horticultural Research Center, Ba Ria-Vung Tau, Vietnam. Email: maivantri2017@gmail.com

Prinya Wongsa, Food Science and Technology Programme, School of Agro-Industry, Mae Fah Luang University, Chiang Rai, Thailand. Email: prinya@mfu.ac.th

Joanna Cho Lee Ying, Post-harvest Handling Programme, Horticulture Research Centre, Malaysian Agricultural Research and Development Institute (MARDI), 43400 Serdang, Selangor, Malaysia. Email: joanna@mardi.gov.my

Preface

The vine-cactus dragon fruit or pitaya is known to have been used for thousands of years by the indigenous people of the Americas. In the mid-19th century, it was introduced by French priests to Indochina, and in 1995, Vietnam was the first country to sell dragon fruit in the world markets, under the name dragon pearl fruit. Long overlooked outside its native range in Central America, it has become an increasingly popular fruit in markets spanning Asia to Europe due to its exotic appearance and potential health benefits. Nowadays, this crop is grown and marketed in over 30 countries.

Dragon fruit has a high economic potential as a fruit crop in arid regions as the plant is exceptionally drought tolerant. In view of the global climate change and the scarcity of water in extensive regions, this species may become a significant option for sustainable dry-land agriculture.

Dragon fruit provides numerous benefits for human health, mostly for control and management of oxidative stress. All parts of the plant (stems, flowers, peels and flesh) contain bioactive compounds. These include betalains, flavonoids, polyphenols, terpenoids and steroids, saponins, alkaloids, tannins and carotenoids, which have been proven to be effective, healthier, safer and sustainable alternatives to synthetic drugs for the prevention and treatment of many human diseases. The unique dragon fruit has a bright future because of its nutraceutical value and high yields, and the fruits can be produced almost year round, which confers huge market advantage. With the changing pattern of global warming, as a high-temperature-tolerant crop, dragon fruit could be an alternative fruit crop in the future.

As a new crop in most of the countries where it is grown, very little research on cultivation and postharvest management was available before 2000. However, in the last 20 years, dragon fruit researchers around the world have selected many good varieties for cultivation in their own country and have standardized the propagation, production and postharvest management technologies. Until now there has been no comprehensive book available on dragon fruit in English. This publication presents the current state of knowledge concerning the history, physiology, culture, postharvest management and value addition of dragon fruit throughout the world. Although it contains a wealth of documented information from literature and extensive personal experience of the authors within their respective disciplines in association with this crop, further research into all aspects is still required.

Sisir Mitra
Kalyani, India, 2023

Acknowledgements

I wish to express my appreciation and gratitude to all the authors who have given their time, without any financial reward, to contribute to this book. I am grateful to the scientists who have provided information, photographs and/or reviewed the chapters: Dr R. Cronje, ARC Institute of Horticultural Research, South Africa; Dr Jonathan Crane, University of Florida, Florida, USA; Mr Faming Zhang, Institute of Tropical and Subtropical Cash Crops, PR China; Dr S. Debnath, Bidhan Chandra Krishi Viswavidyalaya, India; Dr Thingreingam Irenaeus K.S., College of Agricultura, Tripura, India; Dr Elahadi M. Yahia, Autonomous University of Querto, Mexico; Dr Leila Pio, University of Lavras, Brazil; Dr N.M. Chau, SOFRI, Vietnam; Mrs Peyanoot Naka, Horticulture Research Institute, Thailand; Mr Yacob Ahmad, International Tropical Fruits Network, Malaysia; Dr Soesilad Esti Widodo, University of Lampung, Indonesia; Dr Ted Dejong, UC Davis, California, USA; Mr Claude Allen de Mo, Green Cove Springs, Florida, USA; Dr Carlos Cristo, UC Davis, California, USA; Dr Pedro A.M. Guerra, Teguccigaipa, Honduras; Dr Fabio Celape Faleiro, Embrapa, Cerrados, Brazil; Mr Lauren Strever, Amorentia Estate & Nursery, South Africa; Dr Ian Maguire, UF/IFAS, Florida, USA; Dr Jeff Waieleloski, UF/IFAS, Florida, USA; Dr Kusal Roy, Faculty of Agriculture, Bidhan Chandra Krishi Viswavidyalaya, India; Dr J.S. Brar, Punjab Agricultural University, India; Dr Bert Lamerichs, Ines Gangtena y Chillanes, Ecuador; Dr Po-An Chen, Agricultural Technology Research Institute, Taiwan; Dr Dapeng Zhang, Sustainable Perennial Crops Laboratory, Maryland, USA; Dr Long Hai-bo, Chinese Academy of Agricultural Science, PR China; Dr Qin Yonghua, South China Agricultural University, PR China; Dr Taner Bozkurt, Tekfen Agricultural Research, Production and Marketing Institute, Türkiye; Dr Jules Janick, Purdue University, Indiana, USA; Dr M.M. Tomaz de Oliveira, Ben-Gurion University of the Nagev, Israel; Dr Remi Kahane, CIRAD, Montpellier, France; Dr Robert E. Paull, University of Hawaii at Manoa, USA; Dr C.K. Benergama, University of Paradeniya, Sri Lanka; Dr Mai Van Tri, Southern Fruit Research Institute, Tien Giang, Vietnam; Dr H. Lembisana Devi, ICAR, Regional Station, Tripura, India; Dr Nor Elliza Tajidin, Universiti Putra Malaysia, Malaysia; Dr Debalina Maji, Bidhan Chandra Krishi Viswavidyalaya, India; Dr Ricardo Goenaga, US Department of Agriculture, Mayaguez, Puerto Rico; Dr Ben Faber, UC Cooperative Extension, California, USA; and Mr Ronald G. Mangubat, FFTC, Taiwan.

My special thanks to Dr Yosef Mizrahi and Dr Noemi Tel-Zur, Ben-Gurion University of the Negev, Israel, for providing valuable information and photographs, and to Dr P.K. Pathak, West Bengal University of Animal and Fishery Science, India, for his editorial assistance.

The continued support and encouragements received from various editors and assistants of CABI, particularly from Rebecca Stubbs and Emma McCann, are gratefully acknowledged.

On a personal note, I thank my wife, Ipsa, who has unselfishly allowed me to edit this book. Without her sacrifices, I could not possibly have finished.

Sisir Mitra

1 *Selenicereus* (A. Berger) Britt. & Rose (Cactaceae) (including *Hylocereus* (A. Berger) Britt. & Rose)

Nigel P. Taylor*
Depto de Biologia, Universidade Federal São Carlos, Sorocaba, São Paulo, Brazil

1.1 Introduction

The cactus family (Cactaceae) comprises more than 120 genera and 1500 species of mostly spiny succulent plants and is exclusively native to the Americas, with one exception (*Rhipsalis baccifera* (J.S. Muell.) Stearn, which also occurs naturally in the Old World, as well as many introduced, cultivated and sometimes invasive non-native cactus species). In habit, the family is rather diverse, including a few primitively broad-leaved shrubs, trees and climbers, and a majority of leafed or leafless stem-succulent trees, shrubs, cushion-like subshrubs, lianous climbers, epiphytes/lithophytes, and very large to very dwarf, globular single-stemmed species that can sometimes behave as geophytes. Their habitats are equally diverse, ranging from extreme almost desert conditions in coastal northern Chile to perhumid tropical rainforest and high alpine areas in the central Andes (up to 4800 m above sea level), although the majority are found in warm, seasonally dry habitats within 35°N and 35°S of the equator. Morphologically, the family is unique in having three kinds of stem: (i) the conventional stem as in other plants (i.e. the auxoblast), whose nodes subtend the brachyblasts; (ii) brachyblasts (short shoots) known as areoles, often felted and bearing the familiar spines (modified leaves) and flowers/fruit; and (iii) the pericarpel (receptacle stem) into which the flower is inverted and not infrequently also fused inside a hypanthial tube. The fruit of nearly all cacti utilize zoochorous (animal) dispersal, and many are edible, notably those of *Opuntia* Mill., especially *O. ficus-indica* (L.) Mill., *Myrtillocactus* Console, *Stenocereus* (A. Berger) Riccob., *Cereus* Mill. and *Selenicereus* spp., all of which are cultivated for human consumption or harvested from wild stands.

Modern classifications divide the family Cactaceae into four or five subfamilies, of which only the largest, the Cactoideae, concerns us here. The subfamily Cactoideae is as diverse as the family as a whole but is notable for the absence of true leaves (as opposed to leaf-like stems), apart from minute ephemeral and vestigial scale-like structures seen at the apices of actively growing stems in some species. Its division into tribes, now based largely on gene sequence data, is still evolving, but there is general agreement that the plants described in this book belong to the tribe Hylocereeae F. Buxb., a large assemblage of epiphytes and climbers with a centre of diversity in central and northern South America and the Caribbean. The phylogeny (the evolutionary history and relationships) of the Hylocereeae is well understood from molecular studies undertaken in the last 6 years

* Email: nigelcactus.taylor@gmail.com

DOI: 10.1079/9781800623156.0001

(e.g. Korotkova *et al.*, 2017), while the taxonomy of the tribe at the species level was summarized previously by Bauer (2003) and is that followed in the modern standard work on the family, *The New Cactus Lexicon* (Hunt et al., 2006). The work of Korotkova *et al.* (2017) demonstrated that the three commercially cultivated species of dragon fruit, also known as pitahaya or pitaya (pitaia in Brazil), belong in the genus *Selenicereus* together with all other species that were previously classified in the genus *Hylocereus*. Both generic names were established by Britton and Rose in 1909, who took them up from Alwin Berger's 1905 infrageneric classification of the then broadly defined 'hold-all' genus *Cereus* (see section 1.2).

The genus *Selenicereus* as currently circumscribed includes approximately 30 species of large, mostly ribbed or three-angled climbers and epiphytes/lithophytes with or without cylindrical stems, with generally large to very large nocturnal flowers and fleshy fruit, found in nature from Mexico and the Caribbean to as far south as northern Argentina. Some of its species have been known botanically since before the time of Linnaeus, who described and named two in his *Species Plantarum* of 1753: *Cactus grandiflorus* L. (subsequently the nomenclatural type of *Selenicereus*) and *C. triangularis* L. (the type of *Hylocereus*). While the application of these two old names is not in doubt, the same cannot be said for various of the later-described members of the genus, whose stems are often very similar to one another and whose flowers and fruit were frequently unknown or poorly characterized; furthermore, museum specimens were rarely preserved. The geographical origin of some of these taxa was also imperfectly recorded at the time of their description; thus, the most commonly encountered taxon, *Selenicereus undatus*, was originally described from cultivation as having been introduced from China, and its natural range remains uncertain. Bauer (2003) made a valuable attempt to fix the application of such names and his treatment is basically that followed here.

1.2 Taxonomic Summary

Selenicereus (A. Berger) Britt. & Rose, *Contributions from the United States National Herbarium* 12, 429 (1909); *Cereus* subsect. *Selenicereus* A. Berger, Annals of the Missouri Botanical Garden 16, 76 (1905). Type: *S. grandiflorus* (L.) Britt. & Rose.= *Hylocereus* (A. Berger) Britt. & Rose, *Contributions from the United States National Herbarium* 12, 428 (1909); *Cereus* subg. *Hylocereus* A. Berger, *Annals of the Missouri Botanical Garden* 16, 72 (1905). Type: *H. triangularis* (L.) Britt. & Rose.= *Wilmattea* Britt. & Rose, *Cactaceae* 2, 195 (1920). Type: *Cereus minutiflorus* Vaupel.= *Cryptocereus* Alexander, *Cactus and Succulent Journal* 22, 164 (1950). Type: *C. anthonyanus* Alexander.[= *Mediocactus* sensu Britt. & Rose, *Cactaceae* 2, 210 (1920), excl. typ.]= *Selenicereus* subg. *Salmdyckia* D. Hunt, Bradleya 7: 91–92 (1989); *Hylocereus* subg. *Salmdyckia* (D. Hunt) D. Hunt, *Cactaceae Systematics Initiatives* 21, 12 (2006). Type: *Cereus setaceus* DC.

These are climbers, epiphytes or lithophytes, with stems up to 5 m or more, 1.5–10 cm in diameter, segmented or non-segmented, cylindrical with up to 12 ribs or triangular, rarely flattened (*S. anthonyanus*), green to glaucous, freely emitting aerial roots; areoles are conspicuous at first, sometimes subtended by wing-like podaria, bearing from one to many spines, which are fine and bristle-like (especially in seedlings and juvenile growth) to short, stout and conical. The flowers are nocturnal, with anthesis ending shortly after dawn the following day, strongly sweetly fragrant (hawkmoth pollination syndrome), mostly large to very large, approximately 12–40 × 10–35 cm (although only 3–5 × 8 cm in *S. minutiflorus*), and funnel form; the pericarpel is areolate and spiny (including hair spines), or without areoles but bearing large bract scales, or sometimes bearing both spines and bracts, and the hypanthial tube is similar, mostly very long, and rarely short; the outer perianth segments are greenish, brownish, golden or reddish, and narrow, while the inner perianth segments are white, sometimes pink to reddish towards the base or rarely entirely reddish, broadly oblanceolate, and acute to irregularly fimbriate at apex; the stamens are very numerous and long, often in two groups; the style is long and stout, its upper part held above the stamens, variously coloured, with the apex divided into very numerous outspread stigma lobes, which are occasionally bifid; the ovary is multi-loculate with hundreds to more than 1000 ovules. The fruit are large, up to 15 × 12 cm, globose to ovoid, with the pericarp bearing large scales, or if areolate and spiny, these are shed as the fruit ripens, mostly indehiscent, often brightly coloured, filled with sweet, white or pink to purplish-red funicular pulp. There are many seeds, which are brown to brown-black, 1.5–3.8 mm long; the testa may be smooth or not, and a mucilage sheath is sometimes present.

1.3 Key To and Synopsis of Species Commercially Cultivated for Their Fruit

Figure 1.1 shows a comparison of the three species that are cultivated commercially, which can be distinguished using the following key.

1.3.1 Key to the species commercially cultivated for their fruit

1a. Pericarpel and hypanthial tube of flower bearing spiny areoles (*Selenicereus* subg. *Salmdyckia*)........................ ***Selenicereus megalanthus***
1b. Pericarpel and hypanthial tube of flower bearing large adpressed bract scales, spineless (*Selenicereus* subg. *Hylocereus*)............................2
2a. Bract scales on pericarpel and fruit overlapping; fruit pulp white; seedling stem soon becoming triangular......... ***Selenicereus undatus***
2b. Bract scales on pericarpel and fruit few, not overlapping; fruit pulp purplish-red; seedling/juvenile stem five-ribbed for first year or more......................***Selenicereus monacanthus***

1.3.2 *Selenicereus* subgenus *Salmdyckia* D. Hunt

S. megalanthus

S. megalanthus (Vaupel) Moran, *Gentes Herbarum* 8, 325 (1953); *Cereus megalanthus* Vaupel, *Notizblatt des Botanischen Gartens und Museums zu Berlin-Dahlem* 5, 284 (1913); *Hylocereus megalanthus* (Vaupel) Bauer, *Cactaceae Systematics Initiatives* 17, 28 (2003). Type: Peru, San Martín, Tarapoto, *Ule* s.n. (not preserved). Lectotype (Bauer, *Cactaceae Systematics Initiatives* 17, 28 (2003)): Karsten & Schenk, *Vegetationsbilder* 2: table 5 (1904). Epitype (Bauer, *Cactaceae Systematics Initiatives* 17, 28 (2003)): Peru, Huánuco, Tingo Maria, valley of the Río Huallaga, *Johnson* s.n. (US 2906780).

Fig. 1.1. Comparison of the fruit and stems of the three *Selenicereus* species: *S. megalanthus* (left), *S. monacanthus* (centre); *S. undatus* (right). Fruit purchased in Singapore public market; stems grown from seed.

DESCRIPTION. Climber/epiphyte, ascending many metres up trees in habitat, often losing contact with the ground; stem segments/branches up to 200 × 5–7 cm, three-angled/winged, green, emitting aerial roots; areoles on raised stem tubercles, 3–5 cm apart on each angle; spines of mature stems, two to three per areole, 3–5 mm long, broad-based/conical, brown or paler. The flowers are nocturnal but remain open briefly after dawn, and are funnel form, 30–38 × 25–30 cm (Fig. 1.2); the pericarpel is approximately 4 cm long, globose to ovoid, green, with porrect fleshy-based bract scales subtending spiny areoles in their axils; the hypanthial tube is approximately 16 × 3 cm, slightly ridged, green tinged, reddish above, with a few large bract scales increasing in size above, the uppermost reddish; the outer perianth segments are up to around 15 cm, linear, narrow, golden, the inner segments a little shorter but much broader and white. The fruit are ovoid, up to 11 × 6 cm or somewhat larger in cultivated selections, and yellow; the pericarp has large triangular scales/tubercles, which are sometimes slightly reflexed at the apex, bearing areoles with spines 5–10 mm long, which are usually deciduous when the fruit are fully ripe (Fig. 1.3); the funicular pulp is white to pink-magenta. The seeds are brown-black, approximately 3 mm long; the testa is smooth. Seedlings/juveniles are three-angled; areoles have numerous fine bristly spines.

NATURAL RANGE. Colombia to Peru in the eastern drainage of the Andes; 90–1800 m above sea level. Illustrations: Hunt *et al.* (2006, Atlas: pp. 85.3–4). Note: the commercially produced fruit normally has a white pulp, but wild plants can have a pinkish pulp (see illustration cited above).

Fig. 1.2. Flower of *Selenicereus megalanthus*. Photograph courtesy of Sisir Mitra.

1.3.3 *Selenicereus* subg. *Hylocereus* (A. Berger) N.P. Taylor, comb. nov. Basionym: *Cereus* subg. *Hylocereus* A. Berger, *Annals of the Missouri Botanical Garden* 16, 72 (1905)

S. monacanthus

S. monacanthus (Lem.) D. Hunt, *Cactaceae Systematics Initiatives* 36, 33 (2017); *Cereus monacanthus* Lem., *L'Horticulteur Universel* 6, 60 (1845): *Hylocereus monacanthus* (Lem.) Britt. & Rose, *Cactaceae* 2, 190 (1920). Type: not preserved. Neotype (Bauer, 2003: p. 33): Colombia, Magdalena, Parque Tayrona, Playa Arrecifes, 2 m above sea level, 20 March 1999, *Bauer* 46 (ZSS 22705).= *Hylocereus polyrhizus* (F.A.C. Weber ex K. Schum.) Britt. & Rose, *Cactaceae* 2, 185 (1920); *Cereus polyrhizus* F.A.C. Weber ex K. Schum., *Gesamtbeschreibung der Kakteen* 151 (1897). Type: Colombia, cult. Hort. Luxembourg, not preserved. Neotype (Bauer, 2003: 41): Venezuela, Cojedes, Hato Piñero, *Delascio* 14484 (VEN 249248).[= *Hylocereus triangularis sensu* N.P. Taylor, *Gardenwise (Singapore)* 49, 36 (2017), non (L.) Britt. & Rose.]

DESCRIPTION. Climber/epiphyte or lithophyte; stem segments/branches up to 200 × 2–4 cm, three-angled/ribbed, grey-green, without a horny margin, emitting aerial roots; areoles on only slightly raised parts of the ribs, 2–3 cm apart on each angle; the spines of mature stems

Fig. 1.3. Fruit of *Selenicereus megalanthus*, with spiny areoles. Photograph courtesy of Sisir Mitra.

are approximately two per areole, very short, conical, and brown or paler. The flowers are nocturnal, funnel form, approximately 27 × 17 cm; the pericarpel is approximately 3 cm long, globose, green, with large adpressed bract scales naked in their axils; the hypanthial tube is approximately 11 × 3 cm, green, with a few large yellowish semi-adpressed bract scales increasing in size above; the outer perianth segments are linear, narrow and golden; the inner segments are few, with 12 bract scales, much broader, ovate, white, and pink at the base; the style is long and exserted in relation to the stamens. The fruit are broadly ovoid to almost spherical, up to 15 × 12 cm (smaller in wild forms), and red; the pericarp has large triangular adpressed bract scales, which are well spaced and not concealing the pericarp; the funicular pulp is deep purplish-magenta (or white in some wild forms). The seeds are brown-black, approximately 2 mm; the testa is smooth. Seedlings and juveniles are five-ribbed; the areoles have numerous fine bristly spines.

NATURAL RANGE. Nicaragua to Panama, Colombia, Ecuador, Peru, northern Venezuela, Trinidad and Tobago, Guyana and Suriname; 0–2000 m above sea level. Illustrations: Hunt *et al.* (2006, Atlas: pp. 88.1–3). Note: Bauer (2003: pp. 33–41) lists numerous other synonyms, but only *C. polyrhizus* is cited here as the plant in commerce seems to match the protologue of this name in its red fruit (with purplish-red pulp) and five-ribbed juvenile stems, whereas Bauer claims that *S. monacanthus* (as *Hylocereus monacanthus*) has fruit with a white pulp. He states that fruit pulp colour is variable in species of the genus, but at least the plant in commerce has a consistently deeply coloured pulp. The present author (as cited above) misapplied the name *Hylocereus triangularis* to this species, but recently studied wild plants in the Caribbean (Dominican Republic) have pinkish-red fruit with a white pulp (Paul Hoxey, personal communication).

S. undatus

S. undatus (Haw.) D. Hunt, *Cactaceae Systematics Initiatives* 36, 35 (2017); *Cereus undatus* Haw., *Philosophical Magazine* 7, 110 (1830); *Hylocereus undatus* (Haw.) Britt. & Rose, *Flora of Bermuda* 256 (1918). Type: a plant cultivated at the London Horticultural Society, introduced from China, not preserved. Neotype (Taylor, 1995): *Curtis's Botanical Magazine* tabula 1884 (1817) as 'Cactus triangularis'.

DESCRIPTION. Climber/epiphyte or lithophyte; stem segments/branches up to 30 × 4–7.5 cm, three-angled/ribbed, light to dark green, crenate, when older with a horny margin, emitting aerial roots; areoles are in the sinuses of the rib crenations, approximately 4–5 cm apart on each angle; none to three spines of mature stems per areole, 3–6 mm long, conical and grey-brown. The flowers are nocturnal but remain open briefly after dawn and are funnel form, 25–30 × 15–25 cm (Fig. 1.4); the pericarpel is approximately 5 cm long, ovoid, green, with large adpressed and imbricate bract scales naked in their axils; the hypanthial tube is approximately 15 × 4 cm, green, with overlapping yellowish, adpressed to semi-adpressed bract scales increasing in size above; the outer perianth segments

Fig. 1.4. Flower of *Selenicereus undatus*, cultivated in Alvorada de Minas, Minas Gerais, Brazil, 2018.

are linear, narrow and golden; the many inner segments are lanceolate and white throughout. The fruit are globose to oblong, up to 10–15 × 10–12 cm, and yellow to red; the pericarp has large triangular, adpressed, yellowish bract scales, concealing or not the pericarp; the funicular pulp is white (Fig. 1.5). The seeds are brown-black, approximately 1.5 mm; the testa is smooth. The seedlings/juveniles are three-ribbed; the areoles have numerous fine bristly spines.

NATURAL RANGE. Uncertain; cultivated and often escaped throughout the tropics. Illustrations: Hunt *et al.* (2006, Atlas: p. 89.4) and Taylor and Zappi (2004: p. 131, Table 12.1). Note: Bauer (2003: p. 32), as *Hylocereus undatus*, suggests that it may be related to species from Mexico and Central America. However, the gene sequence phylogeny by Korotkova *et al.* (2017) places it as a sister species to *S. triangularis* from the Caribbean (Greater and Lesser Antilles). Thus, it is possible that the cultivated *S. undatus* is derived from this species. The commonly cultivated form has fruit with a yellow pericarp, ripening to pinkish-red, but there are selections that remain yellow when ripe. The pulp of both is white. Mature stems of *S. undatus* can often be recognized for the characteristically corky margin of the exposed angles (see illustration referenced above in Taylor and Zappi, 2004).

Fig. 1.5. Fruit of *Selenicereus undatus*. Photograph courtesy of Sisir Mitra.

1.4 References

Bauer, R. (2003) A synopsis of the tribe Hylocereeae F. Buxb. *Cactaceae Systematics Initiatives* 17, 3–63.

Hunt, D., Taylor, N. and Charles, G. (2006) *The New Cactus Lexicon*. Text and Atlas. DH Books, Milborne Port, UK.

Korotkova, N., Borsch, T. and Arias, S. (2017) A phylogenetic framework for the Hylocereeae (Cactaceae) and implications for the circumscription of the genera. *Phytotaxa* 327, 1–46.

Taylor, N. (1995) The typification of *Hylocereus undatus*. In: Rowley, ed. The Sessé and Mociño cactus Plates – A Postscript. *Bradleya* 13, 119–120.

Taylor, N. and Zappi, D. (2004) *Cacti of Eastern Brazil*. Royal Botanic Gardens, Kew, UK.

2 Origin, Production and History

Sisir Mitra[1]* and Pradyot K. Pathak[2]

[1]*Former Professor, Faculty of Horticulture, Bidhan Chandra Krishi Viswavidyalaya (State Agricultural University), Mohanpur, West Bengal, India;* [2]*KVK, Ashoknagar, West Bengal University of Animal and Fishery Science, Kolkata, West Bengal, India*

2.1 Introduction

Vine cacti of the genus *Selenicereus* are an epiphytic group of plants bearing edible and exotic fruit, known as dragon fruit, pitaya or pitahaya. The flowers are generally nocturnal and remain open for only a few hours. The fruit are medium sweet to sweet, juicy and have black small seeds, with either a spineless or a spiny peel (Morton, 1987). They are crops with high adaptability in tropical and subtropical climates, have great potential as a water-efficient crop and can tolerate temperatures above 38–40°C. These plants belong to the family Cactaceae and so can use Crassulacean acid metabolism pathways to adapt to arid conditions. However, some species prefer rainy areas (annual precipitation 340–3500 mm) and high altitudes of 2750 m above sea level.

The fruit has gained popularity due to its attractive appearance and potential health benefits. The fruit is a source of lycopene, β-carotene and vitamin E, and the seed is rich in essential fatty acids. Red dragon fruit or red pitaya (*Selenicereus polyrhizus*) is the only edible fruit that contains betalains. Betalain pigments have potential benefits in promoting health and preventing diseases in humans due to their antioxidant, anti-inflammatory and chemopreventative activities *in vitro* and *in vivo*. Red dragon fruit peel powder has potential in reducing total cholesterol, triglyceride and low-density lipoprotein cholesterol levels, and in increasing high-density lipoprotein cholesterol levels. Generally, dragon fruit are consumed fresh or processed into juice. The fruit consists of about 37% peel. The peel is a good source of pectin, phenols, antioxidants, betacyanin pigment and total dietary fibre. Details of the flesh, peel and skin composition are given in Chapter 3 (this volume).

Dragon fruit has a non-climatic ripening pattern and should be harvested at full maturity, when quality is at maximum acceptability. Premature fruit is inferior in quality and is susceptible to cell-wall rupture and physiological disorders caused by cellular disorganization; however, overripe fruit rapidly senesce, causing a reduction in both quality and shelf life.

Recent molecular studies on the plastid and nuclear DNA of dragon fruit support the conclusion that the genus *Hylocereus* falls within the genus *Selenicereus* with a single common ancestor to the species in these former two genera (Plume *et al.*, 2013; Korotkova *et al.*, 2017). Morphological and anatomical comparisons reached a similar conclusion (Gómez-Hinostrosa *et al.*, 2014). The reclassification implies fewer species with broader variation in each species and gives greater clarity in the breeding programme

* Email: sisirm55@gmail.com

DOI: 10.1079/9781800623156.0002

(Paull and Chen, 2021). At least six *Selenicereus* (*Hylocereus*) spp. are known to have been used in the development of the varying forms of commercial dragon fruit or pitaya: (i) *Selenicereus undatus*; (ii) *S. monacanthus*; (iii) *S. costaricensis*; (iv) *S. megalanthus*; (v) *S. guatemalensis*; and (vi) *S. ocamponis*. Because the genus *Selenicereus* has already been published (Hunt, 2017), it is regarded as having priority over *Hylocereus*. However, *Hylocereus* has an international market from an economic prospective and is relevant to the Convention on International Trade in Endangered Species of Wild Fauna and Flora (CITES) and trade (Korotkova *et al.*, 2017). In this book, the contributors have used dragon fruit, pitaya or pitahaya, and, in reference to the species in the two genera of climbing vine cacti that are commercially grown, *Hylocereus* (Berger) Britton & Rose and *Selencereus* (Berger) Britton & Rose.

The climbing cacti are often epiphytic with elongated stems that are three angled or three winged in *Hylocereus* spp. and have four or more ribs in *Selenicereus* spp. They are freely branching, with the branches emitting aerial roots. The areoles bear short wool and several short spines, or rarely are spineless. The flowers are very large, funnel form and usually bloom at night. The fruit are spherical to oblong, red or yellow, and fleshy.

2.2 Origin and Distribution

Vine cacti of the genera *Selenicereus* are native to the tropical regions of northern South America, Central America and Mexico. They are currently being cultivated in more than 30 countries of the world. In Central and South America, they are cultivated in Guatemala, Nicaragua, Mexico, Colombia, Costa Rica, Venezuela and Peru. *Selenicereus* (syn. *Hylocereus*) has been introduced for production to the Bahamas, Bermuda, the USA (Florida and California), Brazil, Israel, Australia, Japan, Thailand, Malaysia, Cambodia, China, India, Taiwan, the Philippines, Vietnam, Sri Lanka, Bangladesh, Indonesia and other countries (Ortiz-Hernandez and Carrillo-Salazar, 2012; Wakchaure *et al.*, 2020). The area and production of dragon fruit in major growing countries and the fruit-harvesting periods are presented in Tables 2.1 and 2.2.

The common Spanish names for these fruit are pitaya, pitajaya or pitanga, and cuauhnochtli, in the Nahuatl language, *cierge lizard* and *poire de chardon* in French, and night-blooming cereus, strawberry pear and dragon fruit in English (Paull and Chen, 2021). Many columnar cacti in the genus *Stenocereus* are also called pitaya, pitanaya or pitajaya in Latin America. Bravo-Hollis (1991) believes that the word 'pitaya' is a Quechua word (Antillean) introduced by the Spanish conquerors and refers to any fleshy, juicy and edible fruit from a cactus. In Mexico, 'pitahaya' is used for fruit from epiphytic cacti such as *Hylocereus*, while 'pitaya' is used for fruit from column-like cacti (Ortiz-Hernandez, 1999). *Hylocereus* refers to *Hyle*, Greek for 'forest', while *Cereus* is the name historically used for various kinds of columnar or elongate-cylindrical cacti (Taylor, 2017).

Table 2.1. Area and production of dragon fruit (pitaya) in the major growing countries.

Country	Year	Area (ha)	Production (t)	Reference
China (including Taiwan)	2023	67,000	1,600,000	Zhao (2023)
Vietnam	2021–2022	55,000	1,206,000	Anon. (2022b)
Indonesia	2022	1,580	367,300	Anon. (2022a)
Colombia	2018	397	17,773	Agronet (2021)
Brazil	2017	536	1,459	IBGE (2023)
Japan	2015	22.6	167.2	Nashima *et al.* (2020)
Peru	2022	100	1,500	Hugo Villachica Leon, personal communication
Malaysia	2018	753	9,358	Sarip (2020)
India	2020	4,000	12,000	Wakcahure *et al.* (2020)
Ecuador	2018	1,528	11,613	Anon. (2018)
Thailand	2020	3,495	36,053	Anon. (2023)

Table 2.2. Harvesting time in the major growing countries.

Country	Jan	Feb	Mar	Apr	May	Jun	Jul	Aug	Sep	Oct	Nov	Dec
Australia	✓	✓	✓	✓								
Brazil										✓	✓	✓
China						✓	✓	✓	✓	✓	✓	
Honduras										✓	✓	✓
India							✓	✓	✓	✓		
Indonesia	✓	✓	✓	✓	✓	✓	✓	✓	✓	✓	✓	✓
Israel	✓	✓	✓			✓	✓	✓	✓	✓	✓	✓
Mexico					✓	✓						
Nicaragua						✓	✓	✓	✓	✓		
Philippines				✓	✓	✓	✓	✓	✓	✓	✓	✓
Sri Lanka					✓	✓	✓	✓	✓	✓	✓	✓
Thailand												
Taiwan	✓	✓	✓									
USA	✓	✓										
Vietnam			✓	✓	✓	✓	✓	✓	✓			

Indonesia harvests throughout the year near the equator, from May to October in areas more than 5°N and from November to April in areas more than 5°S. Vietnam harvests year-round using artificial lighting.

However, Suárez (1996) stated that the word 'pitahaya' comes from the Taino language and belongs to the Arawak linguistic family, meaning 'scaly fruit'. The Spanish conquistadors adopted this name, as well as the chroniclers from the earliest historical reference (de Anglería, 1494). Later, the same term was used to name the fruit from columnar cacti and since then there has been confusion, partly remedied by keeping the name 'pitahaya' for scaly fruit and 'pitaya' for fruit from columnar cacti (Rodríguez, 2015). The fruit has a red coat, with protrusions like flames, which reminds people of dragon scales, hence the name 'dragon fruit' (Dr F. Zhang, personal communication).

Historical references to these cacti have been found in chronicles from conquistadors, historians and priests who came to America. They include the outstanding chronicles of Pedro Mártir de Anglería (1494), Gonzalo Fernández de Oviedo y Valdés (1535), Diego de Landa (1560), Tomás López Medel (1565) and Juan López de Melgarejo (1582). Pitahayas have been a source of inspiration for at least 140 artists from 31 countries, while pitayas have been captured by at least 65 artists from 13 countries. The initial contact made by Rodríguez in 2001 with the artistic, historical and literary expressions of the two cacti was in an exposition where Rodríguez brought together 18 works related to pitahayas; they belonged to an emerging personal collection (Rodríguez, 2001). In 2011, he proposed the systematic study of pitahaya visual expressions which resulted in the book *Las Pitahayas en las Artes Plásticas* (Rodríguez, 2011), which included some historical and literary references. His later books, *Las Pitayas en las Artes Plásticas, la Historia y la Literatura* (Rodríguez, 2012) and *Las Pitahayas en las Artes Plásticas, la Historia y la Literatura* (Rodríguez, 2013), included the full literary and historical perspectives.

2.3 Visual Arts

There are many works of art related to pitayas and pitahayas; about 1000 works have been identified, of which 400 were selected and catalogued due to their broad perspective of the plants and their parts, with special emphasis on the fruit; 200 corresponded to pitayas and 200 to pitahayas.

Such visual expressions are, however, varied because they include paintings, drawings, sculptures, woodcarving, crochet, manga and digital art. The painting techniques used are also diverse: murals, oil acrylics, watercolour, pastel, gouache, decoupage and mixed techniques.

Botanical paintings stand out in the selected works. There are works from Mary Emily Eaton that were prepared for the important book *The Cactaceae: Descriptions and Illustrations of Plants of the Cactus Family*, by Britton and Rose (1920) (Fig. 2.1), as well as works by Walter Hood Fitch, published in *Curtis's Botanical Magazine* (1836

Fig. 2.1. *Hylocereus undatus*, watercolour, by Mary Emily Eaton, 1906. Photograph courtesy of C.A. Rodríguez, with permission of the International Society for Horticultural Science.

and 1854, Fig. 2.2), and by Atanasio Echeverría and Juan de Dios Vicente de la Cerda, artists from the Royal Botanical Expedition to New Spain (1786–1803). The botanical illustration tradition was followed by contemporary artists from several countries (Figs 2.3 and 2.4).

Pitayas have been portrayed by at least 65 artists from 13 countries, while pitahayas have inspired at least 140 artists from 31 countries (Rodríguez, 2015). The 205 artists with selected and catalogued art work come from 34 countries: Germany, Argentina, Australia, Brazil, Cambodia, Canada, Colombia, Korea, Cuba, Chile, Spain, the USA, the Philippines, France, Guatemala, Hungary, Ireland, Italy, Japan, Malaysia, Mexico, Nicaragua, the Netherlands, Peru, Poland, Portugal, the UK, Dominican Republic, Singapore, South Africa, Switzerland, Taiwan, Ukraine and Vietnam (Rodríguez, 2015).

Among the most renowned artists that have been inspired by pitayas or pitahayas are Diego Rivera (Mexico, 1886–1957), Frida Kahlo (Mexico, 1907–1954), José María Velasco (Mexico, 1840–1912), Olga Costa (Mexico and Germany, 1913–1993), Manuel Álvarez Bravo (Mexico, 1902–2002), Amelia Peláez (Cuba, 1896–1968) and Robert Ossaye (Guatemala, 1927–1954; Fig. 2.5).

2.4 History

The chroniclers who came to America in the 16th century gave an account of the existence

Fig. 2.2. *Hylocereus triangularis*, watercolour, by Walter Hood Fitch, 1836. Photograph courtesy of C.A. Rodríguez, with permission of the International Society for Horticultural Science.

Fig. 2.3. Pitahayas, watercolour, Rodolfo Navarro Méndez, 1999. Photograph courtesy of C.A. Rodríguez, with permission of the International Society for Horticultural Science).

of pitayas and pitahayas. Pedro Mártir de Anglería (Italy, 1457; Spain, 1526) was the first to report the pitahaya in 1494: 'There is another tree that rises in stone crevices, not in good soil: it is called pythahaya' (de Anglería, 1944). The descriptions of pitayas and pitahayas in 1535 by Gonzalo Fernández de Oviedo y Valdés (Spain, 1478–1557), in 1560 by Diego de Landa (Spain, 1524–1579), in 1565 by Tomás López Medel (Spain, 1520–1582) and in 1582 by Juan López de Melgarejo (Spain, 1546–?) are very illustrative.

This was the way Diego de Landa (1560) described the pitahaya found in Yucatán:

> There are some thorny and ugly thistles and they always grow from vines attached to other trees, scrambled with them. They have a fruit whose skin is red and it is similar to an artichoke, the skin can be easily removed and it does not have any thorns. The flesh inside is white and full of tiny black seeds. It is sweet, delicate and tenderly soft, it melts in the mouth, it is eaten in slices, as oranges, and with salt; the fruits brought by the Indians from the mount can't be enough since the Spanish eat them all.

The oldest works to mention these fruit are three pre-Hispanic archaeological pieces, one from Chavin Peruvian culture (1200–500 BC) and the others from the Olmec (900–500 BC) and Mexican (1250–1500) cultures from Mexico: Hampycamayoc (Fig. 2.6), and Pitaya and Cactus Stela (Fig. 2.7), respectively. Those from the Colonial period include two drawings by the Spanish chronicler Gonzalo Fernández de Oviedo y Valdés in 1535 (Fig. 2.8), and the murals of the Augustinian Convent in Malinalco (Mexico) painted in 1540, containing illustrations of animals and plants including several cacti; and Teonochtli painted in 1552 as part of the Codex de la Cruz-Badiano a Mexican medicine treatise (Fig. 2.9). In the late 18th century, as part of their work in the Royal Botanical Expedition to New Spain, Atanasio Echeverría and Juan de Dios Vicente de la Cerda produced two works (*Cactus triangularis* and *Hylocereus undatus*). From the Independent period, there are three works of art: *Cereus giganteus* and Pápagos (both from 1848; Fig. 2.10) from the American army explorer William Emory, and Candelabro de Oaxaca (1887) from the great Mexican landscape painter José María Velasco. In the 19th century, the British illustrator Walter Hood Fitch produced two works: *Hylocereus triangularis* (1836) and *Hylocereus monacanthus* (1854).

Fig. 2.4. Anhelo y realidad, oil on canvas, by Erasto Celis Aguilar, 2012. Photograph courtesy of C.A. Rodríguez, with permission of the International Society for Horticultural Science.

Fig. 2.5. Pitahayas, oil on canvas, by Roberto Ossaye, 1952. Photograph courtesy of C.A. Rodríguez, with permission of the International Society for Horticultural Science.

Mythology, poetry and narratives about pitayas and pitahayas are diverse. Rodríguez (2015) made a selection of nine stories and 24 poems. The popular culture of Argentina, Peru and Mexico includes legends about pitayas.

Outstanding writers including Nobel laureates of literature such as Gabriela Mistral (Chile, 1889–1957) and Miguel Ángel Asturias (Guatemala, 1899–1974), the Mexican writer Juan Rulfo (1917–1986), the poet and writer André Breton (France, 1896–1966) and the novelist Jack London (USA, 1876–1916) included descriptions of these fruit in their written works. Poets from several nations including Ramón Velez y Herrera (Cuba, 1808–1886), Clemente López Trujillo (Mexico, 1905–1981), Liliana Felipe (Argentina, 1954), Isidoro Arellano Corro (Mexico, 1910–1997), Erasmo Rodríguez Barreto

(Colombia) and Adriano Corrales Arias (Costa Rica, 1958) have written and sung to pitayas and pitahayas. In Yucatán's folk traditions, there are verses called bombas that refer to pitahayas. In Asia's popular culture, the dragon fruit legend was spread, providing an interesting amalgamation of a major Chinese mythological element and a plant that throughout the century was incorporated into its cultural tradition (Rodríguez, 2015)

The poem *Saguaro* by Malinda Leon (1982) expresses the cultural meaning of pitayas or saguaros in indigenous communities from North America:

Fig. 2.6. Hampycamayoc, bas-relief sculpture from Chavin culture (Peru), 1200–500 BC. Photograph courtesy of C.A. Rodríguez, with permission of the International Society for Horticultural Science).

Fig. 2.7. Cactus, carved in basalt, Mexican culture, 1230–1500. Photograph courtesy of C.A. Rodríguez, with permission of the International Society for Horticultural Science).

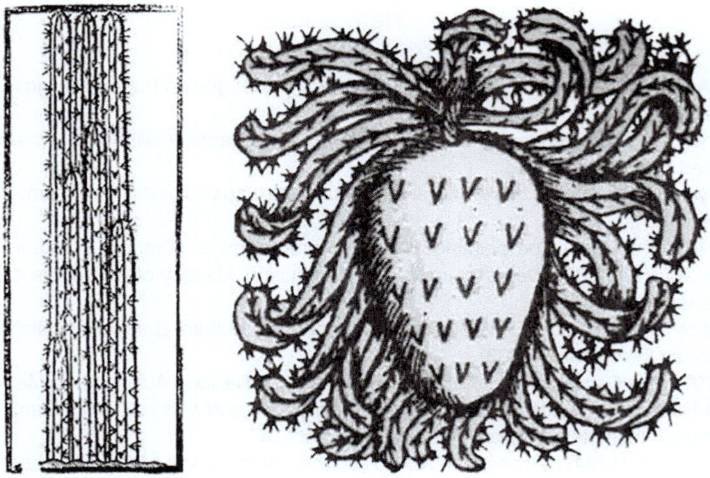

Fig. 2.8. Ink drawings, by Gonzalo Fernández de Oviedo y Valdés, 1535. Photographs courtesy of C.A. Rodríguez, with permission of the International Society for Horticultural Science.

Fig. 2.9. Teonochtli, Codex de la Cruz-Badiano, 1552. Photograph courtesy of C.A. Rodríguez, with permission of the International Society for Horticultural Science.

What do you see in a saguaro cactus?
you see its fruits
you see its syrup
you see its candy
you see its wine
but most of all you see Papago tradition.

Fig. 2.10. Pápagos, lithography, by William Emory, 1848. Photograph courtesy of C.A. Rodríguez, with permission of the International Society for Horticultural Science.

2.5 References

Agronet (2021) Agronet MinAgriculture. Available at: www.agronet.gov.co/Paginas/inicio.aspx (accessed 8 November 2023).

Anon. (2018) Ministry of Agriculture and Livestock (MAG, 2019), Ecuador. Available at: www.agricultura.gob.ec/ (accessed 8 November 2023).

Anon. (2022a) System for Providing Horticultural Statistical Data. https://sipedas.pertanian.go.id/ (accessed 8 November 2023).

Anon. (2022b) Report on estimated planting areas and production of perennial plants in Vietnam, year 2019, 2020, 2021 and 2022. Department of Plant Production, Ministry of Agriculture and Rural Development, Hanoi.

Anon. (2023) Office of Agricultural Economics, Department of Agriculture, Bangkok. Available at. www.oae.go.th (accessed 8 November 2023).

Bravo-Holis, H. (1991) *Las Cactaceas de Mexico*, Vol. II. Universidad Nacional Autónoma de México, Mexico City.

Britton, N.L. and Rose, J.N. (1920) *The Cactaceae: Descriptions and Illustrations of Plants of the Cactus Family*. Carnegie Institution of Washington, Washington, DC.

de Anglería, P.M. (1944) [1494] *Décadas del Nuevo Mundo*. Editorial Bajel, Buenos Aires.

de Landa, D. (1973) [1560] *Relación de las Cosas de Yucatán*, 10th edn. Editorial Porrúa, Mexico City.

Fernández de Oviedo y Valdés, G. (1535) *Historia General y Natural de las Indias, Islas y Tierra-firme del Mar Océano*. Part 1. Available at: www.ems.kcl.ac.uk/content/etext/e026.html (accessed 8 November 2023).

Gómez-Hinostrosa, C., Hernández, H.M., Terrazas, T. and Correa-Cano, M.E. (2014) Studies on Mexican Cactaceae. V. Taxonomic notes on *Selenicereus tricae*. *Brittonia* 66, 51–59.
Hunt, D.R. (2017) Selenicereus. *Cactoaceae Systematics Initiatives* 36, 29–36.
IBGE (Instituto Brasileiro de Geografia e Estatística) (2023) Producao de Pitaia no Brasil. Available at: www.ibge.gov.br/explica/producao-agropecuaria/pitaia/br (accessed 8 November 2023).
Korotkova, N., Borsch, T. and Arias, S. (2017) A phylogenetic framework for the Hylocereae (Cactaceae) and implication for the circumscription of the genera. *Phytotaxa* 327, 1–46.
López Medel, T. (1990) [1565] *De los Tres Elementos: Tratado Sobre la Naturaleza y el Hombre del Nuevo Mundo*. Alianza Editorial, Madrid.
López de Melgarejo, J. (1582) *Relación de la Isla de San Juan de Puerto Rico*. Available at: https://mlab.taik.fi/simultaneous/Text/Rel_Puerto_Rico.html (accessed 8 November 2023).
Morton, J. (1987) Strawberry pear. In: *Fruits of Warms Climates*. Julia F. Morton, Miami, Florida, pp. 347–348. Available at: www.hort.purdue.edu/newcrop/morton/strawberry_pear_ars.html (accessed 8 November 2023).
Nashima, K., Hosaka, F., Shimajiri, Y., Matsumura, M., Tavora, K. *et al.* (2020) SSR marker development and genetic identification of pitaya (*Hylocereus* spp.) collected in Okinawa Prefecture, Japan. *Horticultural Journal* 90, 23–30.
Ortiz-Hernandez, Y.D. (1999) *Pitahaya: Un Nuevo Cultivo Para Mexico*. Editorial Limusa, Mexico City.
Ortiz-Hernandez, Y.D. and Carrillo-Salazar, J. (2012) Pitahaya (*Hylocereus* spp.): a short review. *Communicata Scientiae* 3, 220–237.
Paull, R.E. and Chen, N.J. (2021) Dragon fruit production and global marketing. In: Fullerton, R.A. and Lee, W.L. (eds) *Dragon Fruit from Production to Market*. Food and Fertilizer Technology Center (FFTC) for the Asian and Pacific Region, Taipei, Taiwan, pp. 1–16.
Plume, O., Straub, S.C.K., Tel-Zur, N., Cisneros, A., Schneider, B. and Doyle, J.J. (2013) Testing a hypothesis of intergeneric allopolyploidy in vine Cacti (Cactaceae: Hylocereae). *Systematic Botany* 72, 1022–1026.
Rodríguez, C.A. (2001) *Exposición Internacional: Las Pitahayas en las Artes Plásticas*. Universidad Autónoma Chapingo, Texcoco, Mexico.
Rodríguez, C.A. (2011) *Las Pitahayas en las Artes Plásticas*. Universidad Autónoma Chapingo, Texcoco, Mexico.
Rodríguez, C.A. (2012) *Las Pitayas en las Artes Plásticas, la Historia y la Literatura*. Universidad Autónoma Chapingo, Texcoco, Mexico.
Rodríguez, C.A. (2013) *Las Pitahayas en las Artes Plásticas, la Historia y la Literatura*. Universidad Autónoma Chapingo, Texcoco, Mexico.
Rodríguez, C.A. (2015) Pitayas (*Stenocereus* spp.) and pitahayas (*Hylocereus undatus*) in art, history and literature. *Acta Horticulturae* 1067, 335–341.
Sarip, J. (2020) Dragon fruit exportation practices. Challenges and country experiences: Malaysia. In: *Proceedings of the 3rd DFNet Workshop on Dragon Fruit Value Chain for Global Markets*, Taichung, Taiwan, 22–23 September, pp. 13–26.
Suárez, M.V.M. (1996) *El Español que se Habla en Yucatán*, 3rd edn (corrected by M. Güémez Pineda). Universidad Autónoma de Yucatán, Mérida, Mexico.
Taylor, N.P. (2017) Dragon fruit explained. *Gardenwise* 49, 35.
Wakchaure, G.C., Kumar, S., Meera, K.K., Rane, J. and Pathak, H. (2020) Dragon fruit cultivation in India: scope, marketing, constraints and policy issues. Technical Bulletin No. 46. ICAR-National Institute of Abiotic Stress Management, Maharashtra, India.
Zhao, Y. (2023) China's dragon fruit production jumped to the world's first. *China Fruit News* 40, 52.

3 Composition and Uses

Kundan Kishore[1]* and Ankita Sahu[2]
[1]*Central Horticultural Experiment Station (ICAR-Indian Institute of Horticultural Research), Bhubaneswar, India;* [2]*ICAR-Central Institute for Women in Agriculture, Bhubaneswar, India*

3.1 Introduction

Dragon fruit (also known as pitaya) is known for its nutritional significance. It is a rich source of vitamins, minerals, pigments and bioactive compounds. Moreover, the crop is also rich in antioxidative properties (Jaafar *et al.*, 2009). On the basis of pulp coloration, dragon fruit is categorized into red fleshed (*Selenicereus monacanthus* and *S. costaricensis*) and white fleshed (*S. undatus*). The former possesses the characteristic betalain pigment, which imparts a unique identity to the crop. In contrast, white-fleshed dragon fruit lacks the betalain pigment in its pulp (Fig. 3.1). *S. undatus* (syn. *Hylocereus undatus*), *S. monacanthus* (syn. *H. polyrhizus*) and *S. costaricensis* (syn. *H. costaricensis*) are characterized by red peel, whereas *S. megalanthus* (syn. *H. megalanthus*) is characterized by a knobbly yellow peel with white flesh. The fruit pulp is soft and juicy with numerous tiny edible blackish seeds embedded within the pulp (Le Bellec *et al.*, 2006). The dragon fruit pulp and seeds are edible, whereas the peel or skin is considered waste material (Fig. 3.2). Studies clearly indicate the compositional significance of the pulp, seed and peel. The objective of this chapter is to provide a simplified insight into the composition of dragon fruit and its uses. The proximate composition of dragon fruit is discussed with respect to pulp, seed and peel (Fig. 3.3).

3.2 Fruit Pulp

3.2.1 Fibre

Dietary fibres are plant-based polysaccharides, which are partially broken down by digestive enzymes. The insoluble portion mostly comprises cellulose, hemicellulose and lignin, whereas the soluble fraction is composed of pectins (Anita and Abraham, 2006). Dietary fibres are polymer of monosaccharaides such as fructose, glucose, arabinose, rhamnose, galactose and mannose. Dragon fruit are rich in dietary fibre, but this varies with the genotype. The crude fibre content in the pulp may be 0.88–1.84%, but red-fleshed fruit has a higher dietary fibre content than the white-fleshed clones. Abirami *et al.* (2021) and Liaotrakoon *et al.* (2013a) also reported higher dietary fibre content in red-fleshed dragon fruit compared with white-fleshed fruit. Dragon fruit is also a good source of pectin (Mahattanatawee *et al.* 2006). The presence of a high level of dietary fibre in dragon fruit enhances its nutritional significance, as fibre is considered an important component for heart functioning.

* Email: kkhort12@gmail.com

DOI: 10.1079/9781800623156.0003

Fig. 3.1. Red-fleshed (*Selenicereus monacanthus*) (left) and white-fleshed (*S. undatus*) (right) dragon fruit.

Fig. 3.2. Transverse section of red-fleshed dragon fruit (*Selenicereus monacanthus*).

3.2.2 Carbohydrates and sugar

The pulp of dragon fruit contains a mucilaginous material in which seeds are embedded. This mucilaginous layer is composed of oligosaccharides and polysaccharides (pectin). Glucose and fructose are major oligosaccharides in dragon fruit. However, they vary in proportion. Oligosaccharides in dragon fruit are made up of one glucose unit and two fructose units, one glucose unit and three fructose units (nystose, bifurcose and neobifurcose), and three glucose units and one fructose unit (e.g. stachyose) (Wichienchot *et al.*, 2010). In comparison with other tropical fruits such as banana, mango, jackfruit and pineapple, the carbohydrate content in dragon fruit is relatively low; in the pulp of ripe fruit it varies from 10.4% to 12.3%. The carbohydrate content in white-fleshed dragon fruit is higher than that of the red-fleshed fruit (Liaotrakoon *et al.*, 2013b). Arivalagan *et al.* (2021) also reported a higher content of carbohydrate in white-fleshed dragon fruit (7.32 g 100 g^{-1}) compared with red-fleshed fruit (5.42 g 100 g^{-1}). The starch content in white-fleshed fruit is lower than in red-fleshed fruit: Arivalagan *et al.* (2021) reported that starch content was 0.205 g 100 g^{-1} in white-fleshed and 0.471 g 100 g^{-1} in red-fleshed dragon fruit. The predominant carbohydrates of white- and red-fleshed dragon fruit are

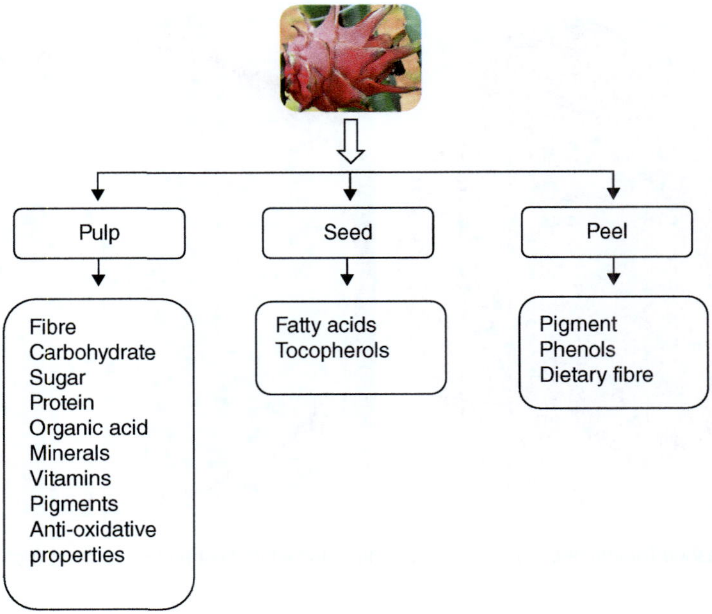

Fig. 3.3. Nutritional composition of the different parts of dragon fruit.

glucose and fructose. Wichienchot *et al.* (2010) determined that the glucose content in white-fleshed fruit (353 ± 0.7 g kg^{-1}) was significantly lower than that of red-fleshed fruit (401 ± 1.27 g kg^{-1}). However, the fructose content of white-fleshed fruit (238 ± 0.84 g kg^{-1}) was significantly higher than that of red-fleshed fruit (158 ± 0.32 g kg^{-1}).

In contrast to the abundance of glucose and fructose in dragon fruit, the content of sorbitol and sucrose is significantly low (Stintzing *et al.*, 2003). Glucose and fructose constitute the major proportion of total sugar content, and the content of glucose may vary from 4.63% to 6.39% (Nomura *et al.*, 2005). Arivalagan *et al.* (2021) reported a higher amount of total sugar, glucose and fructose in white-fleshed dragon fruit of the Andaman white genotype compared with red-fleshed fruit. They reported that in the Andaman white genotype, the glucose content was 2.05%, whereas in the Andaman red genotype, the content was 1.24%. Likewise, the fructose content was also higher in white-fleshed fruit than in red-fleshed fruit. In comparison with glucose and fructose, sucrose is also present but at a much lower level ranging from 0.825 to 2.77 mg g^{-1}. Other simple sugars detected in dragon fruit include mannose, inositol, arabinose, ribose and xylose, and these sugars also vary with genotype. Metabolomic profiling of dragon fruit has demonstrated the presence of 16 types of sugars; among these, glucose (>40%), fructose (>35%) and sorbose (>10%) were the major sugars (Wu *et al.*, 2019). Fruit sugar (i.e. glucose, fructose, sucrose and sorbitol) content increases gradually during fruit maturation and reaches its highest level in the pulp at the mature stage (Hua *et al.* 2018).

3.2.3 Protein

Like banana and jackfruit, dragon fruit has a low protein content. However significant variation exists among genotypes with respect to amino acid and total protein content. Total protein content varies substantially from 0.887% (red flesh) to 1.11% (white flesh). However, Liaotrakoon *et al.* (2013a) reported a higher protein content in red-fleshed dragon fruit pulp than in white-fleshed fruit. A total of 18 types of amino acids have been identified in dragon fruit pulp, among which phenylalanine, arginine, glutamic acid and proline were predominant (Arivalagan *et al.* 2021). Phenylalanine ranged from 183 to 220 mg 100 g^{-1}, followed by arginine (126–149 mg 100 g^{-1}), glutamic acid (101–127 mg 100 g^{-1}) and proline (69.6–116 mg 100 g^{-1}). Metabolomic profiling of dragon fruit demonstrated the presence of 22 amino acids and amines (Wu *et al.*, 2019). The detectable amino acids and amines were alanine, serine, valine, proline, threonine, glutamic acid, glutamine, leucine, isoleucine, acetamide, arginine, methionine, phenylalanine, tyrosine and tryptamine. Amino acids and amines are responsible for fruit quality and nutritional values, but their content decreases with fruit development (Jang *et al.*, 2015). Amino acids are also precursors of betaxanthin and are involved in betalain biosynthesis (Strack *et al.*, 2003). The decrease in content of amino acids with progression of fruit development indicates their conversion into sugars, organic acids and secondary metabolites during fruit development.

3.2.4 Organic acids

The organic acids present in dragon fruit are major components responsible for acidity. The organoleptic properties of fruit such as flavour, colour, aroma and fruity taste are significantly influenced by the organic acid present. Additionally, organic acids contribute towards the postharvest shelf life of the fruit (Al-Farsi *et al.*, 2005). Metabolomic profiling of dragon fruit has demonstrated the presence of several organic acids including malic acid, citramalic acid, succinic acid, fumaric acid, citric acid, quinic acid, α-ketoglutaric acid, hexadecanoic acid and oxalic acid. Wu *et al.* (2020) reported citramalic acid as the dominant acid, accounting for more than two-thirds of the total acid. The profound abundance of citramalic acid in dragon fruit makes it an ideal plant source for this compound. However, Arivalagan *et al.* (2021) quantified organic acids in dragon fruit and found malonic acid and citric acid to be the major ones. Dragon fruit is a low-acid food. Its pH value varies between 4.4 and 5.1, with malic acid being the main acid in the fruit (Nomura *et al.*, 2005). Wu *et al.* (2019) reported that the content of organic acid in mature fruit was higher in the pulp than in the peel. Citramalic acid is the major organic acid in unripe fruit, whereas in the mature pulp, the most dominant organic acid is malic acid (>70%), followed by citramalic acid (23%).

3.2.5 Minerals

Dragon fruit is a wholesome fruit containing essential macro- and micro-minerals. Intake of 100 g of fresh dragon fruit would provide nearly 3.8–6.0% of the recommended dietary allowance (RDA) of phosphorus, 3.3–5.2% RDA of potassium, 9.2–13.5% RDA of magnesium, 3.2–7.6% RDA of calcium, 1.9–2.9% RDA of zinc, 4.1–5.2% RDA of iron, 4.8–6.5% RDA of copper and 1.9–3.1% RDA of manganese (Arivalagan *et al.* 2021). Among the various macro-minerals in dragon fruit, potassium content was higher than the magnesium content, followed by calcium and phosphorus. The content of iron is relatively high among all micronutrients, followed by zinc. The contents of manganese and copper are relatively low. Mineral acquisition varies with genotype, as white-fleshed clones have a higher content of minerals. Gunasena *et al.* (2007) reported that the potassium, phosphorus, sodium and magnesium content in dragon fruit were higher than in mango and pineapple (Stintzing *et al.*, 2003). The content of magnesium is reported to be 38.9 mg 100 g^{-1}, phosphorus 27.5 mg 100 g^{-1}, sodium 8.9 mg 100 g^{-1} and zinc 0.35 mg 100 g^{-1}. Dragon fruit is rich in potassium (398 mg 100 g^{-1}) and calcium (10.2 mg 100 g^{-1}), but the contents are lower than in jackfruit and pineapple. While evaluating the compositional changes during fruit development of dragon fruit, Singh *et al.* (2022) reported that potassium, calcium, magnesium, iron and zinc levels increased during fruit growth and maturation, followed by a slight decrease.

3.2.6 Vitamins

Dragon fruit is considered as a good source of vitamins, containing both water-soluble and fat-soluble vitamins (Hua *et al.*, 2018). The major water-soluble vitamins identified and quantified are vitamin C, niacin, pyridoxine, pantothenic acid, biotin, thiamine and riboflavin. Vitamin C (ascorbic acid) is the predominant water-soluble vitamin found in dragon fruit, ranging from 2.94 to 5.64 mg 100 g^{-1}, followed by pantothenic acid (40.1–49.3 µg 100 g^{-1}) and niacin (7.10–12.55 µg 100 g^{-1}), while lower levels were found of pyridoxine (0.171–0.589 µg 100 g^{-1}), biotin (0.317–1.022 µg 100 g^{-1}), thiamine (1.39–2.43 µg 100 g^{-1}) and riboflavin (0.57–1.95 µg 100 g^{-1}) (Arivalagan *et al.*, 2021). Liaotrakoon *et al.* (2013a) reported that the vitamin C content of the red-fleshed dragon fruit pulp (9.85 mg 100 g^{-1}) was twice that of the white-fleshed pulp. Comparing the vitamin profile of various clones, Arivalagan *et al.* (2021) reported that biotin and riboflavin varied significantly among the clones, and that red-fleshed clones registered the highest biotin content, while riboflavin content was higher in white-fleshed clones. Additionally, consumption of 100 g of fresh dragon fruit can provide about 7.4–14.1% RDA of vitamin C, 1.06–3.41% RDA of biotin, 0.80–0.99% RDA of pantothenic acid, 0.10–0.17% RDA of thiamine and 0.04–0.12% RDA of riboflavin. Fat-soluble vitamins such as vitamins D, E and K have also been detected in dragon fruit, with vitamin E (tocopherol) predominant (75–182 µg 100 g^{-1}), followed by vitamin K and then vitamin D. The levels of vitamins D and K were higher in white-fleshed dragon fruit than in red-fleshed ones. Sandate-Flores *et al.* (2020) reported that yellow dragon fruit (*S. megalanthus*) is a good source of vitamin E. Abirami *et al.* (2021) reported a higher content of carotenoids in *H. costaricensis*. Genotypes with high levels of carotenoids, xanthophylls and vitamin C possess immense potential for the development of nutraceutical products.

3.2.7 Pigments

The red coloration of dragon fruit pulp is attributed to the prevalence of pigments called betalains, which are important pigments in cactus fruit such as *Hylocereus* spp. and *Opuntia* spp. There are two types of betalain: betacyanins (red/purple pigment) and betaxanthins (orange/yellow pigment) (Gibson and Nobel, 1986). Betalains comprise a group of nitrogen-containing, water-soluble pigments, which are present in the pulp of red-fleshed dragon fruit (*H. polyrhizus* and *S. monacanthus*) and in the peel of both red-fleshed and white-fleshed dragon fruit (Moreno *et al.*, 2008). Dragon fruit predominantly contains betanin, phyllocactin, hylocerenin and their isomer forms with a smaller fraction of unidentified betacyanin compounds (Kim *et al.*, 2011). The pigment content (betacyanin) in dragon fruit increases with fruit maturity (Phebe *et al.*, 2009). Pigmentation appears earlier in the pulp than in the peel (Huang *et al.*, 2021). In a mature, red-fleshed dragon fruit, the total betacyanin content varies from 32 to 47 mg 100 g^{-1} (Vaillant *et al.*, 2005), which is comparable to the betacyanin content of red beetroot (*Beta vulgaris*), a commercial source of betacyanin (Azeredo *et al.*, 2007). Thus, red-fleshed dragon fruit is a promising novel source of betacyanin pigment for adding intense pink coloration to food items (Stintzing and Carle, 2004). The characteristic stability of the pigment over a wide pH range aids its application as a natural food colourant of acidic foods. Betacyanins are reported to be potential antioxidants and are considered a new class of dietary antioxidants with potency greater than both catechin and vitamin E. Metabolomic profiling of dragon fruit demonstrated the presence of seven betalains and betalain precursors (Wu *et al.*, 2019). Furthermore, during fruit development and maturation, the decrease in amino acids, accompanied by the increase of sugars and organic acid might contribute to the formation of betalains. Citramalic acid (a potential biomarker metabolite) is also related to betalain formation. According to Wu *et al.* (2019), the betacyanin content in pulp was higher than in peel, while the betaxanthin content was higher in peel than in pulp. However, Suh *et al.* (2014) reported a greater betacyanin and betaxanthin content in peel than in pulp. There exists a significant positive correlation between fructose, glucose, sorbose, glycoside, citric acid and betacyanin content, as well as betaxanthin levels in dragon fruit pulp (Wu *et al.*, 2019).

Liaotrakoon *et al.* (2013a) reported that the total betacyanin content of red-fleshed pulp was higher (16.5 mg 100 g^{-1}) than that of white-fleshed pulp. However, it was slightly lower than that of its peel (20.0 mg 100 g^{-1}). While characterizing dragon fruit genotypes, Arivalagan *et al.* (2021) reported that total betalain content in red-fleshed dragon fruit ranged between 14.4 mg and 23.0 mg of betacyanin equivalent per 100 g of pulp. Yellow dragon fruit has also been reported to be a potential source of pigments (betaxanthins, β-carotene and lycopene) (Huang *et al.*, 2021). As well as betacyanin, pigments such as β-carotene and lycopene have also been reported by others in dragon fruit (Charoensiri *et al.*, 2009).

3.2.8 Phenol and flavonoids

Phenolic compounds are a predominant group of phytochemical secondary metabolites that exhibit strong antioxidant capacities because of the presence of phenolic groups that donate electrons and conjugate with metal ions (Hoyweghen *et al.*, 2012). Phenolic compounds are characterized into different groups such as flavonoids, phenolic acids, stilbenes and lignans, on the basis of the number of carbon molecules present and the structure orientation (Hoda *et al.*, 2019). A total of 80 phenolics comprising phenolic acids, flavonoids, lignans, stilbene and other polyphenols have been identified in dragon fruit (Chen *et al.*, 2021). The major phenolic compounds characterized in both red- and white-fleshed dragon fruit are flavonols, flavanones and hydroxycinnamic acid derivatives (García-Cruz *et al.*, 2017). Additionally, gallic acid, syringic acid, caffeic acid, *p*-coumaric acid, cinnamic acid, quinic acid, quercetin-3-*O*-hexoside, 3,4-dihydroxyvinylbenzene and apigenin have also been categorized in white- and red-fleshed dragon fruit (Luo *et al.*, 2014; Zain *et al.*, 2019; Castro-Enríquez *et al.*, 2020; Lira *et al.*, 2020). Among the cinnamic acids, caffeic acid is predominant, followed by ferulic acid and protocatechuic acid.

The total phenol content differs in white- and red-fleshed dragon fruit, and was found to vary from 24.0 mg gallic acid equivalent (GAE) 100 g^{-1} (white pulp) to 53.2 mg GAE 100 g^{-1} (Arivalagan *et al.*, 2021). Comparative analysis of phenolic contents illustrated that red-fleshed dragon fruit contained comparatively higher levels of phenolic acid than white-fleshed clones, thus resulting in a higher phenol content antioxidant potential in red-fleshed than in white-fleshed dragon fruit. Nurliyana *et al.* (2010) demonstrated that the peel of both *H. undatus* and *H. polyrhizus* contained higher phenolic content than the pulp. The phenolic content in the pulp of *H. undatus* was much lower than that of *H. polyrhizus*, and thus for both *Hylocereus* spp., the peel contains higher radical scavenging activity compared with the pulp. The levels of secondary metabolites in pulp were found to be higher than in peel by Wu *et al.*, (2019). However, Huang *et al.* (2021) reported that the total polyphenols are higher in dragon fruit seeds and peel than pulp. Hua *et al.* (2018) reported that the levels of total phenols, total flavonoids and antioxidant activities are higher in the peel than in the pulp during fruit maturation, suggesting the potential of dragon fruit peel as a good source of natural phenols and flavonoids. It has also been reported that red-fleshed clones have a higher flavonoid content than white-fleshed clones. The total flavonoid content varied among different clones of dragon fruit, with 16.7–37.6 mg 100 g^{-1} in red-fleshed clones (Arivalagan *et al.*, 2021).

3.2.9 Antioxidant potential

The antioxidative properties of dragon fruit could be due to the presence of bioactive compounds such as phenols, flavonoids, pigments such as betacyanin, and vitamins C and E (Nurliyana *et al.*, 2010). Phenolic compounds are the most important antioxidative compounds in plant-based foods. Generally, phenolic compounds exist in varying forms such as simple phenol, phenolic acid and polyphenolic compounds, and provide numerous health benefits (Maestri *et al.*, 2006; Nagasaka *et al.*, 2007). The antioxidative potential of phenolic compounds is mostly attributed to their reduction ability as hydrogen donors and singlet oxygen quenchers. Dragon fruit contains high levels of phenols, potentially making it a fruit with a high level of antioxidative activity (Vaillant *et al.*, 2005; Wu *et al.*, 2006). In addition to phenolic acids, dragon fruit also contains flavonoid compounds such as phloretin-2-*O*-glucoside and myricetin-3-*O*-galactopyranoside, which have antioxidant potential and are at much higher concentration in the peel than in the pulp (Kim *et al.*, 2011). The antioxidative activity of dragon fruit is higher than some tropical fruits such as mango, papaya, lychee and longan (Mahattanatawee *et al.*, 2006).

The betacyanin pigment present in dragon fruit also contributes to the antioxidative properties of the fruit (Delgado-Vargas *et al.*, 2000). The antioxidant role of this pigment is attributed mainly to the chemical structure of betacyanin, as it contains a phenol moiety and a cyclic amine group. In addition, the pigment is reported to be a reducing compound having antioxidative and anti-radical activities. Kanner *et al.* (2001) reported that the pigment is an excellent electron donor with the potential to stabilize free radicals. Additionally, dragon fruit is also rich in other promising natural antioxidative components such as vitamin E (Chemah *et al.*, 2010; Lim *et al.*, 2010) and vitamin C (Choo and Yong, 2011). Vitamin C is regarded as an important cellular antioxidant. It protects the biomembrane against peroxidative damage by scavenging superoxide, hydroperoxide, hypochlorite, hydroxyl radicals, peroxyl radicals and singlet oxygen. Vitamin E exhibits its antioxidative activity by its major biochemical function of reacting with organic peroxyl radicals. Vitamin E also protects cell membranes from lipid peroxidation. Additionally, vitamin C can also protect membranes against peroxidation by enhancing the activity of vitamin E (Sies *et al.*, 1992).

Comparative assessment of antioxidative properties of different species of dragon fruit indicate that red-fleshed fruit have a higher antioxidative potential than white-fleshed fruit (Moreno *et al.*, 2008; Kim *et al.*, 2011). In addition to pulp, the peel of both dragon fruit types is reported to exhibit significant antioxidative potential (Choo and Yong, 2011). It is suggested that the antioxidative activity in red-fleshed dragon fruit is probably related to the presence of betacyanin. Betacyanin contributes to the total phenolic content due to a phenol moiety in its structure (Zainoldin and Baba, 2009). This pigment also exhibits antioxidative potential. These bioactive compounds conferring antioxidant potential make dragon fruit a healthy fruit with a number of nutritional and functional attributes (Liaotrakoon *et al.*, 2013b). Arivalagan *et al.* (2021) also reported considerable variation in antioxidant capacity among the different varieties of dragon fruit. The antioxidant capacity measured in terms of 2,2-diphenyl-1-picrylhydrazyl (DPPH) and 2,2-azino-*bis*-3-ethylbenzothiazoline-6-sulphonic acid (ABTS) radical scavenging potential varied significantly between white- and red-fleshed fruit, but the antioxidative property of red-fleshed fruit was higher than that of white-fleshed fruit.

3.3 Seeds

Dragon fruit contains numerous small black seeds embedded in the pulp. The seeds are consumed along with the pulp. They are a rich source of fatty acids and vitamin E (tocopherol) and hence they are considered a high-value seed. Essential fatty acids, namely linoleic acid and linolenic acid, constitute a significant percentage of the unsaturated fatty acids of the seed oil extract. These essential fatty acids are important fatty acids and constitute necessary substrates in metabolism in the human body and cannot be synthesized *in vivo*. Dragon fruit seeds contain a high level of oil with beneficial properties for health, which is mostly due to the presence of a significant amount of essential fatty acids and vitamin E. Liaotrakoon *et al.* (2013b) reported that both white- and red-fleshed dragon fruit seed contains significant amount of oil (32–34%). The seeds contain about 50% essential fatty acids such as palmitic, stearic, oleic, linoleic and linolenic acids. The major fatty acid is linoleic acid (54–55%), followed by oleic acid (19–24%), palmitic acid (15–18%) and stearic acid (Ariffin *et al.*, 2009). Variation exists among genotypes in the relative content of these fatty acids. Linoleic acid was estimated to be 55.43% and 45.21% in white- and red-fleshed dragon fruit seed oil, respectively. The seed oil in dragon fruit predominantly contains polyunsaturated fatty acids, and these were higher in white-fleshed dragon fruit (55.94%) than in red-fleshed dragon fruit (45.70%). However, the saturated fatty acid and monounsaturated fatty acid (MUFA) content in white-fleshed dragon fruit seed oil was substantially lower in comparison with red-fleshed dragon fruit seed oil. The total amount of unsaturated fatty acids in the white-fleshed dragon fruit seed oil (75.62%) was higher than that of the red-fleshed dragon fruit seed oil (Liaotrakoon *et al.*, 2013a). The amount of essential fatty acids in dragon fruit seed oil is comparable with that present in maize oil (Lemcke-Norojarvi *et al.*, 2001), but is higher than that in olive oil

(Gurdeniz *et al.*, 2010; Krichene *et al.*, 2010) and sesame oil (Elleuch *et al.*, 2007). The predominant triacyglycerols (TAG) in white- and red-fleshed dragon fruit seed oils were found to be triunsaturated TAGs (21.60%) and diunsaturated TAGs (17.89%). However, a moderate amount of monounsaturated TAGs (4.56%) was also present (Liaotrakoon *et al.*, 2013a).

The seeds of dragon fruit are also rich in vitamin E, of which α-tocopherol is the most abundant form. The total vitamin E content in red-fleshed dragon fruit seed oil is significantly higher than in white-fleshed seed oil. In the red-fleshed dragon fruit, the γ- and α-tocopherol contents in seed oil were substantially higher than in the white-fleshed seed oil, while the δ-tocopherol content was equal for both seed oils. The total vitamin E content of the dragon fruit seed oil is also higher than that of olive oil and grape seed oil (Baydar *et al.*, 2007; Krichene *et al.*, 2010). Vitamin E is a potent natural antioxidant and is mostly found in most seed oils, which contain different levels of vitamin E activity. Assessment of antioxidative properties indicated that red-fleshed dragon fruit seeds exhibited higher activity than those of white-fleshed fruit (Chemah *et al.*, 2010). Besides vitamin E and fatty acids, dragon fruit seeds also contain insoluble dietary fibre such as cellulose, hemicellulose and lignin, which may originate from cell wall materials and seed coats (Tarpila *et al.*, 2005). The consumption of dragon fruit has been associated with positive health benefits due to the chemical components present in its pulp and seeds and their bioavailability. A focus on dragon fruit seeds as a prospective source of essential fatty acids, vitamin E and antioxidants may help in the development of valuable food components to aid in nutrition enrichment of human diets.

3.4 Peel

Fruit crops are unique in possessing nutritional value not only in the pulp but also in the peel. The peel is usually considered fruit waste, but due to the presence of bioactive compounds it may also be considered a valuable part of the fruit. Fruit waste causes environmental pollution, and hence converting the waste into wealth by harnessing its valuable compounds is gaining importance. In dragon fruit in particularly, the peel can be considered to have as much nutritional content as the pulp (Gondim *et al.*, 2005). The major functional compounds in dragon fruit peel are betalain pigments, phenolic compounds and dietary fibres, discussed below.

3.4.1 Pigment

Dragon fruit is unique in terms of the presence of betalain pigments, which are categorically present only in members of the order Caryophyllales. Dragon fruit species are reported to be the third largest source of betalains after beetroot and *Amaranthus* spp. (Wybraniec and Mizrahi, 2002; de Azeredo *et al.*, 2009; Silva *et al.*, 2012). Red-skinned dragon fruit such as *H. polyrhizus* and *H. undatus* contain the betacyanin pigment in their peel, while the yellow-skinned dragon fruit (*H. megalanthus*) possess betaxanthin (Cejudo-Bastante *et al.*, 2016). Comparative analysis of pigment content has determined that the total betacyanin content in red-fleshed dragon fruit peel is double the amount in white-fleshed dragon fruit peel (Liaotrakoon *et al.*, 2013b). The major betacyanin components in dragon fruit peel are betanin, phyllocactin, isobetanin and isophyllocactin (Fathordoobady *et al.*, 2016). However, the betacyanin profile of dragon fruit is distinguished from that of beetroot, where only betanin and isobetanin are prevalent (Nemzer *et al.*, 2011).

3.4.2 Phenolic compounds

Dragon fruit peel is a rich source of bioactive constituents such as phenols. It has a higher phenolic content and antioxidant activity than the pulp (Ruzlan *et al.* 2010). A wide variation in phenolic content has been reported across the various species of dragon fruit. Phenolic profiling of dragon fruit peel demonstrated the prominence of gallic acid, tyrosine, *trans*-caftaric acid, hydroxycinnamic acid derivative and myricetin-3-*O*-galactopyranoside. A higher level of phenol content has been reported in the peel of *H. undatus* than that of *H. polyrhizus* (Ruzlan *et al.*, 2010; Kim *et al.*, 2011). Additionally, *H. undatus* peel is also reported to have substantially higher phenol content when compared with the peel of *H. megalanthus* (Ferreres *et al.*, 2017). Abirami *et al.* (2021) reported a comparatively higher phenol and flavonoid content in peel than in pulp, while Fidrianny and Sahar (2014) also reported a high level of total phenol content and flavonoid content in peel extracts of *H. costaricensis*. Phenolic compounds contribute significantly to antioxidant capacity, which is higher in the peel than in the pulp, due to the higher phenol content in the peel. The antioxidative ability of phenolic compounds in dragon fruit peel is attributed to the high radical scavenging and reducing properties. The antioxidant capacity measured in terms of DPPH radical scavenging capacity demonstrated a linear correlation with total phenol content (Kim *et al.*, 2011) and was higher in *H. undatus* peel than in *H. polyrhizus* peel (Ruzlan *et al.*, 2010). The phenolic content in dragon fruit peel is much higher than that in passion fruit, peach and tomato (Contreras-Calderon *et al.*, 2011).

3.4.3 Dietary fibres

Dragon fruit peel is a rich source of dietary fibre. The ratio of insoluble dietary fibre (IDF) to soluble dietary fibre (SDF) is less than 3:1, which provides desirable physiological effects (Nguyen *et al.*, 2020). This indicates that dragon fruit peel is a good source of soluble dietary fibre, unlike certain cereals such as wheat bran and oat bran, which have a high IDF:SDF ratio, indicating a larger content of insoluble dietary fibre (Grigelmo-Miguel and Martin-Belloso, 1999). The dietary fibre in dragon fruit peel is almost comparable with maize bran, lemon peel, orange peel and apple peel, which are established popular sources of dietary fibre (Kahlon and Chow, 2000; Figuerola *et al.*, 2005; Ubando-Rivera *et al.*, 2005). Profiling of the dietary fibre present in dragon fruit peel has indicated the prevalence of rhamnose and galactose as the predominant monosaccharides together with galacturonic acid. The insoluble dietary fraction comprises xylose and galactose with lignin (Zhuang *et al.*, 2012).

Pectin is an important dietary fibre, and in dragon fruit peel, the pectin yield ranges from 10 to 26 g 100 g^{-1}, comparable with other important pectin sources such as orange pulp and apple pomace (Faravash and Ashtiani, 2007). The pectin content in *H. polyrhizus* has been reported to be 10.8–26.4% (Jamilah *et al.*, 2011; Muhammad *et al.*, 2014). In *H. costaricensis*, the pectin content in peel ranges from 18.4% to 23.1% (Tongkham *et al.*, 2017). Pectin also has a number of industrial uses. The pectin from dragon fruit peel exhibits similar textural properties to that in apple and citrus peels, which are common pectin sources commercially (Izalin *et al.*, 2016). Pectin is of particular importance in the food industry, and dragon fruit peel pectin was reported to have a similar property to citrus pectin in terms of oil- and water-holding ability. However, its viscosity is lower than that of apple and citrus pectin. This indicates the practical utility of dragon fruit pectin as a functional ingredient in moderately viscous foods and beverages (Muhammad *et al.*, 2014). The pectin extracted from dragon fruit peel also holds promise for applications in the drug and pharmaceutical industries; in particular, low methoxyl pectin with an esterification degree of 36% is reported to be useful in hydrogel formation, which is a potential drug carrier. Dragon fruit peel also contains several oligosaccharides (raffinose, stachyose and maltotriose); *H. polyrhizus* exhibited higher levels of these than *H. undatus* (Khalili *et al.*, 2014).

3.5 Uses

Dragon fruit is a highly perishable fruit crop species, and more attention is required during harvesting and handling. As the majority of the fruit is used for fresh consumption, there is scope for increased use of processed products (Fig. 3.4). In addition, the peel and seeds can be utilized for different purposes.

3.5.1 Pulp

The fruit is often relished due to its nutritive value, attractive appearance and abundance of bioactive compounds. Dragon fruit is widely commercialized and used extensively in the food industry. It is used in the preparation of juice, jelly, marmalade, jams, wine and beverages. The red-fleshed dragon fruit are mostly used in the preparation of various types of food products

Fig. 3.4. Various uses of dragon fruit.

and beverages because of their unique attractive colour. Fruit juice is a drink prepared by extraction of the natural liquid contained in fruit. Red-fleshed dragon fruit are rich in antioxidants and are used in preparation of fermented fruit juice (Jalgaonkar *et al.*, 2022). Fermented dragon fruit juice is reported to exhibit strong biological activity. A combination of fermented juice (10%) and fresh dragon fruit juice (90%) appeared to have wider consumer acceptability (Muhialdin *et al.*, 2020). Fermented red dragon fruit contains ten times more betacyanins than the fresh juice (Phebe *et al.*, 2009; Choo *et al.*, 2018). The fruit is also used for jelly making. However, dragon fruit has a low pectin content (0.20–1.04%) so pectin needs to be added for the preparation of jelly or jam. A fruit bar can be prepared from dragon fruit in combination with passion fruit. The use of passion fruit with dragon fruit in a bar imparts a unique taste and flavour and in turn improves the acidity, sensory properties and overall consumer acceptability of the final product (Yen *et al.*, 2022). The pulp of dragon fruit is also processed into fruit leather. The moisture content of fresh dragon fruit is very high (>89%), so preparation of fruit leather from the pulp by adding sugar to achieve a higher sugar concentration is an alternative way to prepare a nutritious and appealing product, which can be stored for a relatively longer period. Raj and Dash (2022) reported that dragon fruit leather prepared from dehydrated sheets of the pulp along with a hydrocolloid mixture such as xanthan gum had the maximal sensory attributes. The preparation of fruit spaghetti (carbohydrates with low fat) has also been explored from red-fleshed dragon fruit. Preparation of fruit vinegar from dragon fruit may also be a novel approach, as the fruit contains a number of nutrients. The presence of various bioactive compounds such as betalains, flavonoids and polyphenols has also increased the scope for wine production from dragon fruit. Dragon fruit wine is reported to be moderately clear and a brilliant yellow colour with a discreetly complex detectable aroma, a good balance of several detectable flavours, and a fair texture with a smooth and rich taste that remains in the mouth after drinking. The wine prepared from this exotic fruit has a good blend of sweet/sour/alcoholic taste and has gained unanimous acceptability (Dimero and Tepora, 2018).

The plant pigments (carotenoids, anthocyanin and betacyanin) serve as a natural food colourant. Dragon fruit are primarily popular due to the presence of betalain pigment, which is exclusively found in members of the order Caryophyllales. This pigment has a major advantage over other pigments such as carotenoids, which are fat soluble, as betalains are water soluble. Betalains also exhibit a higher stability over a wide pH range (pH 3–7) and are also stable with food items containing vitamin C. Hence, betalains are emerging as a potential natural colourant in comparison with other pigments. Dragon fruit is processed to produce a colouring powder used as a natural food additive (Moshfeghi *et al.*, 2013). The fruit is also being explored for the production of beauty products. The fruit pigment is used in formulating beauty products such as natural lipstick (Azwanida *et al.*, 2014).

The fruit contains citramalic acid, which is reported to be the dominant organic acid in the fruit. Citramalic acid has the potential to minimize facial skin wrinkles and appears to be promising for use in the cosmetics and medicine industries (Wu *et al.*, 2020).

3.5.2 Seed

Seeds of dragon fruit are a good source of fatty acid and vitamin E. They also contain antioxidant components. Dragon fruit oil may be considered as a 'new' potential oil (Liaotrakoon *et al.*, 2013a). The beneficial effects of the oil are related to its fatty acid composition. Linoleic acid is the most abundant fatty acid present in both white and red-fleshed dragon fruit. The dragon fruit seed oils are high in essential fatty acids particularly in white flesh, whereas seeds of red flesh contain a significant amount of natural antioxidants and vitamin E. Thus, the dragon fruit seed oils can be considered as high-value oils due to their oil composition.

3.5.3 Peel

Dragon fruit contains a substantial proportion of peel (30–35%), which is normally discarded during consumption and processing. The peel is a good source of betacyanins, pectin, dietary fibre and antioxidants. The peel pigments may be blended with other products to enhance their functional qualities. Betalain pigment has been extracted from dragon fruit peel using a microwave (Thirugnanasambandham and Sivakumar, 2017). However, better retention of the red colour is observed in spray drying compared with drum drying of the peel. Peel powder was obtained from freeze-dried peel, which was converted into alcohol-insoluble residues and subjected to rheological measurements. Owing to low values of the degree of esterification and pectin content, dragon fruit peel is not suitable for pectin extraction (Montoya-Arroyo *et al.*, 2014). Chia and Chong (2015) reported that dragon fruit peel powder can be considered a high total dietary fibre food, as it contains 6% fibre. Hence, it can be used as a functional ingredient in food items.

3.6 Conclusion

Dragon fruit is known for its nutritional significance as it possesses minerals, pigments and bioactive compounds. It also has high antioxidative properties. White- and red-fleshed dragon fruit are commonly cultivated species in the world. They differ in their compositional attributes. Red-fleshed dragon fruit is rich in pigments and minerals. The pulp is rich in fibre, minerals, betacyanins, organic acids and proteins and has antioxidative properties. As well as the pulp, the peel and seeds also have commercial significance. The peel is a rich source of betalain pigment and phenolic compounds, whereas the seeds are a rich source of fatty acids and vitamin E, and hence are considered high-value seeds. Essential fatty acids, namely linoleic acid and linolenic acid, constitute a significant percentage of the unsaturated fatty acids of the seed oil extract. In spite of their high nutritional significance, the potential of peel and seeds of dragon fruit are yet to be exploited.

3.7 References

Abirami, K., Swain, S., Baskaran, V., Venkatesan, K., Sakthivel, K. and Bommayasamy, N. (2021) Distinguishing three dragon fruit (*Hylocereus* spp.) species grown in Andaman and Nicobar Islands of India using morphological, biochemical and molecular traits. *Scientific Reports* 11, 1–14.

Al-Farsi, M., Alasalvar, C., Morris, A., Baron, M. and Shahidi, F. (2005) Compositional and sensory characteristics of three native sun-dried date (*Phoenix dactylifera* L.) varieties grown in Oman. *Journal of Agricultural and Food Chemistry* 53, 7586–7591.

Antia, F.P. and Abraham, P. (2006) *Clinical Dietetics and Nutrition*, 4th edn. Oxford University Press, New Delhi.

Ariffin, A.A., Bakar, J., Tan, C.P., Rahman, R.A., Karim, R. and Loi, C.C. (2009) Essential fatty acids of pitaya (dragon fruit) seed oil. *Food Chemistry* 114, 561–564.

Arivalagan, M., Karunakaran, G., Roy, T.K., Dinsha, M., Sindhu, B.C. *et al.* (2021) Biochemical and nutritional characterization of dragon fruit (*Hylocereus* species). *Food Chemistry* 353: 129426.

Azeredo, H.M.C., Santos, A.N., Souza, A.C.R., Mendes, K.C.B. and Andrade, M.I.R. (2007) Betacyanin stability during processing and storage of a microencapsulated red beetroot extract. *American Journal of Food Technology* 2, 307–312.

Azwanida, N.N., Normasarah, N. and Afandi, A. (2014) Utilization and evaluation of betalain pigment from red dragon fruit (*Hylocereus polyrhizus*) as a natural colorant for lipstick. *Jurnal Teknologi* 69, 139–142.

Baydar, N.G., Özkan, G. and Cetin, E.S. (2007) Characterization of grape seed and pomace oil extracts. *Grasas y Aceites* 58, 29–33.

Castro-Enríquez, D.D., Montaño-Leyva, B., Toro-Sánchez, D.C.L., Juárez-Onofre, J.E., Carvajal-Millán, E. *et al.* (2020) Effect of ultrafiltration of pitaya extract (*Stenocereus thurberi*) on its phytochemical content, antioxidant capacity, and UPLC-DAD-MS profile. *Molecules* 25: 281.

Cejudo-Bastante, M.J., Hurtado, N., Delgado, A. and Heredia, F.J. (2016). Impact of pH and temperature on the colour and betalain content of Colombian yellow pitaya peel (*Selenicereus megalanthus*). *Journal of Food Science and Technology* 53, 2405–2413.

Charoensiri, R., Kongkachuicha, R., Suknicom, S. and Sungpuag, P. (2009) Betacarotene, lycopene and alpha-tocopherol contents of selected Thai fruits. *Food Chemistry* 113, 202–207.

Chemah, T.C., Aminah, A., Noriham, A. and Wan Aida, W.M. (2010) Determination of pitaya seeds as a natural antioxidant and source of essential fatty acids. *International Food Research Journal* 17, 1003–1010.

Chen, Z., Zhong, B., Barrow, C.J., Dunshea, F.R. and Suleria, H.A. (2021) Identification of phenolic compounds in Australian grown dragon fruits by LC-ESI-QTOF-MS/MS and determination of their antioxidant potential. *Arabian Journal of Chemistry* 14: 103151.

Chia, S.L. and Chong, G.H. (2015) Effect of drum drying on physico-chemical characteristics of dragon fruit peel (*Hylocereus polyrhizus*). *International Journal of Food Engineering* 11, 285–293.

Choo, K.Y., Kho, C., Ong, Y.Y., Thoo, Y.Y., Lim, R.L.H. *et al.* (2018) Studies on the storage stability of fermented red dragon fruit (*Hylocereus polyrhizus*) drink. *Food Science and Biotechnology* 27, 1411–1417.

Choo, W.S. and Yong, W.K. (2011) Antioxidant properties of two species of *Hylocereus* fruits. *Advances in Applied Science Research* 2, 418–425.

Contreras-Calderon, J., Calderon-Jaimes, L., Guerra-Hernandez, E. and Garcıa-Villanova, B. (2011) Antioxidant capacity, phenolic content and vitamin C in pulp, peel and seed from 24 exotic fruits from Colombia. *Food Research International* 44, 2047–2053.

de Azeredo, H.M.C., Pereira, A.C., de Souza, A.C.R., Gouveia, S.T. and Mendes, K.C.B. (2009) Study on efficiency of betacyanin extraction from red beetroots. *International Journal of Food Science and Technology* 44, 2464–2469.

Delgado-Vargas, F., Jiménez, A.R. and Paredes-López, O. (2000) Natural pigments: carotenoids, anthocyanins, and betalains – characteristics, biosynthesis, processing, and stability. *Critical Reviews in Food Science and Nutrition* 40, 173–289.

Dimero, F.N. and Tepora, T.F. (2018) Processing and development of dragon fruit wine. *International Journal of Environment, Agriculture and Biotechnology* 3, 1943–1947.

Elleuch, M., Besbes, S., Roiseux, O., Blecker, C. and Attia, H. (2007) Quality characteristics of sesame seeds and by-products. *Food Chemistry* 103, 641–650.

Faravash, R.S. and Ashtiani, F.Z. (2007) The effect of pH, ethanol volume and acid washing time on the yield of pectin extraction from peach pomace. *International Journal of Food Science and Technology* 42, 1177–1187.

Fathordoobady, F., Mirhosseini, H., Selamat, J. and Manap, M.Y.A. (2016) Effect of solvent type and ratio on betacyanins and antioxidant activity of extracts from *Hylocereus polyrhizus* flesh and peel by supercritical fluid extraction and solvent extraction. *Food Chemistry* 202, 70–80.

Ferreres, F., Grosso, C., Gil-Izquierdo, A., Valentao, P., Mota, A.T. and Andrade, P.B. (2017) Optimization of the recovery of high-value compounds from pitaya fruit by-products using microwave-assisted extraction. *Food Chemistry* 230, 463–474.

Fidrianny, I. and Sahar, N. (2014). Evaluation of antioxidant activities from various extracts of dragon fruit peels using DPPH, ABTS assays and correlation with phenolic, flavonoid, carotenoid content. *International Journal of Research in Pharmaceutical Sciences* 5, 104–111.

Figuerola, F., Hurtado, M.A.L., Estevez, A.M.A., Chiffelle, I. and Asenjo, F. (2005) Fibre concentrates from apple pomace and citrus peel as potential fibre sources for food enrichment. *Food Chemistry* 91, 395–401.

García-Cruz, L., Dueñas, M., Santos-Buelgas, C., Valle-Guadarrama, S. and Salinas-Moreno, Y. (2017) Betalains and phenolic compounds profiling and antioxidant capacity of pitaya (*Stenocereus* spp.) fruit from two species (*S. pruinosus* and *S. stellatus*). *Food Chemistry* 234, 111–118.

Gibson, A.C. and Nobel, P.S. (1986) *The Cactus Primer*. Harvard University Press, Cambridge, Massachusetts.

Gondim, J.A.M., Moura, M.D.F.V., Dantas, A.S., Medeiros, R.L.S. and Santos, K.M. (2005) Centesimal composition and minerals in peels of fruits. *Food Science and Technology* 25, 825–827.

Grigelmo-Miguel, N. and Martin-Belloso, O. (1999) Comparison of dietary fibre from by-products of processing fruits and greens and from cereals. *LWT – Food Science and Technology* 32, 503–508.

Gunasena, H.P.M., Pushpakumara, D.K.N.G. and Kariyawasam, M. (2007) Dragon fruit *Hylocereus undatus* (Haw.) Britton and Rose. *Underutilized Fruit Trees in Sri Lanka* 1, 110–141.

Gurdeniz, G., Ozen, B. and Tokatli, F. (2010) Comparison of fatty acid profiles and midinfrared spectral data for classification of olive oils. *European Journal of Lipid Science Technology* 112, 218–226.

Hoda, M., Hemaiswarya, S. and Doble, M. (2019) Phenolic phytochemicals: sources, biosynthesis, extraction, and their isolation. In: *Role of Phenolic Phytochemicals in Diabetes Management*. Springer Nature, Singapore, pp. 13–44.

Hoyweghen, L.V., Beer, T.D., Deforce, D. and Heyerick, A. (2012) Phenolic compounds and anti-oxidant capacity of twelve morphologically heterogeneous bamboo species. *Phytochemical Analysis* 23, 433–443.

Hua, Q., Chen, C., Zur, N.T., Wang, H., Wu, J., *et al.* (2018) Metabolomic characterization of pitaya fruit from three red-skinned cultivars with different pulp colors. *Plant Physiology and Biochemistry* 126, 117–125.

Huang, Y., Brennan, M.A., Kasapis, S., Richardson, S.J. and Brennan, C.S. (2021) Maturation process, nutritional profile, bioactivities and utilisation in food products of red pitaya fruits: a review. *Foods* 10: 2862.

Izalin, M.N., Kharidah, M., Jamilah, B. and Noranizan, M. (2016) Functional properties of pectin from dragon fruit (*Hylocereus polyrhizus*) peel and its sensory attributes. *Journal of Tropical Agriculture and Food Science* 44, 95–101.

Jaafar, R.A., Rahman, A.R., Mahmod, N.Z. and Vasudevan, R. (2009) Proximate analysis of dragon fruit (*Hylecereus polyhizus*). *American Journal of Applied Sciences* 6: 1341.

Jalgaonkar, K., Mahawar, M.K., Bibwe, B. and Kannaujia, P. (2022) Postharvest profile, processing and waste utilization of dragon fruit (*Hylocereus* spp.): a review. *Food Reviews International* 38, 733–759.

Jamilah, B., Shu, C., Kharidah, M., Dzulkily, M. and Noranizan, A. (2011). Physico-chemical characteristics of red pitaya (*Hylocereus polyrhizus*) peel. *International Food Research Journal* 18, 279–286.

Jang, Y.K., Jung, E.S., Lee, H.A., Choi, D. and Lee, C.H. (2015) Metabolomic characterization of hot pepper (*Capsicum annuum* "CM334") during fruit development. *Journal of Agricultural and Food Chemistry* 63, 9452–9460.

Kahlon, T. and Chow, F. (2000) *In vitro* binding of bile acids by rice bran, oat bran, wheat bran, and corn bran. *Cereal Chemistry* 77, 518–521.

Kanner, J., Harel, S. and Granit, R. (2001) Betalains – a new class of dietary cationized antioxidants. *Journal of Agricultural and Food Chemistry* 49, 5178–5185.

Khalili, R.M.A., Abdullah, A.B.C. and Manaf, A.A. (2014) Isolation and characterization of oligosaccharides composition in organically grown red pitaya, white pitaya and papaya. *International Journal of Pharmacy and Pharmaceutical Sciences* 6, 131–136.

Kim, H., Choi, H.K., Moon, J.Y., Kim, Y.S., Mosaddik, A. and Cho, S.K. (2011) Comparative antioxidant and antiproliferative activities of red and white pitayas and their correlation with flavonoid and polyphenol content. *Journal of Food Science* 76, C38–C45.

Krichene, D., Allalout, A., Mancebo-Campos, V., Salvador, M.D., Zarrouk, M. and Fregapane, G. (2010) Stability of virgin olive oil and behaviour of its natural antioxidants under medium temperature accelerated storage conditions. *Food Chemistry* 121, 171–177.

Le Bellec, F., Vaillant, F. and Imbert, E. (2006) Pitahaya *(Hylocereus* spp.): a new fruit crop, a market with a future. *Fruits* 61, 237–250.

Lemcke-Norojarvi, M., Kamal-Eldin, A., Appelqvist, L.A., Dimberg, L.H., Ohrvall, M. and Vessby, B. (2001) Corn and sesame oils increase serum γ-tocopherol concentrations in healthy Swedish women. *Journal of Nutrition* 131, 1195–1201.

Liaotrakoon, W., de Clercq, N., van Hoed, V. and Dewettinck, K. (2013a) Dragon fruit (*Hylocereus* spp.) seed oils: their characterization and stability under storage conditions. *Journal of the American Oil Chemists' Society* 90, 207–215.

Liaotrakoon, W., de Clercq, N., van Hoed, V., van de Walle, D., Lewille, B. and Dewettinck, K. (2013b) Impact of thermal treatment on physicochemical, antioxidative and rheological properties of white-flesh and red-flesh dragon fruit (*Hylocereus* spp.) purees. *Food and Bioprocess Technology* 6, 416–430.

Lim, H.K., Tan, C.P., Karim, R., Ariffin, A.A. and Bakar, J. (2010) Chemical composition and DSC thermal properties of two species of *Hylocereus* cacti seed oil: *Hylocereus undatus* and *Hylocereus polyrhizus*. *Food Chemistry* 119, 1326–1331.

Lira, S.M., Dionísio, A.P., Holanda, M.O., Marques, C.G., da Silva, G.S. *et al.* (2020) Metabolic profile of pitaya (*Hylocereus polyrhizus* (FAC Weber) Britton & Rose) by UPLC-QTOF-MSE and assessment of its toxicity and anxiolytic-like effect in adult zebrafish. *Food Research International* 127: 108701.

Luo, H., Cai, Y., Peng, Z., Liu, T. and Yang, S. (2014) Chemical composition and *in vitro* evaluation of the cytotoxic and antioxidant activities of supercritical carbon dioxide extracts of pitaya (dragon fruit) peel. *Chemistry Central Journal* 8: 1.

Maestri, D.M., Nepote, V., Lamarque, A.L. and Zygadlo, J.A. (2006) Natural products as antioxidants. In: Imperato, F. (ed.) *Phytochemistry: Advances in Research*. Research Signpost, Kerala, India, pp. 105–135.

Mahattanatawee, K., Manthey, J.A., Luzio, G., Talcott, S.T., Goodner, K. and Baldwin, E.A. (2006) Total antioxidant activity and fiber content of select Florida-grown tropical fruits. *Journal of Agricultural and Food Chemistry* 54, 7355–7363.

Montoya-Arroyo, A., Schweiggert, R.M., Pineda-Castro, M., Sramek, M., Kohlus, R. *et al.* (2014) Characterization of cell wall polysaccharides of purple pitaya (*Hylocereus* sp.) pericarp. *Food Hydrocolloids* 35, 557–564.

Moreno, D.A., García-Viguera, C., Gil, J.I. and Gil-Izquierdo, A. (2008) Betalains in the era of global agri-food science, technology and nutritional health. *Phytochemistry Reviews* 7, 261–280.

Moshfeghi, N., Mahdavi, O., Shahhosseini, F., Malekifar, S. and Taghizadeh, S.K. (2013) Introducing a new natural product from dragon fruit into the market. *International Journal of Recent Research and Applied Studies* 15, 269–272.

Muhammad, K., Zahari, N.I.M., Gannasin, S.P., Adzahan, N.M. and Bakar, J. (2014) High methoxyl pectin from dragon fruit (*Hylocereus polyrhizus*) peel. *Food Hydrocolloids* 42, 289–297.

Muhialdin, B.J., Kadum, H., Zarei, M. and Hussin, A.S.M. (2020) Effects of metabolite changes during lacto-fermentation on the biological activity and consumer acceptability for dragon fruit juice. *LWT – Food Science and Technology* 121: 108992.

Nagasaka, R., Chotimarkorn, C., Shafiqul, I.M., Hori, M., Ozaki, H. and Ushio, H. (2007) Anti-inflammatory effects of hydroxycinnamic acid derivatives. *Biochemical and Biophysical Research Communications* 358, 615–619.

Nemzer, B., Pietrzkowski, Z. and Sporna, A. (2011) Betalainic and nutritional profiles of pigment-enriched red beet root (*Beta vulgaris* L.) dried extracts. *Food Chemistry* 127, 42–53.

Nguyen, S.N., Tu Ngo, T.C., Tran, T.T., Nguyet, T.N.M. and Man, L.V.V. (2020) Pasta from cellulase-treated wheat bran and durum semolina: effects of vital gluten addition and/or transglutaminase treatment. *Food Bioscience* 38: 100782.

Nomura, K., Ide, M. and Yonemoto, Y. (2005) Changes in sugars and acids in pitaya (*Hylocereus undatus*) fruit during development. *Journal of Horticultural Science and Biotechnology* 80, 711–715.

Nurliyana, R., Osman, I.S., Zahir, I.S., Suleiman, K.M., Aisyah, M.R. and Rahim, K.K. (2010) Antioxidant study of pulps and peels of dragon fruits: a comparative study. *International Food Research Journal* 17, 367–375.

Phebe, D., Chew, M.K., Suraini, A.A., Lai, O.M. and Janna, O.A. (2009) Red-fleshed pitaya (*Hylocereus polyrhizus*) fruit colour and betacyanin content depend on maturity. *International Food Research Journal* 16, 233–242.

Raj, G.B. and Dash, K.K. (2022) Development of hydrocolloids incorporated dragon fruit leather by conductive hydro drying: characterization and sensory evaluation. *Food Hydrocolloids for Health* 2: 100086.

Ruzlan, N., Idid, S.O., Idid, S.Z., Koya, M.S., Rehan, M.A. and Kamarudin, K.R. (2010). Antioxidant study of pulps and peels of dragon fruits: a comparative study. *International Food Research Journal* 17, 367–375.

Sandate-Flores, L., Rodríguez-Rodríguez, J., Velázquez, G., Mayolo-Deloisa, K., Rito-Palomares, M. *et al.* (2020) Low-sugar content betaxanthins extracts from yellow pitaya (*Stenocereus pruinosus*). *Food and Bioproducts Processing* 121, 178–185.

Sies, H., Stahl, W. and Sundquist, A.R. (1992) Antioxidant functions of vitamins: vitamins E and C, beta-carotene, and other carotenoids. *Annals of the New York Academy of Sciences* 669, 7–20.

Silva, L.R., Valentao, P. and Faria, J. (2012) Phytochemical investigations and biological potential screening with cellular and non-cellular models of globe amaranth (*Gomphrena globosa* L.) inflorescences. *Food Chemistry* 135, 756–763.

Singh, A., Swami, S., Panwar, N.R., Kumar, M., Shukla, A.K. *et al.* (2022) Development changes in the physicochemical composition and mineral profile of red-fleshed dragon fruit grown under semi-arid conditions. *Agronomy* 12: 355.

Stintzing, F.C. and Carle, R. (2004). Functional properties of anthocyanins and betalains in plants, food, and in human nutrition. *Trends in Food Science and Technology* 15, 19–38.

Stintzing, F.C., Schieber, A. and Carle, R. (2003) Evaluation of colour properties and chemical quality parameters of cactus juices. *European Food Research and Technology* 216, 303–311.

Strack, D., Vogt, T. and Schliemann, W. (2003) Recent advances in betalain research. *Phytochemistry* 62, 247–269.

Suh, D.H., Lee, S., Heo, D.Y., Kim, Y.S., Cho, S.K. *et al.* (2014) Metabolite profiling of red and white pitayas (*Hylocereus polyrhizus* and *Hylocereus undatus*) for comparing betalain biosynthesis and antioxidant activity. *Journal of Agricultural and Food Chemistry* 62, 8764–8771.

Tarpila, A., Wennberg, T. and Tarpila, S. (2005) Flaxseed as a functional food. *Current Topics in Nutraceutical Research* 3, 167–188.

Thirugnanasambandham, K. and Sivakumar, V. (2017) Microwave assisted extraction process of betalain from dragon fruit and its antioxidant activities. *Journal of Saudi Society of Agricultural Sciences* 16, 41–48.

Tongkham, N., Juntasalay, B., Lasunon, P. and Sengkhamparn, N. (2017) Dragon fruit peel pectin: microwave-assisted extraction and fuzzy assessment. *Agriculture and Natural Resources* 51, 262–267.

Ubando-Rivera, J., Navarro-Ocana, A. and Valdivia-Lopez, M. (2005) Mexican lime peel: comparative study on contents of dietary fibre and associated anti-oxidant activity. *Food Chemistry* 89, 57–61.

Vaillant, F., Perez, A., Davila, I., Dornier, M. and Reynes, M. (2005) Colorant and antioxidant properties of red-purple pitahaya (*Hylocereus* sp.). *Fruits* 60, 3–12.

Wichienchot, S., Jatupornpipat, M. and Rastall, R.A. (2010) Oligosaccharides of pitaya (dragon fruit) flesh and their prebiotic properties. *Food Chemistry* 120, 850–857.

Wu, L.C., Hsu, H.W., Chen, Y.C., Chiu, C.C., Lin, Y.I. and Ho, J.A. (2006) Antioxidant and antiproliferative activities of red pitaya. *Food Chemistry* 95, 319–327.

Wu, Y., Xu, J., He, Y., Shi, M., Han, X., Li, W., Zhang, X. and Wen, X. (2019) Metabolic profiling of pitaya (*Hylocereus polyrhizus*) during fruit development and maturation. *Molecules* 24: 1114.

Wu, Y., Xu, J., Shi, M., Han, X., Li, W., Zhang, X. and Wen, X. (2020) Pitaya: a potential plant resource of citramalic acid. *CyTA-Journal of Food* 18, 249–256.

Wybraniec, S. and Mizrahi, Y. (2002). Fruit flesh betacyanin pigments in *Hylocereus* cacti. *Journal of Agricultural and Food Chemistry* 50, 6086–6089.

Yen, T.T., Quan, T.H., Nhung, H.T.H., Tram, G.P.N., Karnjanapratum, S. and Benjakul, S. (2022) Development of antioxidative red dragon fruit bar by using response surface methodology for formulation optimization. *Applied Food Research* 2: 100173.

Zain, N.M., Nazeri, M.A. and Azman, N.A. (2019) Assessment on bioactive compounds and the effect of microwave on pitaya peel. *Jurnal Teknologi* 81, 11–19.

Zainoldin, K.H. and Baba, A.S. (2009) The effect of *Hylocereus polyrhizus* and *Hylocereus undatus* on physicochemical, proteolysis, and antioxidant activity in yogurt. *World Academy of Science, Engineering and Technology* 60, 361–366.

Zhuang, Y., Zhang, Y. and Sun, L. (2012) Characteristics of fibre-rich powder and antioxidant activity of pitaya (*Hylocereus undatus*) peels. *International Journal of Food Science and Technology* 47, 1279–1285.

4 Processing and Products

Prinya Wongsa[1]* and Sisir Mitra[2]
[1]*Food Science and Technology Programme, School of Agro-Industry, Mae Fah Luang University, Chiang Rai, Thailand;* [2]*Former Professor, Bidhan Chandra Krishi Viswavidyalaya (State Agricultural University), Mohanpur, West Bengal, India*

4.1 Introduction

Ripe dragon fruit (also known pitaya) is recognized as a source of bioactive compounds including polyphenols, pectin and dietary fibres, and provides various health benefits. However, the fruit are highly perishable and their shelf life is short under ambient conditions. In many countries, postharvest losses of fruit have been a major produce loss. Micro-, small- and medium-scale food-processing enterprises play an important role in processing a diversity of food products to reduce postharvest losses and food waste, extend the shelf life of food, and generate local and national incomes (Jalgaonkar *et al.*, 2020; Owino and Ambuko, 2021). Dragon fruit is usually harvested at full maturity stage. The fruit are widely consumed fresh as a dessert fruit because of their sensory attributes including appearance, texture and flavour. As well as consumption as a dessert fruit, the fruit flesh is used to prepare dried products. The ripe pulp is also used to prepare juice, fermented drinks, wine, jams, jellies and fruit leathers. The inedible parts of dragon fruit including the peel may be utilized for extraction of pectin, colourants and other products (Jalgaonkar *et al.*, 2020). Several of the products produced from dragon fruit are shown in Fig. 4.1.

4.2 Products from the Edible Parts of Dragon Fruit

4.2.1 Fresh-cut fruit

Fresh-cut dragon fruit is often available in markets in South-east Asia. The fresh produce is packed in a plastic tray and wrapped with plastic film. Sim *et al.* (2013) reported that *Salmonella* spp. could grow on fresh-cut dragon fruit under inappropriate storage conditions. Therefore, fresh-cut dragon fruit could be a potential vehicle for salmonellosis. Fresh-cut dragon fruit should be stored at 4°C to ensure the safety of the product, as well as to extend its shelf life. Moreover, the fresh produce should be consumed as soon as possible, especially when cold storage is not readily available. Trial shipments of peeled dragon fruit from Israel to Europe demonstrated the feasibility of exporting the fresh-cut product to distant markets (Goldman *et al.*, 2005).

The application of natural preservative and non-thermal treatments including green tea extract (tea:water ratio of 2.5 to 10), and cold plasma treatment (20 and 40 W) to the fresh-cut dragon fruit to protect against bacterial pathogens (*Escherichia coli*, *Salmonella typhimurium* and *Listeria monocytogenes*) was studied (Matan *et al.*, 2015). The change in pathogens

* Email: prinya@mfu.ac.th

DOI: 10.1079/9781800623156.0004

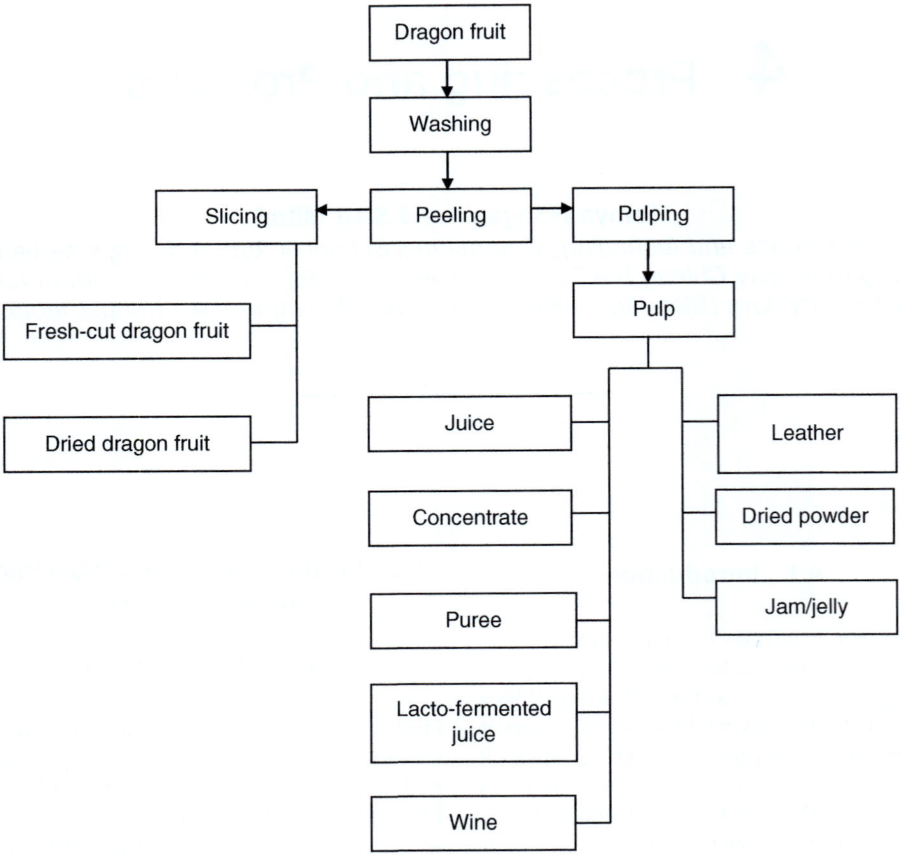

Fig. 4.1. Processing of dragon fruit and its products.

after plasma treatment of fresh-cut dragon fruit treated with green tea at 4 ± 1°C was also investigated. The results showed that atmospheric-pressure radiofrequency (RF) plasma at 40 W could extend protection against all pathogen growth on the surface of fresh-cut dragon fruit treated with 5.0% green tea for at least 15 days. Without the plasma treatment, green tea at 2.5–10.0% could not inhibit all bacterial growth. Therefore, the combination of green tea extract and atmospheric-pressure RF plasma treatment could inhibit the growth of pathogens on fresh-cut dragon fruit and extend its shelf life (Matan *et al.*, 2015).

An alternative method tested peppermint oil to inhibit surface mould and prolong the shelf life of dragon fruit during storage. Peppermint oil-adsorbed activated carbon at different concentrations (100–1000 ml l^{-1}) was placed with the dragon fruit in a storage box at 25 ± 2°C and 75 ± 5% relative humidity for 21 days. The result revealed that peppermint oil-adsorbed activated carbon at 700 ml l^{-1} could provide 100% inhibition of surface mould and fungi for more than 14 days of storage (the control started decaying at day 7) (Chaemsanit *et al.*, 2018).

4.2.2 Drying

Drying is the oldest food-preservation method aimed at removing moisture from food material. However, drying using heat results in inferior product quality due to degradation of colour and nutrients. Freeze-drying or lyophilization is a drying technique that gives dried products a superior quality due to the low-temperature processing (Ratti, 2001). During freeze-drying of foods, water is eliminated at low temperature under vacuum conditions by sublimation from a solid phase into a vapour phase (Marques *et al.*, 2007).

Dragon fruit can be successfully freeze-dried. Liaotrakoon *et al.* (2012) showed that freeze-dried pulp and peel of red-fleshed dragon fruit were a potential source of betacyanins, with a content of up to 360.14 mg 100 g^{-1}. The visual colour of freeze-dried dragon fruit after rehydration was similar to their original colour. Freeze-dried pulp of red-fleshed dragon fruit can be used as a natural colourant due to its relatively high content of betacyanin, its water solubility, its vitamin C (ascorbic acid) content and its colour stability under low-acid condition. The vacuum freeze-drying technology of dragon fruit was studied by Yong *et al.* (2017). Orthogonal experiments were designed to study the effects of different slice thicknesses, sublimation temperature and analytical temperature. Through comprehensive analysis, it was concluded that the optimal conditions for drying of fruit were: a thickness of 10 mm, a sublimation temperature of 25°C and an analytical temperature of 50°C. The product showed a sensory score of 95, a vitamin C content of 74 mg 100 g^{-1}, a protein content of 9 g 100 g^{-1}, a total dietary fibre content of 22.5 g 100 g^{-1} and a sugar:acid ratio of 64:37.

In contrast, thermal processing, based on the inactivation of microorganisms and enzymes, normally results in nutritional and sensory quality losses and rehydration is often poor. Hot-air drying is a conventional process used to preserve foods in which the solid to be dried is exposed to a continuously flowing hot stream of air, allowing moisture to evaporate (Nordin *et al.*, 2014).

Microwave-assisted air drying is a method whereby hot-air drying is combined with microwave heating in order to enhance the drying rate. Microwave-assisted air drying increases the drying rate and enhances the rehydration capacity of dried products and also overcomes shrinkage problems (Chandrasekaran *et al.*, 2013). The addition of microwave energy to the hot-air drying process enabled a reduction in drying time of red-fleshed pitaya slices by over 50% compared with simple hot-air drying; however, microwave vacuum drying reduced the drying time by 83% compared with simple hot-air drying. Microwave vacuum drying also retained 44.6% of the vitamin C content in the red-fleshed pitaya. Other than visual appearance, rehydration ratio and vitamin C content, no significant differences in dried product quality were observed among the samples dried under hot air, microwave/hot air and microwave vacuum drying (Nordin *et al.*, 2014). Several dried products (Fig. 4.2) and dried flower products and tea from the stigma and anthers of dragon fruit flowers (Fig. 4.3) have been developed.

Bhagya Raj and Dash (2020) prepared dragon fruit slices using microwave vacuum

Fig. 4.2. Dried products from dragon fruit. Photograph courtesy of Dr Mai van Tri.

drying of white-fleshed fruit. The slices were pre-treated with citric acid (0.5–1.5%) and then dried in a microwave vacuum dryer at different combinations of microwave power (200–600 W) and vacuum level (3–9 kPa). The optimized conditions for microwave power, vacuum level and citric acid concentration were found to be 450 W, 9 kPa and 1.35%, respectively. The predicted response under optimum conditions for total phenolic content, drying efficiency, colour change and rehydration ratio of dried dragon fruit slices was found to be 8.32 mg gallic acid equivalents (GAE) g^{-1} of dry weight, 59.10%, 14.58 and 3.46, respectively. Consequently, Bhagya Raj and Dash (2022) also investigated the effect of intermittent microwave convective drying (IMCD) on the quality of dried dragon fruit. Three levels of microwave power (200–600 W) and a temperature of 600°C for the hot air were examined. They observed that the total phenolic content of the dragon fruit slices obtained by IMCD ranged from 5.75 to 6.58 mg GAE g^{-1} of dry weight. Within the experimental range of process variables under IMCD conditions, the drying efficiency, colour change and rehydration ratios of the dragon fruit slices were 15.29–51.93%, 18.64–24.84 and 19.08–3.24, respectively. The results revealed that the IMCD method produced dried fruit slices with reduced colour changes and higher total phenolic content and rehydration ratio values.

Fig. 4.3. Tea prepared from dragon fruit stigma and anthers. Photograph courtesy of Dr Mai van Tri.

The drying process also plays an important role in the production of pitaya powder. Dried pitaya slices can be ground into a powder, which extends the shelf life. Yusof *et al.* (2011) reported that the initial moisture content of sliced pitaya was 85.16%, while the moisture content decreased to 5.77% after it had been made into a powder. The colour of the powder was redder than the pitaya slices. The water activity of the powder was 0.45, while the fruit had a value of 0.96. Thus, pitaya powder has the potential to be a natural colouring agent and health supplement.

4.2.3 Crisps

Dragon fruit crisps are quite popular because of their texture, colour, crunchiness and nutrition (Fig. 4.4). Yong *et al.* (2016) studied the effects of microwave (1.0, 2.0 and 4.0 W g^{-1}) and explosion puffing combination drying (MDx-EPD) on the physical and physicochemical properties of pitaya crisps. Compared with conventional hot-air explosion puffing, MDx-EPD-dried crisps showed faster rehydration rates, as well as decreasing the glass transition temperature (T_g). In addition, freeze-drying of dragon fruit crisps

Fig. 4.4. Dragon fruit crisps.

followed by explosion puffing drying (FD-EPD) showed better crisp quality in terms of texture, colour and nutrition (Yong *et al.*, 2017). However, less discoloration due to browning was observed when hot-air-dried explosion puffing drying was carried out. The improvements in texture and colour in FD-EPD dried crisps were confirmed by sensory evaluation.

4.2.4 Dried powder

Dried dragon fruit powder was prepared by mixing dragon fruit juice with different maltodextrin concentrations (20% and 30%) and at different inlet air temperatures of 145°C, 155°C, 165°C and 175°C, respectively (Tze *et al.*, 2012). An increase in inlet air temperature resulted in a decrease in moisture content and water activity at the higher concentration of maltodextrin. The best condition for spray-drying in terms of betacyanin content was an inlet air temperature of 155°C and a maltodextrin concentration of 20%. From a nutritional point of view, the dragon fruit powder was found to be rich in protein, fat, ash, fibre and antioxidants.

Yusof *et al.* (2020) standardized a method for spray-drying of white-fleshed dragon fruit powder using maltodextrin as the wall material. The inlet air temperatures (140°C, 150°C and 160°C), outlet air temperatures (75°C, 80°C and 85°C) and maltodextrin concentrations (20%, 25% and 30%) were tested as independent variables. Process yield, moisture content, water activity, solubility, hygroscopicity and bulk density of the powders were analysed as responses. The process yield increased significantly ($P < 0.05$) with increasing inlet air temperature and decreasing maltodextrin concentration. The combination of an increase in outlet air temperature and a decrease in maltodextrin concentration significantly ($P < 0.05$) reduced the moisture content and water activity of the white-fleshed dragon fruit powder. Powder solubility and hygroscopicity increased significantly ($P < 0.05$) as inlet and outlet air temperature increased. Bulk density values decreased as the inlet and outlet air temperature increased. The optimum conditions for spray-drying of white dragon fruit powder that would produce a high yield, low moisture content, low water activity, high solubility, low hygroscopicity and high bulk density were determined to be 150–153°C inlet temperature, 80–82°C outlet temperature and 18–20% maltodextrin (Yusof *et al.*, 2020). Dragon fruit powder produced from dragon fruit juice has potential use as a natural colouring agent and health supplement. Dragon fruit colouring powder has been used as a natural food additive to colour rice, milk, yoghurt, juice and pastry (Moshfeghi *et al.*, 2013).

Anthocyanins are one of the commonly used natural colourants for beverages, confectionary, fruit preparation and ice cream. The instability of anthocyanins is influenced by light, pH, temperature and oxygen during processing. Zaidel *et al.* (2015) carried out experiments to encapsulate the anthocyanins extracted from red-fleshed dragon fruit using maltodextrin as the wall material and the microwave-assisted technique with different drying times. The encapsulation drying time varied from 1 to 3 min with fixed power at 330 W. The results indicated that 2 min encapsulation drying time showed the highest yield of encapsulation, whereas 3 min showed the highest encapsulation efficiency, followed by 2 min with 4% differences. From this study, it was observed that encapsulation at 2 min drying time was the most feasible encapsulation process. Thermal studies at 60°C and 80°C showed that anthocyanins degraded faster at the higher temperature (80°C) for all capsule types. Dragon fruit peel (40% peel extract) may be used for salt fortification, as well as iodized salt. Fortification of salt with dragon fruit peel has the ability to bind vitamin C, which can be used as an alternative source of vitamin C (Maflahah and Asfan, 2020).

4.2.5 Biscuits

Biscuits (Fig. 4.5) were prepared by mixing pitaya powder at 30%, 40%, 50% and 60% (w/w) with wheat flour and using the traditional creaming method (Pawde *et al.*, 2020). Biscuits fortified with 50% pitaya powder showed an increased fibre content of about fivefold (7.81%), increased minerals (0.91%) and an improved spread ratio (8.76%) when compared with wheat flour biscuits. A 100 g daily consumption of fortified biscuits may meet the 20% recommended daily intake requirement for dietary fibre. The effects of pitaya peel supplementation

Fig. 4.5. Biscuits prepared using dragon fruit powder. Photograph courtesy of Dr Mai van Tri.

in the biscuit recipe on the nutritional quality, *in vitro* glycaemic index and antioxidant release from enriched fibre cookies were investigated (Mai *et al.*, 2023). The higher the ratio of pitaya peel powder in the recipe, the greater the dietary fibre, betacyanin and phenolic contents, and antioxidant activities of the product. Biscuits supplemented with 10–25% pitaya peel powder were classified as having a low glycaemic index, ranging from 51.9 to 45.7, compared with the glucose reference. The release of betacyanins, phenolics and antioxidant activities from the biscuit sample was recorded in the salivary, gastric, intestinal and colon steps of *in vitro* digestion, and gastric digestion showed the highest release of antioxidant content and activity. The increased pitaya peel powder ratio in the biscuit recipe improved the antioxidant activities of the aqueous fractions at the four digestive steps.

4.2.6 Juice

Fresh dragon fruit juice can be prepared by conventionally pulping flesh and seeds together using a kitchen blender. In addition, commercially, dragon fruit juice has been produced from fresh fruit using commercial enzymes such as Pectinex Ultra SP-L and Pectinex CLEAR. There are differences in sensory quality and chemical composition in fresh fruit pulp and juice produced with and without enzyme treatment. Dragon fruit juice produced after enzyme treatment and pasteurization did not shown marked changes in chemical parameters such as moisture, ash, fat, carbohydrate and energy. Enzyme treatment using Pectinex CLEAR increased the protein content by 2.2% (w/w) from 0.23% (w/w) in raw juice (Nur 'Aliaa *et al.*, 2011). The vitamin C content in the treated samples was reduced slightly because of the heat treatment. However, the phenolics content was slightly higher, up to 7%, suggesting that dragon fruit juice is richer in antioxidant capacity than the unprocessed fruit. Therefore, the application of pectinases affected several properties of red-fleshed dragon fruit juice and promotes the potential of dragon fruit juice as a functional beverage (Nur 'Aliaa *et al.*, 2011).

A high extraction yield (86.35%) and reduced sugar content in the juice (28.87 mg ml^{-1}) from red-fleshed dragon fruit pulp were reported at 40°C for 120 min and with a combination of Pectinex Ultra SP-L and Viscozyme L (70:30, v:v), while extraction without enzyme treatment resulted in a lower yield (54.04%) and less of a reduction in sugar content in the juice (20.63 mg ml^{-1}) (Phuong and Tuan, 2016). The application of enzymes also increased the total acidity (from 0.47 to 0.75 g 100 ml^{-1}) and vitamin C content (from 27.94 to 32.29 mg 100 g^{-1}).

Red-fleshed pitaya juice is attracting increasing attention due to its nutritional functionalities, special purple-red colour, and soft, bright and fresh taste (Zhu *et al.*, 2021). Thermosonication (TS) is an energy-efficient technology combining ultrasonication with moderate heat. TS preserved the colour of red-fleshed pitaya juice better than heat treatment alone (Liao *et al.*, 2020). Zhu *et al.* (2021) studied the effects of TS on the microbial safety and quality of red-fleshed pitaya juice during storage. Freshly prepared red-fleshed pitaya juices were thermosonicated at 475 and 560 W for 20 min. Following TS processing, native microbiota including aerobic bacteria, yeasts and moulds were reduced to fewer than 10 colony-forming units ml^{-1}. Growth during storage was slow and equal to heat-processed (83°C, 15 min) samples. During storage at 4°C for 28 days, the soluble solid content, pH, activities of polyphenol oxidase and peroxidase, and browning degree remained unchanged. A visible colour decay was observed in TS-processed juice at day 10, mainly resulting from decomposition of betacyanins and the growth of residual native microbiota. Compared with heat-treated juice, better colour retention was obtained by TS treatment. One of the industrial challenges in red-fleshed pitaya juice production is to enhance the extraction of functional compounds such as phytochemicals from the fruit flesh (Tan *et al.*, 2023). Radziejewska-Kubzdela *et al.* (2020) observed that pretreatment of the flesh with ultrasound, enzymes or heat often results in the extraction of these compounds in the juice. The levels of organic acids, phenolics and anthocyanin in red-fleshed pitaya juice increased after 60 min of ultrasonic pretreatment (Tan *et al.*, 2023).

A pitaya suspension beverage can be prepared using pitaya pulp, citric acid, white sugar, xanthan gum, sodium carboxymethyl cellulose (CMC-Na) and carrageenan. The best composition was suggested as 15% pitaya pulp, 6% sugar, composite stabilizer (0.15% sodium carboxymethyl cellulose, 2% xanthan gum) and 0.12% citric acid. The product had moderate fruit grain, a good suspension effect, a sweet and sour taste, and the aroma of pitaya (Huajing *et al.*, 2020).

Lacto-fermented juice

Fermentation of dragon fruit juice was carried out using lactic acid bacteria such as *Lactobacillus plantarum* FBS05 for 48 h at 37°C (Muhialdina *et al.*, 2020). The antibacterial activity of the fermented dragon fruit juice was 95.52%, 95.63%, 96.05% and 93.38% towards *E. coli*, *Salmonella typhimurium*, *Pseudomonas aeruginosa* and *Staphylococcus aureus*, respectively. However, the antioxidant activity results showed no significant differences for fermented and non-fermented juice. The lacto-fermentation of dragon fruit juice extended the shelf life for 3 months (Muhialdina *et al.*, 2020). Ong *et al.* (2012) applied phenotypic and genotypic methods to confirm the identity of lactic acid bacteria present in fermented red-fleshed dragon fruit beverages. A total of 21 members of the genus *Enterococcus* were isolated and characterized based on their biochemical characteristics. Choo *et al.* (2018) studied the effect of storage temperature and duration on the stability of fermented red-fleshed dragon fruit drink on its betacyanin content, physicochemical and microbiological qualities and its sensory acceptability. The findings showed that both storage temperature and duration have a significant effect on the betacyanin content and physicochemical properties. There was a 15.73% loss of betanin (from 16.53 to 13.93 g l^{-1}) at 4°C with no significant changes in the physicochemical properties from week 2 onwards compared with a 56.32% betanin loss (from 16.53 to 7.22 g l^{-1}) at 25°C. At week 8, fermented red-fleshed dragon fruit drink stored at 4°C still contained 13.93 g l^{-1} of betanin with a pH value of 3.46 (Choo *et al.*, 2018).

Carbonated juice

A carbonated drink can be produced from red-fleshed dragon fruit (Fig. 4.6). Red-fleshed dragon fruit was hydrolysed using pectinase enzymes at different temperatures (35°C, 40°C, 45°C and 50°C), hydrolysis times (1, 2, 3 and 4 h), concentrations of pectinase enzyme (0.4%, 0.6%, 0.8% and 1.0%) and enzyme pectinase content (0.2%, 0.4%, 0.6% and 0.8%) (Mai *et al.*, 2022). Carbonated dragon fruit juice provided a higher polyphenol content, vitamin C content and 2,2-diphenyl-1-picrylhydrazyl (DPPH) scavenging activity. These results are expected to aid in

Fig. 4.6. Dragon fruit juice from white-fleshed (left) and red-fleshed (right) dragon fruit.

diversification of products from dragon fruit raw materials to meet the increasing demand of consumers. Jalgaonkar *et al.* (2018) prepared a dragon fruit-based ready-to-serve drink using dragon fruit (60–80%, v:v), grape juice (0–10%, v:v) and sugar syrup (2–6% v:v). The juice concentrations were optimized using response surface methodology. The optimum juice percentages obtained for the best blend formulation were 70% dragon fruit, 5% grape juice and 3% sugar syrup. The beverage was found to be acceptable in terms of its nutritional value and overall acceptability (Jalgaonkar *et al.*, 2018).

4.2.7 Wine

Dragon fruit is widely used to produce a value-added wine and liqueur from the flesh of the fruit (Fig. 4.7). Dam (2013) suggested inoculation of *Saccharomyces ovifomis* 50% (v:v) and *S. vini* 50% (v:v). The produced wine had 12% (v:v) ethanol content, pH 3.48 and 6.27°Brix. Wine has also been fermented from red-fleshed dragon fruit juice by alcoholic fermentation with different yeasts including *Saccharomyces cerevisiae* EC-1118, *Torulaspora delbrueckii* Biodiva and *Lachancea thermotolerans* Concerto (Jiang *et al.*, 2020a). There were similar alcoholic fermentation rates in terms of sugar utilization and ethanol production (8–9%, v:v) by these three yeast strains. The yeast *T. delbrueckii* produced a significantly lower amount of glycerol and acetic acid, while *L. thermotolerans* produced more lactic and succinic acids. In terms of volatile compounds, *S. cerevisiae* produced the highest levels of esters, while *T. delbrueckii* produced more alcohols, isoamyl acetate and terpenes. The alcoholic fermentation caused significant degradation of the betacyanin pigments and total phenolic compounds but better retention of antioxidant activity and colour stability for wines fermented using *L. thermotolerans* and *T. delbrueckii* compared with *S. cerevisiae.* These results indicated that it is possible to use pure non-*Saccharomyces* yeast to produce red-fleshed dragon fruit wine for the market.

The high pectin content in red-fleshed dragon fruit presents a challenge in industrial processing. Pretreatment with pectinase of red-fleshed dragon fruit wine fermented with *T. delbrueckii* is considered necessary (Jiang *et al.*, 2020b). Pectinase enzyme (Pectinex Ultra SP-L), added at 0.1% (v:v) was applied after pasteurization of the juice followed by fermentation with *T. delbrueckii* at 20°C for 14 days. Pectinase pretreatment did not affect yeast growth or the production of ethanol (8%, v:v) and glycerol (6 g l^{-1}) but significantly increased the juice/wine yield by 16% (v:v). Moreover, fermented pectinase-treated samples possessed aroma compound profiles with enriched esters and terpenes but decreased higher alcohols. In addition, pectinase

Fig. 4.7 (A) Wine and (B) liquor from dragon fruit flesh. Photograph courtesy of Dr Mai van Tri.

treatment increased the total phenolic content but decreased the betacyanin content and colour intensity compared with the control sample.

4.2.8 Concentrate

The effects of heat pasteurization, pH adjustment and ascorbic acid (vitamin C) addition as well as storage under agitation and under light or dark conditions on betacyanin content in red-fleshed dragon fruit juice and concentrate have been studied (Wong and Siow, 2015). The concentrate was produced by concentrating clarified red-fleshed dragon fruit juice in a rotary evaporator at 40°C. Addition of 0.25% ascorbic acid, adjusting the pH to 4.0 and pasteurization at 65°C for 30 min were the best processing conditions to retain the betacyanin content in red-fleshed dragon fruit juice. The ascorbic acid stabilized the betacyanin in both the juice and concentrate under agitated and non-agitated conditions, while light degraded the betacyanin content in both the juice and concentrate (Wong and Siow, 2015).

4.2.9 Puree

White-fleshed and red-fleshed dragon fruit purees were subjected to heat treatment for 0–60 min at temperatures between 50°C and 90°C (Liaotrakoon *et al.*, 2013a). The physicochemical values and antioxidative properties of the unheated red-fleshed purees were significantly higher than those of the unheated white-fleshed purees. The betacyanin content of the red-fleshed puree decreased to 32.35% after treatment at 90°C for 60 min, based on the initial puree. Thus, heated dragon fruit purees, particularly from red-fleshed fruit, offer the possibility of application in foodstuffs due to their interesting physicochemical properties and nutritional attributes after thermal treatment.

4.2.10 Fruit leather

The development of dragon fruit leather with different hydrocolloids including gellan gum, xanthan gum and guar gum was studied (Jalgaonkar *et al.*, 2020). The prepared fruit puree mixed with hydrocolloid was dried by conductive hydro-drying at three different temperatures of water (70°C, 80°C and 90°C) to form the fruit leather. The total phenolic content and antioxidant activity of the leather formed at 90°C were in the range of 4.09–5.03 mg GAE g^{-1} of dry weight and 73.68–82.47%, respectively. Based on the preferences of the sensory panellists, the average preference score suggested that dragon fruit leather prepared with xanthan gum was preferred over the other hydrocolloids.

4.2.11 Jelly and jam

Dragon fruit jelly was prepared using a standard formulation with 0.05%, 0.1% and 1.5% pectin (Islam *et al.*, 2011). The products were stored at ambient temperature (27–34°C) for a period of 6 months. During storage, changes in colour, flavour, odour, total soluble solids and pH were studied. No special change in ingredients was found and little change of pH was observed during the first 4 months of storage. Jelly prepared using 1.5% pectin showed the highest score for colour, flavour, turbidity and overall acceptability. Panchal *et al.* (2018) reported that dragon fruit jelly prepared from 1000 ml of fruit extract, 550 g of sugar and 11 g of pectin showed the highest scores on sensory quality. The prepared jelly was bottled in pasteurized polyethylene terephthalate (PET) bottles (500 ml capacity) and stored at ambient (30 ± 2°C) and refrigerated (5 ± 2°C) temperatures for up to 90 days. During storage, the jelly total soluble solids, titratable acidity, reducing sugars and total sugars increased, but the moisture content, ascorbic acid and organoleptic quality slightly decreased with increased storage period. Microbial growth was slightly increased under both storage conditions but was within the specification limit at the end of the storage period. Therefore, the prepared jelly was safe and suitable for consumption for up to 3 months. The shelf life of red-fleshed dragon fruit jam was predicted using the Arrhenius model based on the accelerated shelf life test (Zaidiyah *et al.*, 2019). The fruit jam was prepared using pectin as the stabilizer and with 10 min of cooking time. The zero-order kinetics model was the most appropriate model to represent the change of total plate count values, and the first-order model was best for the change in antioxidant activity. According to the chosen models, the shelf life of red dragon fruit jam during storage at 28°C was estimated to be 99 days.

Jam (Fig. 4.8) is a popular food in many countries and makes it possible to consume a particular fruit out of season. The addition of peel to dragon fruit jam showed the potential to produce a product with no waste (Castro *et al.*, 2021). As well as the advantage of reducing the amount of waste from personal consumption and industries, the beneficial properties of dragon fruit peel can result in a better taste. It is also a considerable source of antioxidant compounds such as vitamins B_1, B_5 and B_6, as well as providing sensory features that were considered pleasant by consumers within food safety and quality standards.

4.2.12 Yoghurt

Fruit yoghurt is increasingly becoming popular because it provides probiotics and prebiotics, as well as nutritious value. Yoghurts have been produced from white-fleshed and red-fleshed dragon fruit, using mashed pulp at different concentrations (10%, 20% and 30%) (Zainoldin and Baba, 2009). The probiotic admixture produced by Chr. Hansen (Denmark) (consisting of *Lactobacillus thermophilus*, *L. casei*, *L. bifidus* and *L. acidophilus*) and Biolife Advanced Multiblend PROBIOtix (consisting of *Lactobacillus acidophilus*, *L. casei*, *L. bulgaricus*, *L. rhamnosus*, *Bifidobacterium infantis*, *B. longum* and *B. bifidum*) along with yoghurt starter culture were used. The pH of the white-fleshed dragon fruit decreased from 4.03 to 3.98 and that of red-fleshed dragon fruit from 4.01 to 3.95 by increasing the fruit concentration from 10% to 30% (v:v) in the yoghurt. In both white- and red-fleshed yoghurt, the taste, texture and flavour development of the yoghurt was due to proteolysis. Jayasinghe *et al.* (2015) prepared white-fleshed yoghurt using starter cultures from Chr. Hansen and 10.0% fruit, 10% sugar and 8% gelatin. The pH of the yoghurt decreased from 4.06 to 3.9. The developed yoghurt comprised 23.58% total solids, 9.64% non-fat solids and 3.2% fat. Santana *et al.* (2015) added quinoa (*Chenopodium quinoa*) and sucralose along with the white-fleshed dragon fruit pulp. The resultant yoghurt had a pH of 4.99, 14.4°Brix, 0.8% acidity (lactic acid equivalent), 88% moisture, 5% protein, 1.88% fibre, 1.31% ash, 10.55% carbohydrate and 2.38% lipids. In addition, it constituted a functional prebiotic because of the high content of dietary fibre.

Probiotic yoghurt has good healthcare functions and is also rich in nutrients. Sun *et al.* (2018) fermented dragon fruit yoghurt with a single strain of *Lactobacillus plantarum*. After evaluation, they reported that the optimal formulation of probiotic dragon fruit yoghurt was 12% dragon fruit juice, 11% white sugar and 2% inoculum of *L. plantarum* with a fermentation time of 19 h at 37°C.

Dragon fruit flesh is also used for flavouring ice cream (Fig. 4.9), which is very popular in Vietnam.

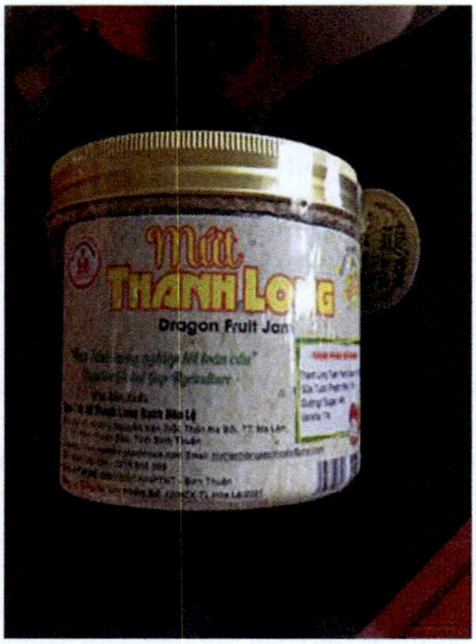

Fig. 4.8. Dragon fruit jam. Photograph courtesy of Dr Mai van Tri.

4.3 Products from Non-edible Parts of Dragon Fruit

Processing of the non-edible parts of dragon fruit is also important to produce value-added products. Jam can be prepared from the skin of dragon fruit by adding 0.3% pectinase and 0.2% pectin (w:w) (Dam, 2013). The product had a good structure, colour and taste. The stem of the

Fig. 4.9. Dragon fruit flavoured ice cream. Photograph courtesy of Dr Mai van Tri.

dragon fruit plant can be used as raw material for making soft drinks (raw material and water at a ratio of 1:4, w:w) by incubation with 0.25% Pectinex Ultra SP at 45–55°C, pH 4.5, for 2 h and then blending at a ratio of 10.4% sugar, 0.1% citric acid, 0.03% kiwi flavour, followed by pasteurization at 95°C for 10 min. Tea has also been produced from the flower remnants of dragon fruit, using 10% dried dragon fruit flower, 0.2% liquorice, 0.08% citric acid, 0.03% aroma, 9% sugar and 80.69% water. The tea infusion was a brownish-yellow colour and had a better taste and aroma than conventional tea made using *Camellia sinensis*.

4.3.1 Food additives

The addition of red-fleshed dragon fruit peel extracts to beef sausages was tested (Manihuruka *et al.*, 2017). The fruit peel extracts were obtained by maceration using solvent at pH 5. The results showed that antioxidant and antimicrobial activities were associated with high levels of phytochemical compounds and total phenols in the extract. Red dragon fruit peel extracts at various percentages (0%, 20%, 30% and 40%) were added to beef sausages, and their physicochemical characteristics, nutrients, antioxidant activity and microbiological profile analysed. The results showed that the addition of red dragon fruit peel extracts significantly reduced the texture values but increased the intensity of red colour. It was concluded that red dragon fruit peel extract containing phytochemical compounds was effective as an antibacterial agent and natural antioxidant. The addition of red dragon fruit peel extracts was effective in increasing the antioxidant activity and decreasing thiobarbituric acid reactive substances.

Dragon fruit peel powder has been produced and used as an antioxidant dietary fibre at concentrations of 1.5% and 3.0% to alter the quality and susceptibility to lipid oxidation of chicken nuggets during 20 days in refrigerated storage (Madane *et al.*, 2020). The dragon fruit peel contained dietary fibre (56.91%) with high levels of insoluble dietary fibre and phenolics (36–39 mg GAE 100 g^{-1}) and possessing good radical-scavenging activity. High-performance liquid chromatography analysis revealed the presence of high concentrations of gallic and ferulic acid. Incorporation of dragon fruit peel in nuggets caused a decrease in pH but improved emulsion stability and cooking yield, and resulted in higher protein and ash and lower fat content. The treated nuggets had significantly higher dietary fibre and total phenolics content than the controls. Dragon fruit peel significantly decreased lipid peroxidation, odour scores and microbial load in the chicken nuggets during 20 days of storage. Therefore, antioxidant dietary fibre can be used in animal foods without affecting the quality and acceptability of products.

Ground dragon fruit peel was produced using drum drying at 1 rpm with 2 bars of steam pressure (Chia and Chong, 2015). The betacyanin content was found to be twofold higher in drum-dried powder (41.55 mg g^{-1} of dry weight) than in fresh sample (80.21 mg g^{-1} of dry weight). The drying technique was found to preserve up to 98.62% of the total phenolic content, and there was a reduction of 3.32 mg Trolox equivalents g^{-1} of dry matter in the radical-scavenging activity. The density of the powder was 0.1315 g ml^{-1}, with 51.44% soluble in water. The dried dragon fruit peel had a water-holding capacity of 2.52 g g^{-1} of sample, an oil-holding capacity of 3.56 g of oil g^{-1} of sample and a swelling capacity of 6.23 ml g^{-1}. Therefore, dried dragon fruit peel can be considered a potential functional ingredient.

4.3.2 Food colourant

Dragon fruit skin contains pigments including betacyanins and anthocyanins. Dragon fruit skin waste can be utilized to produce food colourant that contains bioactive compounds. Utilization of agricultural waste has enabled additional profit for dragon fruit farmers and the juice industry. Dragon fruit peel extract powder can be produced by mixing the skin extract with maltodextrin (10–90%, w:w) before freeze-drying (Shofinita *et al.*, 2020). An increase in maltodextrin concentration in the freeze-dried feed resulted in a reduction in the bioactive contents in the dried samples. The moisture content of the powder with the addition of 90% maltodextrin was found to be 9.3 ± 3.0%, compared with 26.5 ± 6.0% for powder without maltodextrin addition. High recoveries of bioactive compounds during freeze-drying were observed (>60%).

Alginate microspheres loaded with betacyanins from the peel of red dragon fruit have been studied for their storage and heat stability (Fathordoobady *et al.*, 2013). The investigation of simulated gastrointestinal release showed that the alginate-loaded betacyanin extract contained higher amounts of antioxidants. Both the half-life and total betacyanin retention in the microspheres during storage were improved in comparison with non-capsulated extract and commercial betanin solution (control). Alginate microspheres loaded with betacyanins are effective models for food colourant additives and oral delivery purposes.

4.3.3 Pectin

Dragon fruit peel from the juice industry is an alternative source of pectin for application in jam processing. Hunter L^*, a^* and b^* values of the dragon fruit peel pectin were determined as 39.95, 34.87 and 20.18, respectively (Nur Izalin *et al.*, 2016), and it was light red in colour compared with commercial apple and citrus pectin. The water-holding capacity of the pectin was similar to that of apple pectin (5.50 and 5.45 g g^{-1}, respectively). Its oil-holding capacity and swelling capacity also showed no significant difference from that of citrus pectin ($P > 0.05$). It was successfully used in pineapple jam but at a high percentage of up to 2% to make it set well. In terms of sensory attributes, no significant differences were observed between the mean scores of the jam produced using dragon fruit peel pectin and apple pectin except for the colour attribute ($P < 0.05$). Therefore, pectin obtained from dragon fruit peel can be used as a thickener in food products such as low-viscosity foods and beverages.

Different contents of pectin are extracted from the different fractions of red-fleshed dragon fruit peel. The highest pectin yield (26.38% on a dry-weight basis) was obtained from the fresh inner layer of the peel by extraction using 1% citric acid at 73°C for 67 min, and a sample:citric acid ratio of 1:4 (w:v). This pectin also demonstrated the highest degree of esterification (63.74%) when compared with pectin from other fractions of the dragon fruit peel. The calculated degree of esterification confirmed that the extracted pectin was high in methoxyl pectin. The monosaccharide composition was determined using high-performance liquid chromatography and revealed that the pectin was predominantly composed of galacturonic acid (39.11%), followed by moderate concentrations of mannose, rhamnose, galactose and glucose, and minor amounts of xylose and arabinose.

Chua *et al.* (2020) studied the pretreatment and extraction process of pectin from red-fleshed dragon fruit skin. The optimum oven-drying temperature was suggested as 45°C with the highest pectin yield (6.27%) and the lowest ash content (8.34%). The degree of esterification of the pectin pre-optimization (36.36%) and post-optimization (33.27%) characterized it as a low-methoxyl pectin. Comparison of the Fourier transform infrared spectra with commercial grade pectin showed that pectin extracted from dragon fruit peels has high potential in food applications (Chua *et al.*, 2020).

Microwave-assisted extraction has also been used at different intensities of 300, 450, 600 and 800 W to extract pectin from dragon fruit peel (Rahmatia *et al.*, 2019). The optimal extraction conditions were pH 2.07, an extraction time of 65 s and a solid:liquid ratio of 66.57 (w:v). The maximum extraction yields at 300, 450, 600 and 800 W were 11.8%, 14.9%, 17.2% and 18.5%, respectively. The microwave-assisted extraction process was found to be effective in obtaining high-quality dragon fruit peel pectin, with 59.73–69.68% D-galacturonic acid, 45.82–46.95% degree of esterification and 15.32–21.22% crystallinity.

4.3.4 Seed oil

Dragon fruit contains plenty of grainy seeds, and these seeds are rich in polyunsaturated fatty acids. Because of the high lipid content and functional properties, dragon fruit seed oil is considered a high-value product (Liaotrakoon *et al.*, 2013b). The oil has anti-ageing properties, as it smoothens and tightens the skin. With its high levels of antioxidants and unsaturated fatty acids, it is a great addition to creams, lotions and moisturizers (Boyapati *et al.*, 2023).

Red pitaya seed oil is rich in fatty acids, and about 30% oil yield has been reported (Ariffin *et al.*, 2009). Analysis of hot-extracted seed oil indicated the presence of palmitic acid (15.72%), oleic acid (24.89%) and linoleic acid (49.81%), while cold-extracted seed oil exhibited marginally higher contents of palmitic acid (15.78%) and linoleic acid (49.81%) (Murugesu *et al.*, 2013). The results showed that different seed extraction techniques influence the oil yield and the seed oil properties.

4.4 Conclusion

Dragon fruit is an emerging fruit crop in subtropical and tropical countries in Asia and North and South America, and the area of production is increasing. Although the availability of fresh fruit is currently very seasonal and localized, the potential for the domestic and international market is high. Dragon fruit contains nutritional and health-benefit properties. Recently, several processing technologies have been developed to utilize dragon fruit as well as its by-products for food processing and value addition. These processing technologies help to reduce postharvest losses and production costs, and achieve a better quality of end products. Products such as minimally processed fresh-cut fruit, jam, jelly, beverages and leathers have great potential to attract consumers and increase the marketability of the fruit. In addition, processing of by-products (i.e. peel) will not only enhance profitability but also provide new products for the food, cosmetic and pharmaceutical industries. At present, these processed products of dragon fruit rarely appear in the market and more research is required to improve their trading opportunities. Therefore, the key players in the dragon fruit supply-chain including farmers, food processors, traders and consumers will be extensively benefited by effective dragon fruit plantations and fruit processing.

4.5 References

Ariffin, A., Bakar, J., Tan, C., Rahman, R., Karim, R. and Loi, C. (2009) Essential fatty acids of pitaya (dragon fruit) seed oil. *Food Chemistry* 114, 561–564.

Bhagya Raj, G.V.S.B. and Dash, K.K. (2020) Microwave vacuum drying of dragon fruit slice: artificial neural network modelling, genetic algorithm optimization, and kinetics study. *Computers and Electronics in Agriculture* 178: 105814.

Bhagya Raj, G.V.S.B. and Dash, K.K (2022) Effect of intermittent microwave convective drying on physico-chemical properties of dragon fruit. *Food Science and Biotechnology* 31, 549–560.

Boyapati, T., Rana, S.S. and Ghosh, P. (2023) Microwave-assisted extraction of dragon fruit seed oil: fatty acid profile and functional properties. *Journal of the Saudi Society of Agricultural Sciences* 22, 149–157.

Castro, J.C., dos Anjos, M.M., Hoinatski, A., Klosaski, S.J., Mikcha, J.M.G. *et al.* (2021) Dragon fruit (*Hylocereus undaters* Haw.) jam: use full, development and characterization. *Research, Society and Development* 10: e6510716255.

Chaemsanit, S., Matan, N. and Matan, N. (2018) Effect of peppermint oil on the shelf-life of dragon fruit during storage. *Food Control* 90, 172–179.

Chandrasekaran, S., Ramanathan, S. and Basak, T. (2013) Microwave food processing – a review. *Food Research International* 52, 243–261.

Chia, S.L. and Chong, G.H. (2015) Effect of drum drying on physico-chemical characteristics of dragon fruit peel (*Hylocereus polyrhizus*). *International Journal of Food Engineering* 11, 285–293.

Choo, K.Y., Kho, C., Ong, Y.Y., Thoo, Y.Y., Lim, R.L.H. *et al.* (2018) Studies on the storage stability of fermented red dragon fruit (*Hylocereus polyrhizus*) drink. *Food Science and Biotechnology*, 27, 1411–1417.

Chua, B.L., Tang, S.F., Ali, A. and Chow, Y.H. (2020) Optimisation of pectin production from dragon fruit peels waste: drying, extraction and characterization studies. *SN Applied Sciences* 2, 621–633.

Dam, S.M. (2013) Development of different processed products from the edible and inedible parts of the dragon fruit (*Hylocereus undatus*). *Acta Horticulturae* 989, 271–278.

Fathordoobady, F., Jarzębski, M., Pratap-Singh, A., Guo, Y. and Abd-Manap, Y. (2013) Encapsulation of betacyanins from the peel of red dragon fruit (*Hylocereus polyrhizus* L.) in alginate microbeads. *Food Hydrocolloids* 113: 106535.

Goldman, G., Vinokur, Y., Horev, B., Lurie, S., Rodov, V. and Liguori, G. (2005) Fresh-cut products from cactus species. *Acta Horticulturae* 682, 1961–1966.

Huajing, S., Yuan, H.X. and Jin, K. (2020) Development of pitaya suspended beverage. *Storage and Process* 18, 112–117.

Islam, M.Z., Khan, M.T.H., Hoque, M.M. and Rahman, M.M. (2011) Studies on the processing and preservation of dragon fruit (*Hylocereus undatus*) jelly. *Agriculturists* 10, 29–35.

Jalgaonkar, K., Mahawar, M.K., Kale, S., Kale, P.N., Bibwe, B. *et al.* (2018) Response surface optimization for development of dragon fruit based ready to serve drink. *Journal of Applied and Natural Science* 10, 272–278.

Jalgaonkar, K., Mahawar, M.K., Bibwe, B. and and Kannaujia, P. (2020) Postharvest profile, processing and waste utilization of dragon fruit (*Hylocereus* spp.): a review. *Food Reviews International* 38, 733–759.

Jayasinghe, O., Fernando, S., Jayamanne, V. and Hettiarachchi, D. (2015) Production of a novel fruit-yoghurt using dragon fruit (*Hylocereus undatus*). *European Scientific Journal* 11, 208–215.

Jiang, X., Lu, Y. and Liu, S.Q. (2020a) Effects of different yeasts on physicochemical and oenological properties of red dragon fruit wine fermented with *Saccharomyces cerevisiae*, *Torulaspora delbrueckii* and *Lachancea thermotolerans. Microorganisms* 8: 315.

Jiang, X., Lu, Y. and Liu, S.Q. (2020b) Effects of pectinase treatment on the physicochemical and oenological properties of red dragon fruit wine fermented with *Torulaspora delbrueckii. LWT – Food Science and Technology* 132: 109929.

Liao, H., Zhu, W., Zhong, K. and Liu, Y. (2020) Evaluation of colour stability of clear red pitaya juice treated by theromosonication. *LWT – Food Science and Technology* 121, 108997.

Liaotrakoon, W., de Clercq, N., Lewille, B. and Dewettinck, K. (2012) Physicochemical properties, glass transition state diagram and colour stability of pulp and peel of two dragon fruit varieties (*Hylocereus* spp.) as affected by freeze-drying. *International Food Research Journal,* 19, 743–750.

Liaotrakoon, W., de Clercq, N., Hoed, V.V., de Walle, D.V., Lewille, B. and Dewettinck, K. (2013a) Impact of thermal treatment on physicochemical, antioxidative and rheological properties of white-flesh and red-flesh dragon fruit (*Hylocereus* spp.). *Pure Food Bioprocess Technology* 6, 416–430.

Liaotrakoon, W., de Clercq, N., Hoed, V.V. and Dewettinck, K. (2013b) Dragon fruit (*Hylocereus* spp) seed oils: their characterization and stability under storage conditions. *Journal of the American Oil Chemical Society* 90, 207–215.

Madane, P., Das, A.K., Nanda, P.K., Bandyopadhyay, S., Jagtap, P. *et al.* (2020) Dragon fruit (*Hylocereus undatus*) peel as antioxidant dietary fibre on quality and lipid oxidation of chicken nuggets. *Journal of Food Science and Technology* 57, 1449–1461.

Maflahah, I. and Asfan, D.F. (2020) Quality characteristics of salt fortification with dragon fruit peel in plastic packaging during storage period. *Jurnal Tecknologi dan Manajemen Agroindustri* 9, 73–81.

Mai, H.T.Q., Pham, V.T., Do, V.L., Phuc, T.B., Truc, T.T. and Ngan, T.T.K. (2022) Production process of a carbonated drink from red flesh dragon fruit (*Hylocereus polyrhizus*). *Materials Science Forum* 1048, 514–523.

Mai, T.H.A., Tran, T.T.T. and Le, V.V.M. (2023) Effects of pitaya peel supplementation on nutrition quality, overall sensory acceptance, *in vitro* glycemic index, and antioxidant release from fiber-enriched cookies. *Journal of Food Quality* 2023: 3166524.

Manihuruka, F.M., Suryatib, T. and Arief, I.I. (2017) Effectiveness of the red dragon fruit (*Hylocereus polyrhizus*) peel extract as the colourant, antioxidant, and antimicrobial on beef sausage. *Media Peternakan* 40, 47–54.

Marques, L.G., Ferreira, M.C. and Freire, J.T. (2007) Freeze-drying of acerola (*Mapighia glabra* L.). *Chemical Engineering and Processing* 46, 451–457.

Matan, N., Puangjinda, K., Hothisuwan, S. and Nisoa, M. (2015) Combined antibacterial activity of green tea extract with atmospheric radio-frequency plasma against pathogens on fresh-cut dragon fruit. *Food Control* 50, 291–296.

Moshfeghi, N., Mahdavi, O., Shahhosseini, F. and Malekifar, S. (2013) Introducing a new natural product from dragon fruit into the market. *International Journal of Recent Research and Applied Studies* 15, 269–272.

Muhammad, K., Zahari, N.I.M., Gannasin, S.P., Adzahan, M.N. and Bakar, J. (2014) High methoxyl pectin from dragon fruit (*Hylocereus polyrhizus*) peel. *Food Hydrocolloids* 42, 289–297.

Muhialdina, B.J., Kadumb, H., Zareic, M. and Hussind, A.S.M. (2020) Effects of metabolite changes during lacto-fermentation on the biological activity and consumer acceptability for dragon fruit juice. *LWT – Food Science and Technology* 121: 108992.

Murugesu, S., Ariffin, A.A., Ping, T.C. and Chern, B.H. (2013) Physicochemical properties of oil extracted from the hot and cold extracted red pitaya (*Hylocereus polyrhizus*) seeds. *Journal of Food Chemistry and Nutrition* 1, 78–83.

Nordin, M.F.M., Puspasari, I., Tarsirin, S.M., Daud, W.R.W. *et al.* (2014) Quality changes of red pitaya (*Hylocereus undatus*) slices dried in hot air, microwave-hot air and microwave-vacuum dryers. *Iranica Journal of Energy & Environment* 5, 313–322.

Nur 'Aliaa, A.R., Mazlina, M.K.S. and Taip, F.S. (2011) Effects of commercial pectinases application on selected properties of red pitaya juice. *Journal of Food Process Engineering* 34, 1523–1534.

Nur Izalin, M.Z., Kharidah, M., Jamilah, B. and Noranizan, M.A. (2016) Functional properties of pectin from dragon fruit (*Hylocereus polyrhizus*) peel and its sensory attributes. *Journal of Tropical Agriculture and Food Science* 44, 95–101.

Ong, Y.Y., Tan, W.S., Rosfarizan, M., Chan, E.S. and Tey, B.T (2012) Isolation and identification of lactic acid bacteria from fermented red dragon fruit juices. *Journal of Food Science* 77, M560–M564.

Owino, W.O. and Ambuko, J.L. (2021) Mango fruit processing: options for small-scale processors in developing countries. *Agriculture* 11: 1105.

Panchal, J.B., Gaikwad, R.S., Dhemre, J.K. and Chavan, U.D. (2018) Studies on preparation and storage of jelly from dragon fruit (*Hylocereus undatus*). *Journal of Pharmacognosy and Phytochemistry* 7, 2648–2655.

Pawde, S., Talib, M.I. and Parate, V.R. (2020) Development of fiber-rich biscuit by incorporating dragon fruit powder. *International Journal of Fruit Science* 20, S1620–S1628.

Phuong, T.N.M. and Tuan, D.Q. (2016) Application of hydrolytic enzymes for improvement of red dragon fruit juice processing. *Asia Pacific Journal of Sustainable Agriculture, Food and Energy* 4, 1–4.

Radziejewska-Kubzdela, E., Szwengiel, A., Ratajkiewicz, H. and Nowak, K. (2020) Effect of ultra sound, heating and enzymatic pre-treatment on bioactive compounds in juice from *Berberis amurensis* Rupr. *Ultra Sinochem* 63: 104971

Rahmatia, S., Abdullaha, A. and Ka, O.L. (2019) Effects of different microwave intensity on the extraction yield and physicochemical properties of pectin from dragon fruit (*Hylocereus polyr*hizus) peels. *Bioactive Carbohydrates and Dietary Fibre* 18: 100186.

Ratti, C. (2001) Hot air and freeze-drying of high-value foods: a review. *Journal of Food Engineering* 49, 311–319.

Santana, A.T.M.C., Bachiega, P., Assis, R.Q, Rios, A.O. and Souza, E.C. (2015) Chemical, physical and nutritional profile of dragon fruit (*Hylocereus undatus*) yoghurt enriched with quinoa (*Chenopodium quinoa*) and sucralose. *Revista Brasileira de Produtos Agroindustriais* 17, 285–292.

Shofinita, D., Bindar, Y., Samadhi, T.W., Jaelawijaya, A.A. and Fawwaz, M. (2020) Reducing the stickiness of dragon fruit skin extract powder as food colourant by addition of maltodextrin during freeze drying. *AIP Conference Proceedings* 2219: 070011.

Sim, H.L., Hong, Y.-K., Yoon, W.B. and Yuk, H.-G. (2013) Behavior of *Salmonella* spp. and natural microbiota on fresh-cut dragon fruits at different storage temperatures. *International Journal of Food Microbiology* 160, 239–244.

Sun, X.-J., Wang, Y.-R., Ding, R.-X., Ming, L.-Y., Yue, X.-Q. and Wu, J.-R. (2018) Development of probiotic dragon fruit yoghurt. *Food Research and Development* 39, 99–103.

Tan, C.X., Lin, S.W., Tan, S.S. and Tan, S.T. (2023) Characterization of juice extracted from ultrasonic treated red pitaya flesh. *Horticulturae* 9: 92.

Tze, N.L., Han, C.P., Yusof, Y.A., Ling, C.N., Talib, R.A. *et al.* (2012) Physicochemical and nutritional properties of spray-dried pitaya fruit powder as natural colourant. *Food Science and Biotechnology* 21, 675–682.

Wong, Y.-M. and Siow, L.-F. (2015) Effects of heat, pH, antioxidant, agitation and light on betacyanin stability using red-fleshed dragon fruit (*Hylocereus polyrhizus*) juice and concentrate as models. *Food Science and Technology* 52, 3086–3092.

Yong, Y.J., Yan, Z.L., Feng, B.J., Xuan, L., Quin, C.Q. and Wu, X. (2016) Influences of microwave pre-drying and explosion puffing drying induced cell wall polysaccharide modification on physicochemical properties, texture, microstructure and rehydration of pitaya fruit chips. *Food Science and Technology* 70, 271–279.

Yong, Y.J., Lyu, J., Feng, B.J., Yan, Z.L. and Zhou, M. (2017) Hot air drying and freeze-drying pre-treatments coupled to explosion puffing drying in terms of quality attributes of mango, pitaya and papaya fruit chips. *Journal of Food Processing and Preservation* 41: e13300.

Yusof, N., Adzahan, N.M. and Muhammad, K. (2020) Optimization of spray drying parameters for white dragon fruit (*Hylocereus undatus*) juice powder using response surface methodology (RSM). *Malaysian Journal of Applied Sciences* 5, 45–56.

Yusof, Y.A., Mohd Salleh, F.S., Chin, N.L. and Talib, R.A. (2011) The drying and tabletting of pitaya powder. *Journal of Food Process Engineering* 35, 763–771.

Zaidel, D.N.A., Makhtara, N.A., Jusoha, Y.M.M. and Muhamad, I.I. (2015) Efficiency and thermal stability of encapsulated anthocyanins from red dragon fruit (*Hylocereus polyrhizus* (Weber) Britton & Rose) using microwave-assisted technique. *Chemical Engineering Transactions* 43, 127–132.

Zaidiyah, S.H., Anwar, S.Q. and Nasution, A.I.S. (2019) Shelf-life estimation of red dragon fruit jam using accelerated shelf-life testing (ASLT) method. *IOP Conference Series: Earth and Environmental Science* 365: 12029.

Zainoldin, K.H. and Baba, A.S. (2009) The effect of *Hylocereus polyrhizus* and *Hylocereus undatus* on physicochemical, proteolysis, and antioxidant activity in yogurt. *World Academy of Science Engineering and Technology* 3, 884–889.

Zhu, W., Ai, Y., Fang, F. and Liao, H. (2021) Application of thermosonication in red pitaya juice processing: impacts on native microbiota and quality properties during storage. *Foods* 10: 1041.

5 Breeding and Varieties

Sisir Mitra*
Former Professor, Bidhan Chandra Krishi Viswavidyalaya (State Agriculture University), Mohnapur, West Bengal, India

5.1 Breeding

Dragon fruit is considered an exotic vine-cactus species with characteristic fruit. It has great commercial potential worldwide due to consumer preference for new, exotic, nutrient-rich fruits and its adaptability to new environments with tolerance to abiotic stresses such as drought and temperature extremes (Morillo *et al.*, 2022), although extremely high temperatures over 45°C cause stem damage in *Hylocereus undatus* (Nobel and de la Barrera, 2005) and high summer temperatures inhibit flowering in *H. undatus* and *H. monacanthus* (Nerd *et al.*, 2002).

Dragon fruit has a Crassulacean acid metabolism (CAM), which gives it drought tolerance, and it grows primarily in tropical dry and low-land deciduous forests. In Mexico, it can be found in environments ranging from 2 to 2750 m above sea level with an annual rainfall varying from 340 to 3500 mm (Calix de Dios, 2004).

The vine-cactus dragon fruit (also known as pitaya) of the family Cactaceae (subfamily Cactoideae, tribe Hylocereeae) is known to have been used for thousands of years by the indigenous people of the Americas (Ortiz-Hernandez and Carrillo-Salazar, 2012). In the mid-19th century, it was introduced by French priests to Indochina (at that time the name of Vietnam, Laos and Cambodia). In 1995, Vietnam was the first country to sell dragon fruit in world markets, under the name dragon pearl fruit. Many countries obtained their clones from Vietnam, although these are not good in terms of quality and are quite bland (Mizrahi, 2014).

Ben-Gurion University (Negev, Beer-Sheva, Israel) established a large, live gene bank of *Hylocereus* genotypes by collecting genetic resources from around the world (Nerd and Mizrahi, 1997). The collection, established in 1884, includes more than 200 genotypes from ten taxonomically identified *Hylocereus* (now renamed *Selenicereus*) spp. (Tel-Zur *et al.*, 2011b). Yosii Mizrahi, Noemi Tel-Zur and their co-researchers have contributed significantly to our understanding of the cytogenetics and breeding of pitaya with their numerous publications.

Among the 15 *Hylocereus* spp. (Bauer, 2003), the species that were cultivated from their native areas as fruit crops were *H. undatus* (white flesh and red peel), *H. monocanthus* (purple flesh and red peel) and *H. megalanthus* (white flesh, yellow spiny peel and a superior taste) (Weiss *et al.*, 1994). *H. monocanthus* and *H. undatus* are diploid ($2n = 2x = 22$) and *H. megalanthus* is tetraploid (Tel-Zur *et al.*, 2004). The crosses between the tetraploid *H. megalanthus* and various *Hylocereus* genotypes yielded diploid, triploid,

* Email: sisirm55@gmail.com

DOI: 10.1079/9781800623156.0005

tetraploid, pentaploid and hexaploid hybrids (Tel-Zur *et al.*, 2004, 2005, 2011a,b; Cisneros *et al.*, 2013). In Israel, two commercial triploid hybrids were developed ('5-75' and '12-14'), which have good taste, a long storage life and ripen in late autumn to midwinter, differing from all other *Hylocereus* genotypes (Mizrahi, 2014). Most of the cultivated species are usually found in their geographical origin, except for those in South-east Asia. *H. polyrhizus* and *H. costaricensis* have been cultivated in Central America (Nicaragua), *H. megalanthus* has been cultivated primarily in South America (Colombia and Ecuador), and *H. undatus* has been cultivated in South-east Asia (Vietnam, Thailand, Taiwan, Sri Lanka, Malaysia and the Philippines) and in some areas of southern Mexico (Castillo *et al.*, 2005).

5.1.1 Breeding objectives

Hylocereus spp. bear fruit in a variety of shapes, sizes, peel and flesh colours, and tastes. The wide genetic variability present among these species can be used as a valuable platform for genetic improvement. The diploid *Hylocereus* spp. produce large and attractive fruit, but with an inferior taste, which is a major disadvantage that limits their marketability, whereas the tetraploid *H. megalanthus* bears spiny fruit with a superior taste but is inferior to the diploid *Hylocereus* spp. in yield, size and appearance (Tel-Zur, 2013). The long-term breeding goal by Tel-Zur and his group in Israel was initially to improve fruit quality and yield and to prolong the harvesting season (Tel-Zur, 2022). In Brazil, Embrapa Cerrados, a unit of the Brazilian Agricultural Research Corporation located in Planaltina, initiated hybridization work to obtain hybrids of *Selenicereus undatus* (syn. *H. undatus*) and *S. costaricensis* (syn. *H. costaricensis*) that combined the characteristics of high productivity, self-compatibility and larger fruit size present in *S. undatus* with the red flesh of *S. costaricensis*. This hybrid was successfully obtained and random amplified polymorphic DNA (RAPD) molecular markers were used to confirm hybridization (de Lima *et al.*, 2013).

The goal of the pitaya breeding programme in Taiwan is to develop varieties with a combination of pleasant flavour, spineless fruit, attractive flesh colour, large fruit size and long shelf life (Lee and Liu, 2021).

5.2 Selection

Selection is the basis of plant breeding, and must be applied in different ways according to the method of reproduction of the plant. Selection does not create genetic variability but merely acts on the genetic variability already available. In most dragon fruit orchards, the growers multiply their material by cuttings, which root easily. This guarantees the uniformity of the orchard and precocity of production. Thus, these orchards show practically no genetic variability, which is undesirable in terms of plant genetic improvement. Seed propagation may be a resource for introducing variability. Plants obtained by sexual propagation may be similar to one of the parents, to both or to neither (da Silva *et al.*, 2017). Most of the present dragon fruit varieties in different countries are selections made by growers/researchers from naturally occurring genotypes. A genotype introduced from another country was also directly utilized as a new variety when it was found to be promising in performance trials.

Commercial plantations of *H. undatus* were established in Central America with plants collected from home gardens and, to a lesser extent, wild plants, which has resulted in fruit that have a high heterogeneity in characteristics such as colour, size and shape (Fig. 5.1) (Castillo *et al.*, 2005). In Nicaragua, Gonzalez and Alvarado (2004) evaluated 50 qualitative and quantitative characteristics capable of measuring the phenotypic variability of *H. undatus*. The authors proposed a detailed guide of morphological descriptors and recognized 14 characteristics that could be used to differentiate materials. Castillo *et al.* (2005), however, concluded that the reproductive characteristics were the most important for separating genotypes. In Israel, Tel-Zur *et al.* (2011b) conducted morphological and genetic characterization of 64 accessions of members of the genera *Hylocereus*, *Selenicereus* and *Epiphyllum* and found high genetic variability among accessions and therefore excellent prospects for domestication and conservation of these genetic resources.

Fig. 5.1. Genetic variability in dragon fruit. Photograph courtesy of Dr J. Mouyal.

However, some of the different characteristics of genetic diversity of the population and identification of species based on morphological characteristics are strongly influenced by the environment, and therefore molecular identification has become the main tool for characterization and management of germplasm collections of many species of fruit trees (Dotor *et al.*, 2016).

A large collection of dragon fruit germplasm accessions is curated at the University of California South Coast Research and Extension Centre (Irvine, California). The collection was first established in 2005 and consists of 378 individual plants, potentially representing 54 varieties and seven different species introduced from Nicaragua, Mexico, Florida and San Diego (Pagliaccia *et al.*, 2015). Close examination of various accessions indicated significant similarities among plants from different varieties and significant differences among plants within the same variety. This suggested that some plants within each of the designated varieties were different and that some of the designated cultivars with different names may be the same clone. Pagliaccia *et al.* (2015) used amplified fragment length polymorphism (AFLP) to determine whether there was redundancy in some of the commonly named varieties, as well as to access the overall diversity of the collection. A total of 51 markers were scored from 230 accessions. Seven distinct clades were found with 126 putative clones, but only three clades had high bootstrap support (>80%) while one had moderate support (60%). Some of the differently named varieties were identical, but there was also genotypic diversity within putatively named varieties. Tel-Zur *et al.* (2004) used RAPD markers that differentiated 13 species within the genera *Hylocereus* and *Selenicereus* and were also able to differentiate unique genotypes among different accessions of *H. polyrhizus*, *H. undatus*, *S. grandiflorus* and *S. megalanthus* (syn. *H. megalanthus*).

In Colombia, the germplasm bank of yellow pitaya and its wild relatives is located in the National University of Colombia, Palmira. It has 300 introductions, both cultivated and wild, of yellow pitaya (*S. megalanthus*, 238 accessions) and red pitahaya (*H. undatus*, *H. costaricensis* and other *Hylocereus* spp., 36 accessions). Morillo-Coronado *et al.* (2021) evaluated the morpho-agronomic characteristics of the yellow pitaya genotype from five municipalities in Boyaca, Colombia. The results revealed that cladode characteristics such as cladode width, distance between areoles, number of spines, length of areole, margin ribs of the cladode and waxiness could be used to identify *S. megalanthus* clones. Characteristics such as fruit weight, size, colour of peel and flesh, as well as total soluble solids (TSS) and titratable acidity (TA) were highly variable among the genotypes. Morphological, biochemical and molecular characterization of four dragon fruit genotypes (*Hylocereus* spp.) grown in Andaman and Nicobar Islands revealed the presence of a considerable amount of genetic variation among them, which could be used as key traits for distinguishing three of the species (*H. undatus*, *H. megalanthus* and *H. costaricensis*) (Abirami *et al.*, 2021). Cluster analysis distinguished the three different *Hylocereus* spp. on the basis of geographical origin and flesh colour by forming separate groups. Key traits identified for distinguishing the three different species were: flesh/peel colour of the fruit, number of spines and length of areoles in the cladode (Abirami *et al.*, 2021).

Embrapa Cerrados set up the first pitaya Brasil Active Germplasm Bank (BAG) in 1996, with the objective of characterizing the agronomic and commercial potential of the main species of pitaya. The established pitaya BAG collected more than 200 germplasm accessions from different regions of Brazil. After evaluation (productivity, physical and chemical characteristic of the fruit, self-compatibility and disease resistance), six superior accessions of *S. undatus*, seven of *Selenicereus setaceus* (syn. *H. setaceus*), four of *S. costaricensis* and four of *S. megalanthus* were selected. The morphological characteristics of the plant, flowers and fruit from different accessions showed the existence of genetic variability for size and shape of the fruit, flesh colour, soluble solids and acid content of the fruit (Faleiro and Junqueira, 2021). Junqueira *et al.* (2010) conducted a study of the genetic diversity of 13 pitaya accessions kept in the active bank of the BAG using RAPD markers. More than 95% showed polymorphic markers, from which distances were estimated, and genetic differences between accessions and cluster analysis were determined to identify groups of genetic similarity. The 13 accessions were divided into at least seven groups of genetic similarity, of which two groups were in the species *S. setaceus*, two groups were within *S. undatus*, one group was within *S. megalanthus* and two groups had unknown taxonomy. Genetic improvement of the different pitaya species together with the agronomic validation work in different regions and production systems led to the development of five varieties: 'BRS Lua do Cerrado' ('BRS LC'), 'BRS Luz do Cerrado' ('BRS LZC'), 'BRS Mini-pitaya do Cerrado' ('BRS MPC'), 'BRS Granada do Cerrado' ('BRS GC') and 'BRS Amber do Cerrado' ('BRS AC') (Faleiro and Junqueira, 2021).

In Puerto Rico, eight selections ('NOI-13', 'NOI-14', 'NOI-16', 'N97-15', 'N97-17', 'N97-18', 'N97-20' and 'N97-22') were introduced from the US Department of Agriculture/Agricultural Research Service (USDA-ARS) Tropical Plant Genetic Resources and Disease Research Institute (Hilo, Hawaii) and four varieties ('American Beauty', 'Halley's Comet', 'Purple Haze' and 'Cosmic Charlie') were introduced from a nursery in Miami, Florida, and planted in 2009. Initial observations showed that 'NOI-13', 'NOI-17', 'NOI-20' and 'N97-22' were high yielders. Except for 'NOI-13', all the other varieties were white fleshed (Goenaga *et al.*, 2020). Ramirez (1999) evaluated 40 descriptors in four Mexican *Hylocereus* accessions and found high phenotypic variation in the shoot, flower and fruit. Fruit weight varied from 100 to 1200 g, and the ripening period from 27 to 120 days (Ramirez, 1999; Ortiz-Hernandez, 2000).

5.3 Hybridization

Hylocereus spp. hybridize quite easily, and since the late 1980s and early 1990s, amateur breeders in the USA, Israel and South-east Asia have developed a number of hybrids (Thomson, 2002). These hybrids generally include crosses of *H. guatemalensis* or *H. megalanthus* with *H. undatus* as the other parent (Pagliaccia *et al.*, 2015). The Faculty of Agrarian and Veterinary Science of Jaboticabal, Brazil (FCAV/UNNSER), initiated hybridization work in 2010. The evaluated progeny were obtained from interspecific cross-pollination of *H. undatus* × *H. polyrhizus* and *H. setaceus*. A total of 45 progenies and six parental lines were evaluated, aiming to evaluate the genetic diversity of the progeny for future use in breeding programmes. Based on cladode characteristics, six characteristics were assessed: length and diameter of the stem, distance between areoles, arch height, and number and size of spines or areoles. Using Euclidean distance matrix analysis, they constructed a dendrogram using the unweighted pair group method with arithmetic mean (UPGMA). There was great variability among the hybrids, and eight showed promise for use in breeding programmes (da Silva *et al.*, 2017).

In Vietnam, the hybrid 'Long Dinh 1' was created by crossing the 'Binh Thuan' variety (*H. undatus*) as the female parent and the red-fleshed 'Colombia' clone (*H. polyrhizus*) as the male. The hybrid is self-compatible, has red-purple flesh and is good quality (Hang *et al.*, 2014). This hybrid was released in 2005. Another hybrid, 'Long Dinh 5', was developed by crossing 'Long Dinh 1' and the 'Cho Gao' variety as the pollen donor. It is also self-compatible, and has attractive fruit traits and good tolerance to some diseases (Hang *et al.*, 2014). Plant and Food Research in New Zealand and the Southern Horticultural Research Institute (SOFRI) in Vietnam have been working together since 2013 to breed new varieties of dragon fruit with excellent grower and consumer characteristics. Three new varieties from the joint dragon fruit breeding programme are set to be launched on the global market. All of the new varieties are red skinned, with a sweeter, more aromatic flavour than the current varieties and with a crisp dense texture. The varieties to be released vary in flesh colour, one having the traditional white flesh most commonly seen, one being a unique pink-fleshed variety and the other being a dark, appealing, red-fleshed variety. The varieties are believed to be the first canker-tolerant dragon fruit to be commercialized (Fresh Plaza, 2023).

A number of hybrids were developed recently in China. They include 'Hongshuijing No. 6' ('Baishujing' × 'Guangminghong'), 'Jingdu 1' (*H. costaricensis* × *H. polyrhizus*), 'Daqiu 4' ('Guangminghong' × 'Baishijing'), 'Yuehong No. 5' ('Dahong' × Common white flesh) and 'Meilong 1' (*H. costaricensis* × *H. undatus*). All of these hybrids are self-compatible (fruit setting >90%) and have good-quality fruit. The hybrid 'Yuehong No. 5' has strong resistance to pitaya canker disease (Su *et al.*, 2017; Huang *et al.*, 2021; Dai *et al.*, 2022; Li *et al.*, 2022; Wang *et al.*, 2022).

An Israeli breeding programme carried out at Ben-Gurion University has focused on improving fruit traits (Nerd *et al.*, 2002). The first results obtained from this breeding programme indicated a close genetic relationship among the *Hylocereus* spp. as they inter-cross easily. Interspecific homoploid crosses among diploid *Hylocereus* spp. resulted in diploid hybrids with a mix of their parental morphological traits, several of them with improved fruit quality (Tel-Zur *et al.*, 2004). Reciprocal crosses among four diploid *Hylocereus* spp. (*H. costaricensis*, *H. polyrhizus*, *H. undatus* and other *Hylocereus* spp.) yielded viable diploid hybrids with regular chromosome pairing (Tel-Zur *et al.*, 2003). Reciprocal crosses between these *Hylocereus* spp. and *H. megalanthus* yielded viable triploid, pentaploid, hexaploid and aneuploid hybrids. Morphological and phenological traits confirmed the hybrid origins. *In situ* detection of ribosomal DNA sites was in accord with the ploidy status of the species and hybrids. Genomic *in situ* hybridization results indicated that the overall sequence composition of *H. megalanthus* is similar to that of *H. ocamponis* and *S. grandiflorus*. A high sequence similarity was also found between the parental genomes of *H. polyrhizus* and *H. megalanthus* in one triploid hybrid (Tel-Zur *et al.*, 2004). The fruit of the triploid and 3x aneuploid hybrids combined the taste quality of *H. megalanthus* and the attractive appearance of *H. polyrhizus*. The harvesting season of *Hylocereus* spp. in Israel occurs during summer/autumn, while *H. megalanthus* ripens during winter/spring. Triploid hybrids ripen during autumn/winter, between the *Hylocereus* spp. and *H. megalanthus* seasons, making it possible to supply the markets with fresh fruit throughout much of the year (Tel-Zur, 2013). Tel-Zur *et al.* (2005) observed partial fertility in all of the triploid hybrids studied, in contrast to the complete sterility frequently observed in triploids. They noted that different proportions of functional male and female gametes were produced by the various hybrids. Wide ranges of percentage of fruit set and viable seeds were observed (Tel-Zur *et al.*, 2005). Among the triploids studied, two hybrids (S-75 (Fig. 5.2) and 12-14) that were judged superior in terms of fruit quality and yields were released to the farmers (Tel-Zur, 2022). Crosses between triploid hybrid S-75 and diploid *Hylocereus* spp. were reported to be fully self-compatible and exhibited similar fruit weight and similar total seed numbers following cross- or self-pollination. Flow cytometric analysis revealed that all of the self-compatible hybrids were tetraploids, suggesting a possible relationship between self-compatibility and ploidy level (Tel-Zur *et al.*, 2012b).

Fig. 5.2. Red-yellow dragon fruit F$_1$ hybrids. Photograph courtesy of Dr N. Tel-Zur.

5.4 Intergeneric Hybridization

Originally, *Hylocereus* and *Selenicereus* were considered two separate genera. A number of researchers developed hybrids using these two genera, considering them as separate genera. A geographic overlap exists among the native habitats of *Hylocereus*, *Selenicereus* and *Epiphyllum* spp. (Barthlott and Hunt, 1993; Hunt, 2006), and hence natural hybridization may be expected (Tel-Zur *et al.*, 2012a). Natural hybridization occurs in the wild in Mexico, resulting in intergeneric hybrids (Ortiz-Hernandez, 2000). Tel-Zur *et al.* (2012a) obtained synthetic intergeneric hybrids by controlled hand cross-pollination among the genera *Hylocereus*, *Selenicereus* and *Epiphyllum*. Hybrids between *Hylocereus* and *Selenicereus* produced viable seeds, a finding that showed the lack of post-zygotic barriers. Partial to complete sterility was observed in the intergeneric *Epiphyllum* × *Selenicereus* hybrids. Homoploid crosses between the diploid *H. undatus* as the maternal parent and the diploid *S. grandiflorus* as the parental parent produced diploid hybrids. Intergeneric interploidy crosses between *S. grandiflorus* as the maternal parent and *H. megalanthus* as the parental parent produced triploid hybrids (Tel-Zur *et al.*, 2012a). The close genetic relationship between the two genera (*Hylocereus* and *Selenicereus*) allows gene flow and the development of novel and unique plant material for breeding purpose, with the resulting intergeneric hybrids offering a possible bridge for the introgression of desirable traits (Tel-Zur, 2022). Intergeneric hybrids of *H. undatus* × *S. megalanthus* (Fig. 5.3), *H. polyrhizus* × *S. megalanthus* (Fig. 5.4), *S. megalanthus* × *H. undatus* (Fig. 5.5) and interspecific hybrid of *S. grandiflorus* × *S. megalanthus* (Fig. 5.6) have been developed by Mizrahi and co-researchers in Israel and now are now being evaluated (Y. Mizrahi personal communication).

5.5 Autopolyploidization

Autopolyploids may originate from the failure of cytokinensis during a mitotic cell cycle, thus doubling the chromosome number in somatic cells, or from meiotic polyploidization (i.e. $2n$ gametes) (Dhooghe *et al.*, 2011). Polyploidy – whole genome duplication either as a hybrid between species (allopolyploidy) or autopolyploids not due to hybridization – has seemingly occurred in dragon fruit (Lichtenzveig *et al.*, 2000; Tel-Zur *et al.*, 2004; Plume *et al.*, 2013). Artificial autopolyploid plants (produced by treatment with colchicine and oryzalin) were obtained from

Fig. 5.3. Intergeneric hybrid of *H. undatus* × *S. megalanthus*. Photograph courtesy of Dr Y. Mizrahi.

Fig. 5.4. Intergeneric hybrid of *H. polyrhizus* × *S. megalanthus*. Photograph courtesy of Dr Y. Mizrahi.

the donor lines of the diploid *H. polyrhizus*, the tetraploid *H. megalanthus* and the interspecific triploid hybrid S-75 (Tel-Zur, 2013). Morphological differences were observed between the donor plant and its autopolyploid, and a significant reduction in stomatal density, fruit size, number of viable seeds and pollen viability along with an increase in pollen size (Tel-Zur *et al.*, 2011a;

Fig. 5.5. Intergeneric hybrid of *S. megalanthus* × *H. undatus*. Photograph courtesy of Dr Y. Mizrahi.

Fig. 5.6. Intergeneric hybrid of *S. grandiflorus* × *S. megalanthus.* Photograph courtesy of Dr Y. Mizrahi.

Cohen and Tel-Zur, 2012; Cohen *et al.*, 2013). Flow cytometric and chromosome counts confirmed polyploidization in 14 autotetraploid *H. monacanthus*, one autohexaploid S-75 hybrid and 14 auto-octaploid *H. megalanthus* (Tel-Zur *et al.*, 2011a). Autotetraploid *H. monacanthus* exhibited smaller fruit size and fruit weight and lower pollen viability compared with the donor plant but had larger pollen grains; only aborted fruit were obtained from these lines. The most valuable change observed was the breakdown of the self-incompatibility system in the *H. monacanthus* autotetraploid lines (Cohen and Tel-Zur, 2012). Tel-Zur (2022) concluded that autopolyploidization is not a suitable breeding method to induce the production of large fruit in *Hylocereus* spp. and may be exploited to disrupt the self-incompatibility of *Hylocereus* spp.

5.6 Androgenesis and Gynogenesis

In vitro androgenesis in plants was first performed almost 60 years ago (Guha and Maheshvari, 1964). This process involves the development of androgenic embryos from the cells of the male gametophytic pathway, the microspores. The development pathway change from a gametophyte to a sporophyte results in the formation of an androgenic embryo (Touraev *et al.*, 2001). After chromosome doubling, these plants become double haploid, which means that they can reasonably be expected to be homozygous at every genetic locus. Double-haploid plants are an attractive material for a variety of research and breeding programmes, including the development of new cultivars and varieties, genetic mapping and localization of quantitative trait loci (Devaux and Pickering, 2005).

Pre-embryonic and embryonic stages and seed development in the diploid *H. monacanthus* and *H. undatus* and the tetraploid *H. megalanthus* were studied by Cisneros *et al.* (2011) (Fig. 5.7). Embryo development time was significantly more prolonged in *H. megalanthus*. The seedcoat was formed by one layer of sclerenchymatous cells, but was more lignified in *H. megalanthus*. Total seed number and seed viability were significantly lower in *H. megalanthus* than in the other two taxa. Embryos at different developmental stages were observed in aborted *H. megalanthus* seeds (Cisneros *et al.*, 2011).

Hylocereus spp. have large hermaphrodite flowers with a huge number of megaspores and microspores per flower and a prolonged flowering season, all of which facilitate the development of haploid protocols (Tel-Zur, 2022). Benega Garcia *et al.* (2009a) successfully developed

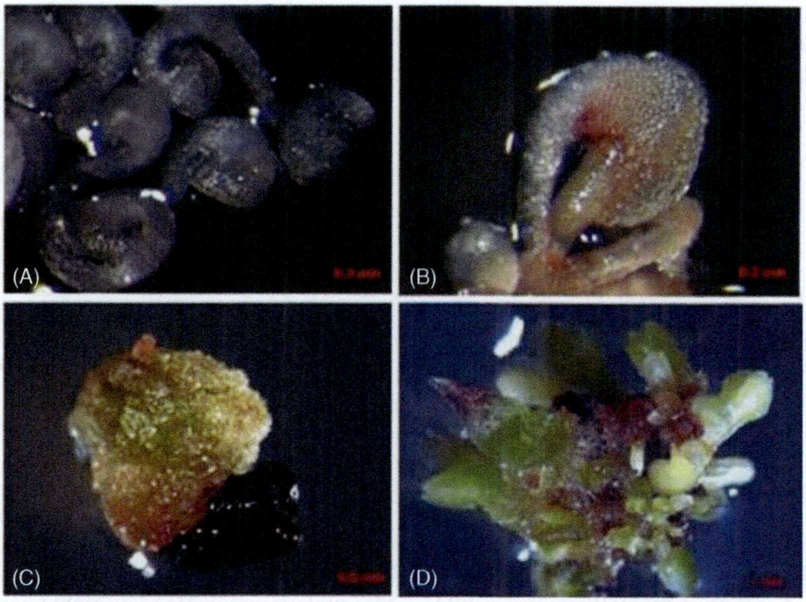

Fig. 5.7. *In vitro* development of isolated ovules in species of vine cacti. (A) Plated ovules. (B) Ovule enlargement. (C) Gynogenic compact green callus formation. (D) Direct gynogenic embryoid formation. Photographs courtesy of Dr N. Tel-Zur and Dr A. Cisneros.

dihaploid plants of *H. megalanthus* using the ovule culture technique. Haploid lines of *H. monacanthus* were developed via direct embryogenesis. This was the first report on the production of monoploid, haploid and dihaploid plants in members of the Cactaceae (Benega Garcia *et al.*, 2009b).

Li *et al.* (2018) studied the putative gamete-derived progenies from two *Hylocereus* spp., the diploid *H. monacanthus* and the tetraploid *H. megalanthus* to confirm their gamete origin and to evaluate their potential use as a genetic resource. Gamete origin was proved in all five *H. monacanthus* lines obtained and in 49 of the 70 *H. megalanthus* lines using flow cytometry and simple sequence repeat markers. The five double-haploid (2x) *H. monacanthus* lines showed low vigour and abnormal flower development, with malformed ovules and aborted pollen grains. Only one flower set fruit, giving sexual viable seeds. For *H. megalanthus*, both abnormal ovules and defective anthers were observed in the dihaploid (2x) and double dihaploid (4x) lines. Among the 46 dihaploid lines, only 14 set fruit. Another 13 dihaploid lines formed flower buds that abscised before anthesis or soon after pollination. The severe sterility of the double-haploid *H. monacanthus* lines and the reduced fertility of all of the dihaploid and double dihaploid *H. megalanthus* lines can be linked to their reduced heterozygosity, which drastically affected the development of normal female and male organs. The researchers concluded that chromosome doubling, as occurred spontaneously, is not sufficient to restore fertility in *Hylocereus* spp.

5.7 Embryo Rescue

Embryo rescue is one of the oldest and most successful *in vitro* tissue culture techniques used to nurture immature or weak embryos to produce plants and is commonly used for producing plants from hybridizations that produce aborted seeds (Sage *et al.*, 1999; Reed, 2004). Embryo culture, in which embryos are excised and placed directly on to culture media, is the most widely used embryo rescue procedure (Miyajima, 2006). Studies carried out by Weiss *et al.* (1994) and Tel-Zur *et al.* (2005) observed that in *H. monacanthus* and *H. undatus*, the embryo reaches the final stage (i.e. the cotyledon) 17 days after pollination (DAP), while in *H. megalanthus* 70 days are needed. The number of ovules in each ovary is very high in *Hylocereus* spp. In *H. undatus*, it is approximately 7200, in *H. monacanthus* around 5300 and in *H. megalanthus* approximately 2000 ovules per flower.

Fertilized ovules carrying embryos at very early pro-embryonic stages were excised from ovaries at 5 DAP and placed on half-strength basal Murashige and Skoog medium containing 680 μM glutamine, 0.55 μM α-naphthalene-acetic acid, 0.45 μM thidiazuron and various concentrations of sucrose (Cisneros and Tel-Zur, 2010). After 30 days in culture, ovules were isolated from the surrounding tissue and transferred to the same fresh medium. The cross between *H. polyrhizus* × *H. undatus* with a sucrose concentration of 0.09 M showed the best response in terms of increased ovule size and callus formation. In terms of embryo conversion, polyembryony and the number of regenerated plants, the highest responses were observed on culture medium supplemented with 0.17 M sucrose. All tested plants were found to be diploid by flow cytometric analysis. Fluorescent AFLP confirmed the hybrid origin of the regenerated plants (Figs 5.8 and 5.9). In a subsequent study by Cisneros *et al.* (2013), embryo rescue was performed following interploidy interspecific crosses of *H. megalanthus* × *H. monacanthus*/*H. undatus*. The rescued embryos were at very early stages of development at 10 or 30 DAP (i.e. the pro-embryo and globular stages, respectively). Some embryos developed into normal plantlets, but others only developed callus and died. The resulting putative hybrids showed triploid, tetraploid and other ploidy levels.

5.8 Abiotic Stress Tolerance

Various aspects of the physiology of the genus *Hylocereus* have been documented. This group of plants utilizes the CAM photosynthetic pathway and has an efficient system for retaining water by closing their stomata during the day (Nobel and de la Barrera, 2005). Total net carbon dioxide uptake is maximal at day/night temperatures of 30°C/20°C but is inhibited at day/night temperatures of 42°C/32°C (Nobel and de la Barrera, 2005). The maximum temperature

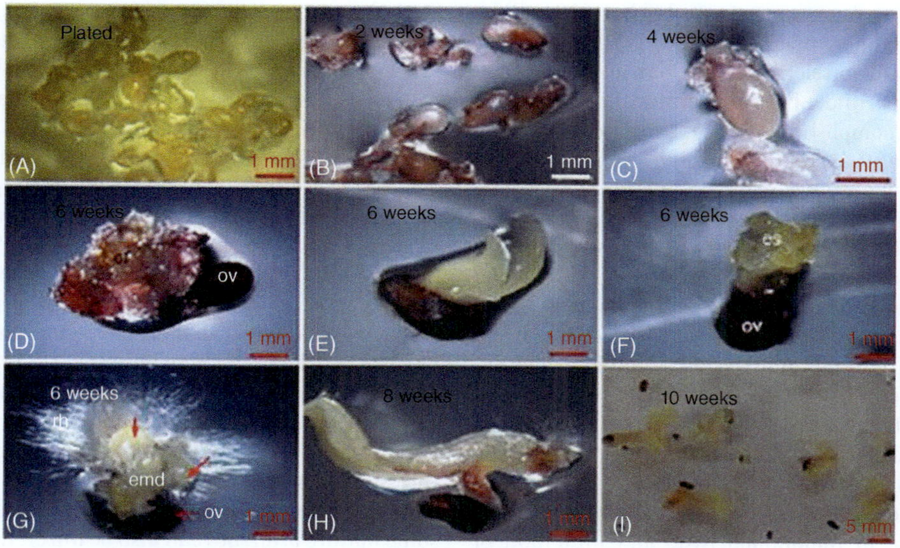

Fig. 5.8. Embryo development following *H. polyrhizus* × *H. undatus* cross. (A) Plated ovules five days after pollination. (B) Two weeks in culture. (C) Isolated ovules, 4 weeks in culture. (D)-(G) Embryo after 6 weeks in culture. (D) Formation of friable callus. (E) Normal embryo development. (F) Multiple embryogenic structures developed from a single embryo. (G) Multiple embryogenic structures with radicular hair. (H) Embryo separated from the ovule coat, 8 weeks in culture. (I) Embryos with whole cotyledonary leaves and radicle, 10 weeks in culture. *emd* embryoids, *ov* ovules, *rh* radicular hair. Photographs courtesy of Dr N. Tel-Zur and Dr A. Cisneros.

Fig. 5.9. Plant regeneration and acclimation. (A-F) Embryogenic structures in regeneration medium. (A) Green embryonic structures, 2 weeks in culture. (B) Multiple embryoids formation, 3 weeks in culture. (C) Different embryoid shapes, 4 weeks in culture. (D) Appearance of yellow shoots, 6 weeks in culture. (E) Multiple plants regenerated from one embryo, 6 weeks in culture. (F) Regenerated plantlets, 8 weeks in culture. (G) Complete plant regeneration (left: *H. polyrhizus* × *H. undatus*, right: *H. undatus* × *H. polyrhizus*) showing shoots and root development, 10 weeks in culture. (H) Plantlets hardening-off in pots. (I) Two month plants. Photographs courtesy of Dr N. Tel-Zur and Dr A. Cisneros.

for *H. undatus* was estimated to be 39°C and the minimum temperature for this species is 7°C (Mizrahi and Nerd, 1999). Cultivation of *Hylocereus* spp. under high temperature conditions can be accomplished by two strategies: (i) using hybrids that can tolerate high temperature; and (ii) using grafted plants (Tomaz de Oliveira and Tel-Zur, 2022).

Tomaz de Oliveira *et al.* (2020) investigated the degree of heat tolerance (45°C/35°C, day/night) in the diploids *H. undatus* and *H. monacanthus* and the tetraploid *H. megalanthus*, and nine of their interspecific interploid hybrids. The tetraploid *H. megalanthus* (classified as heat sensitive) showed a 65% decrease in quantum efficiency of photosystem II (F_v/F_m) and severe visual stem damage, along with a marked reduction in total chlorophyll content, a large increase in malondialdehyde and inhibition of catalase activity. The diploid species *H. undatus* and *H. monacanthus* (classified as low-tolerance species) exhibited slight stem 'liquification'. The allotetraploid hybrid Z-16 showed the best performance under heat stress (21% decrease in F_v/F_m) and the absence of sun damage, coupled with a small decrease in total chlorophyll content, a slight increase in malondialdehyde, high antioxidant activity and proline accumulation progressing with time (Tomaz de Oliveira *et al.*, 2020). Most hybrids performed better than their parental species, and these hybrids are suitable for cultivation in heat-challenging areas (Fig. 5.10) (Tel-Zur, 2022).

High summer temperatures (40°C/30°C, day/night) cause poor fruiting, reduced fruit/seed weight and delayed fruit development in the 'Da-Hong' pitaya (*H. polyrhizus*), a commercial variety grown in Taiwan. Chu and Chang (2022) stated that at high temperature, incomplete fertilization occurs due to inferior pollen viability, which was considered the main cause of poor fruiting. High temperature also caused less regreening of cladodes, but this did not seem to impact fruit production (Chu and Chang, 2020).

Tomaz de Oliveira *et al.* (2021) further studied the effect of plants grafted on to *Hylocereus* spp. on tolerance to high temperature. They grafted the diploid *H. undatus*, the tetraploid *H. megalanthus*, a gamete-derived dihaploid *H. megalanthus* line and a self-compatible allotetraploid (Z-10) on to *H. undatus*, and grafted plants were compared with non-grafted plants. Self-grafted, grafted and non-grafted plants were acclimatized for 1 week and then exposed to heat stress (45°C/35°C, day/night) for 3 days, followed by a 1-week recovery period under optimal temperatures (30°C/22°C, day/night). Physiological and biochemical determinations and evaluation of expression of genes encoding heat-shock proteins showed that self-grafted and grafted plants recovered faster and suffered less stem damage than the non-grafted plants, suggesting that the grafting process can induce 'cross-tolerance' (i.e. the plants exhibited improved performances when exposed to more than one stress).

Low-temperature chilling or freezing damage in pitaya may occur when the temperature goes below 0°C (Fig. 5.10). The minimum threshold that most pitaya cultivars and varieties can tolerate is 0°C. Buds, twigs and even some mature branches can also be frozen when the temperature is lower than 5°C. When the temperature is lower than –2°C for more than 12 h, pitaya plants will be seriously injured, and when the temperature is lower than –4°C, the plants will die (Ortiz-Hernandez and Carrillo-Salazar, 2012). Gong *et al.* (2019) carried out a comparative study on the cold-resistance physiology of red- and white-fleshed pitaya. The red-fleshed pitaya variety 'Taiwan No. 2' and the white-fleshed variety 'White Crystal' were used as experimental materials and were cold stressed at the low temperature of 3°C for 48 or 96 h, or recovered at room temperature after 96 h of cold stress as a control. The relative conductivity, malondialdehyde content, soluble sugar content, soluble protein content, proline content, and activity of superoxide dismutase (SOD) and catalase (CAT) were determined and the cold resistance was analysed. The results showed that, after cold stress, the soluble protein, proline content, and SOD and CAT activities of 'White Crystal' pitaya were higher than those of 'Taiwan No. 2' pitaya, and the cold resistance of 'White Crystal' was stronger than that of 'Taiwan No. 2'.

5.9 Varieties

The named varieties and hybrids of pitaya available in different countries are shown in Table 5.1 and described below.

Fig. 5.10. Heat damage and chilling injury. Photographs courtesy of Dr Y. Mizrahi.

Table 5.1. Named varieties and hybrids of dragon fruit available in different countries.

Country	Varieties
Vietnam	TLA, TL5, LD1, LD5, LD17, LD 18, DF 18, DF 14, DF 16, Binh Thuan, Than Long
China	Hongshuijing, Hongguan No. 1, Hongguan No. 2, Shuangse No. 1, Guanhuahong, Baiyulong, Zilong, Dalong, Guanhubai, Fenhong No. 1, Hongshuijing No. 6, Shuangse 1, Jingdu 1, Daqiu 4, Yuehong No. 5, Meilong 1, Zilong, Dahong
Taiwan	Kin-Yun 1-MI-Bao, Da-Hong, HI-Xianghong, Tricolor Dragon
Malaysia	HU1, HU2, Vietnamese White, Townsend Pink, Asunta, Cebra, Purple Haze, Maria Rosa, Physical Graffiti, Ax Hybrid, S8 Sugar
USA	American Beauty, Dark Star, Haley's Comet, Natural Mystic, Physical Graffiti, Purple Haze, Vietnamese Jaina, Zamorano, Harpua, Cebra, Rosa, Orejona, Lisa, Sin Espinas, San Ignacio, Mexicana, Colombiana, Delight, Seoul Kitchen, Armando, Valdivia Roja, Bien Hoa Red, Bien Hoa White, Red Jiana, El Grullo
Israel	Omer, Sweety, Nili, Golden, triploid S-75, tetraploid (S-75 × *H. undatus*)
Brazil	BRS Lua do Cerrado, BRS Luz do Cerrado, BRS Mini-pitaya do Cerrado, BRS Ambar do Cerrado, BRS Granada do Cerrado
Ecuador	Pichincha, Palora
Honduras	Cebra, Lisa, Orejona, Rosa, Chocoya
Nicaragua	Cebra, Lisa, Sin Espinas, Orejano, San Ignacio, Chocoya
Colombia	Colobiana
Philippines	Saniata 1 (Imee), Saniata 2 (Imelda), Saniata 3 (Miriam), Saniata 4 (Edit), Saniata 5 (Mimi), Saniata 6 (Elena)
Mexico	Mexicana, Vladivia Roja, El Grullo

5.9.1 China

'Hongshuijing'

This is a self-incompatible, red-fleshed variety with thorns on the peel. The fruit shape index is 0.9 ± 0.1, the average fruit weight is 370.4 ± 30.9 g and the edible percentage is 73.6 ± 3.1%. The fruit is medium in quality (TSS 13.9 ± 1.4°Brix, total sugar 9.7 ± 1.0%, TA 0.21 ± 0.04%) (Chen *et al.*, 2018).

'Dalong'

'Dalong' is a red-fleshed variety selected by the Hainan Academy of Agricultural Science (Luo *et al.*, 2018). The fruit is oval to long oval, with a fruit shape index of 1.34. The average weight of a single fruit is 510.38 g and the maximum fruit weight is 825 g. It has a soft, purple-red pericarp, 2.60 mm thick. The flesh is bright coloured, purple to red, and with TSS of 15.18°Brix, total sugar of 10.42%, total acid of 0.25% and vitamin C content of 86.70 mg kg^{-1}. The variety is self-fruitful (>90%).

'Hongguan No. 1'

This variety has a red skin and red flesh and was selected from 'Hongshuijing' dragon fruit (*H. monocanthus*) through seedling selection (Chen *et al.*, 2018). It is a self-pollinating variety (83.7 ± 10.7%), with an average fruit weight of 390.3 ± 27.8 g, a fruit shape index of 1.0 ± 0.1%, and good-quality fruit with TSS of 16.5 ± 2.3°Brix, total sugar of 10.9 ± 0.9% and TA of 0.18 ± 0.03%. It has better marketability in terms of visual appearance of the fruit and TSS content during storage than 'Hongshuijing'.

'Shuangse No. 1'

This variety was also selected from 'Hongshuijing' dragon fruit through seedling selection (Chen *et al.*, 2018). It is a self-pollinating variety (80.5 ± 9.5%). It has a red skin and the flesh is red without and white within. The fruit shape index is 1.1 ± 0.1. The fruit is larger than that of 'Hongguan No. 1' (445.7 ± 42.4 g). It has a good-quality fruit with a TSS of 15.4 ± 1.3°Brix, total sugar of 10.7 ± 1.1% and TA of 0.20 ± 0.03%. It also has a better shelf life than 'Hongshuijing'.

'Hongguan No. 2'

This variety was selected from 'Hongshuijing' dragon fruit through seedling selection (Ma *et al.*, 2023). It is a self-pollinated variety, with spherical fruit, green–red scales and an average fruit weight of 393.5 g. The flesh is red, compact, crispy and sweet with a TSS of 14.4°Brix, total sugar of 10.9% and TA of 0.17%. It has long shelf life. The mature fruit can be stored for 7–10 days at room temperature and for more than 30 days at 7°C with 70% relative humidity.

'Fenhong No. 1'

This is an excellent dragon fruit variety selected from 'Hongshuzjing'. It has strong self-pollination ability and vigorous growth. The average fruit weight is 301.3 g. The flesh is pink, crispy and juicy with a 70.1% edible rate. The fruit quality is good (TSS 14.6°Brix, TA 0.3%). The mature fruit can be stored for 7–10 days at room temperature and for more than 30 days at 7°C with 70% relative humidity (Zhang *et al.*, 2023).

'Hongshuijing No. 6'

This is a new cultivar with outstanding quality, adaptability and consistent yield. The seedling was derived from a cross between 'Baishujing' and 'Guangminghong' (Li *et al.*, 2022). The fruit skin is red and 0.28 cm thick, and the flesh is purplish-red, crispy and sweet, with an excellent flavour and consistent quality. The average fruit weight is 295.0 g. The TSS and TA are 14.2°Brix and 0.19%, respectively. It has a very good shelf life of 7–9 days at room temperature and 20–30 days under refrigeration.

'Guanhubai'

'Guanhubai' is a new, white-fleshed cultivar selected from 'Hongshuijing' derived by seedling selection (Ye *et al.*, 2018b). The fruit is a long oval. The average fruit weight is approximately 300 g. The flesh is white with an average TSS content of 15.2°Brix and TA of 0.13%.

'Shuangse 1'

This is also a seedling selection of 'Hongshuijing'. The fruit is nearly spherical in shape and the peel is dark red. The average fruit weight is 350.7 g but can be up to 740.5 g. The flesh is red without and white within, and is smooth with a special flavour. The fruit quality is medium (TSS 13.5°Brix, TA 0.23%). The cultivar has strong self-pollination ability and a long shelf life (Qin *et al.*, 2018).

'Guanhuahong'

'Guanhuahong' is a red-fleshed dragon fruit cultivar selected from 'Hongshuijing' through seedling selection (Ye *et al.*, 2018a). The average fruit weight is 376.7 g, and it is nearly oval to round in shape. The flesh is red, with an average TSS of 14.5°Brix and TA of 0.17%. This variety has strong self-compatibility.

'Jingdu 1'

This variety was selected from the hybrid offspring between *H. costaricensis* and *H. polyrhizus* (Su *et al.*, 2017). The fruit is short and oval, with a purplish-red background colour. The average fruit weight is about 572.5 g, with an edible rate of 78.8%. The flesh is delicate and sweet (21.2°Brix) and has a rose fragrance. It is self-compatible, a high yielder and has excellent quality fruit.

'Daqiu 4'

This variety was derived from the cross between 'Guangminghong' as the female parent and 'Baishuijing' as the male (Wang *et al.*, 2022). It is a vigorous plant. The fruit is nearly circular with an average weight of 3216.0 g, and an edible rate of 73.2%. The fruit peel has many bracts and is light red. The flesh is purplish-red, smooth, refreshing and sweet (TSS 14.24°Brix, TA 0.191%).

'Yuehong No. 5'

This was selected from the progeny of a cross between 'Dahong' × Common white flesh (Dai *et al.*, 2022). The fruit is oval and medium in size (310.0 g). After full ripening, the peel is pink and the flesh is white, compact, crispy, juicy and sweet (TSS 14.14°Brix, TA 0.172%). The fruit setting rate from natural pollination is 92%. This variety has strong resistance to dragon fruit canker disease.

'Meilong 1'

This was selected from hybrid offspring of a Costa Rican native species (*H. costaricensis*) and 'Baiyulong' (*H. undatus*). The fruit has a long oval shape, a bright green squam and red peel. The average fruit weight is 525 g and the edible rate is 76%. The flesh is red, delicate, sweet (TSS 20.1°Brix) and slightly fragrant. The flower has high self-compatibility and gives a high and stable yield (Huang *et al.*, 2021).

'Zilong'

This selection was bred by researchers in Hainan Province. It has red skin and red flesh, and can flower and bear fruit 12 months after planting. The fruit is medium in size (380 g), oval to round in shape and has a red skin. The flesh is purplish-red in colour, with a delicate juicy texture and a TSS of 16–19%. It has a rich flavour and a smooth feel in the mouth.

The 'Dahong' (Fig. 5.11) and 'Zilong' (Fig 5.12) varieties were developed by the Tropical Crops Genetic Resource Institute (Chinese Academy of Tropical Agricultural Science), the Institute of Tropical Fruit Trees (Hainan Academy of Agricultural

Fig. 5.11. 'Dahong' variety.

Fig. 5.12. 'Zilong' variety.

Science) and the Fruit Tree Institute, Fujian Academy of Agricultural Sciences.

5.9.2 USA

A number of dragon fruit varieties/selections have been registered (Gasic *et al.*, 2016) and are maintained at San Diego, California. Commercial nurseries in California and Florida also maintain several varieties. Brief descriptions of a few of these are given below.

'Cebra' (H. polyrhizus)

This originated in Nicaragua. The fruit is oval and large (~500 g) with red flesh and a thick rind, and has a TSS of 15.75°Brix. The fruit has harvested from late June to late December or in January in Southern California.

'Rosa' (H. polyrhizus)

This originated in Nicaragua. The fruit is round and medium in size (~400 g), with a thin peel, and is susceptible to cracking when mature. It has an acceptable quality, with TSS of 16.05°Brix. The fruit is harvested from late June to late December or January.

'Orejona' (H. polyrhizus)

This originated in Nicaragua. It has red-fleshed fruit with average flavour and size (~450 g), a red, thick skin and a TSS of 15.78°Brix. It is harvested from late June to late December.

'Lisa' (H. polyrhizus)

This originated in Nicaragua. The fruit is oval in shape, medium to large (465 g), with red skin and dark red flesh, which is sweet (17.02°Brix). It is harvested from late June to late December.

'Sin Espinas' (H. polyrhizus)

This originated in Nicaragua. It has red-fleshed fruit that is round in shape, medium in size (~400 g) and glossy, with pinkish-red skin and of good quality (TSS 16.5°Brix). It is harvested from early June to December.

'San Ignacio' (H. polyrhizus)

This originated in Nicaragua. The fruit is large (~550 g) and oval in shape with red flesh. It has an average flavour, a thick red skin and acceptable fruit quality (TSS 15.6°Brix). It is harvested from late June to late December or January.

'Mexicana' (H. undatus)

This originated in Mexico. It has white-fleshed, oblong-shaped fruit, which are large (~550 g) with an attractive appearance, a relatively thin pinkish-red skin and acceptable quality (TSS 14.04°Brix). In Southern California, it is harvested from mid-July to mid-October.

'Colombiana' (H. megalanthus)

This originated in Colombia. It has white-fleshed, oblong-shaped fruit, which are very small in size (160 g), with a yellow skin and a thick and spiny rind. The flavour is excellent and extremely sweet (TSS 21°Brix). In Southern California, it is harvested from November to February.

'Valdivia Roja' (H. ocamponis)

This originated in Mexico. The fruit is oblong, red-fleshed, firm and small (250 g), with a greenish-red skin with very few short bracts, and is excellent quality (TSS 17.9°Brix). It is harvested in Southern California from July to late December or January.

'Bien Hoa Red' (H. guatamalensis)

This originated in Guatemala. It has round, medium-sized fruit (~350 g). The flesh is fuchsia coloured and juicy. It has a reddish-green, thin skin with scattered compact bracts. The fruit is of exceptional quality (TSS 18.9°Brix). In Southern California, it is harvested from June to November.

'Bien Hoa White' (H. undatus)

This originated on Bien Hoa Farms, Fallbrook California. It has white-fleshed, oblong-shaped fruit, which is medium in size (~400 g). The skin is pinkish-red and relatively thin with long bracts, and is poor quality (TS 11.85°Brix). It is harvested in Southern California from mid-July to mid-October.

'Delight' (Hylocereus *sp.*)

This was bred and selected by Paul Thomson in San Diego County in the early 1980s. It has a pinkish-white- or light pink-fleshed oval fruit, which is medium in size (~370 g). The rind is greenish-red and has bracts that are short and scattered. It has excellent fruit quality (TSS 18.08°Brix). In Southern California, it is harvested from late June to mid-December.

'American Beauty' (H. guatemalensis)

The parentage of this variety is unknown. It was introduced from Guatemala to Florida, and then to California. It has round-shaped fruit of medium size (380 g), with fuchsia-coloured flesh. The skin is greenish-red, and the flesh is juicy with an excellent flavour and quality (TSS 18.51°Brix). It is harvested in Southern California from June to November.

'Haley's Comet' (Hylocereus *sp.*)

This was developed by Paul Thomson, in San Diego County, California, in the early 1980s, It has pink-fleshed fruit with excellent flavour; the fruit is round and large (~480 g), and the rind is bright red (Fig. 5.13) and thin, with an intermediate number of bracts. The fruit is good quality (TSS 16.7°Brix) and rated high in sensory evaluations. It is harvested in Southern California from July to December.

'Physical Graffiti' (Hylocereus *sp.*)

This was also developed by Paul Thomson in the early 1980s. It is a very productive variety, with pink-fleshed, oblong-shaped, medium fruit (~375 g). The rind is bright red (Fig. 5.14) and relatively thin with few bracts. The fruit is excellent quality (TSS 17.93° Brix). It is harvested in Southern California from July to December.

'Vietnamese Jaina' (H. undatus)

This originated in Florida. It has attractive, white-fleshed, oval-shaped fruit, which is medium in size (~340 g). The skin is pinkish-red (Fig. 5.15) and relatively thin, and the flesh is very firm and white, with numerous small, black seeds. It is excellent quality (TSS 15.6°Brix), and is a self-pollinating variety and a consistent producer.

'Red Jiana' (H. polyrhizus)

This variety has dark red-fleshed fruit of 300–500 g (Fig. 5.16). It requires a pollinator and has very large flowers (30–40 cm). It is a high yielder.

'Seoul Kitchen' (H. undatus)

This originated in Florida. It has white-fleshed, round, pinkish-red-skinned fruit (Fig. 5.17), which is large (~520 g) and of marginal quality (TSS 12.18°Brix). It is a vigorous plant, and a strong and prolific bearer. It is harvested from mid-July to December in Southern California.

'Armando' (H. polyrhizus/costaricensis)

This was collected by Ramiro Lobo in Las Manos, near the border of Honduras and Nicaragua. It has round fruit, medium in size (~400 g), with firm, red flesh. The skin is purplish-red, of medium thickness and with numerous bracts. It is good quality (TSS 16.11°Brix). In Southern California, it is harvested from mid-June to January.

'Zamorano' (H. polyrhizus)

This variety was introduced from Honduras. The fruit is medium in size (350–450 g) with a dark red flesh and dusky red exterior (Fig. 5.18). It is self-fertile. It has a mild sweet flavour.

'El Grullo' (H. ocamponis)

This came from an unnamed selection collected by Ramiro Lobo in El Grullo State, Mexico. The

Fig. 5.13. 'Haley's Comet' variety. Photograph courtesy of Dr P. Wangsa.

Fig. 5.14. 'Physical Graffti' variety. Photograph courtesy of Dr P. Wangsa.

fruit is large (~490 g), oval and red fleshed, with a greenish-red skin. The rind is thick and hard with numerous compact bracts. The fruit is excellent quality (TSS 18.11°Brix). It is harvested from mid-July to December in Southern California.

5.9.3 Taiwan

'Kin-Yun 1-MI-Bao'

The variety was registered in 2012. The fruit is oval or spherical in shape, and the scales are shorter than most white-fleshed varieties. It has a shallow calyx depth and is not easily cracked. It has good-quality fruit (TSS 18°Brix), with good keeping quality.

Fig. 5.15. 'Vietnamese Jaina' variety. Photograph courtesy of Ian Maguire, UF/IFAS.

'Da-Hong'

This variety has large fruit (400 g), with dark red flesh and a thin peel. It is self-compatible, and its scales are wide, short and thin. The fruit has excellent quality (TSS 20°Brix). This variety accounts for about 60% of red-fleshed varieties in Taiwan.

'HI-Xianghong'

A strong aroma is the most prominent feature of this variety. The fruit is long and oval shaped. It has longer rewinding scales and does not crack easily. It is self-compatible and the fruit is good quality (TSS 18°Brix).

'Tricolor Dragon'

This is a hybrid of the variety 'Xiang-Long' from Central and South America and a white-fleshed variety. Its flesh is bicolor and varies with temperature. The flesh colour is white between May and July, turns pink near the peel and white in the centre between July and September, and becomes red near the peel between October and November, showing clear red and white colours. After November, the flesh colour turns pink. It has a very good taste.

5.9.4 Malaysia

Only two varieties have been registered with the Department of Agriculture (Malaysia). These

Fig. 5.16. 'Red Jiana variety'. Photograph courtesy of Ian Maguire, UF/IFAS.

Fig. 5.17. 'Seoul Kitchen' variety. Photographs courtesy of Dr P. Wangsa.

Fig. 5.18. 'Zamorano' variety. Photograph courtesy of Ian Maguire.

are 'HU1' ('Pink Dragon Sunlike') and 'HU2' ('Iguana') (Fig. 5.19A, B), and are widely planted throughout the country. Several other varieties have been introduced but not registered, such as 'Vietnamese White', 'Townsend Pink', 'Asunta', 'Cebra', 'Purple Haze', 'Maria Rosa', 'Physical Graffiti', and 'S8 Sugar' (Sarip, 2020).

5.9.5 Vietnam

'Binh Thua'

This cultivar/clone was introduced by the French in the late 1800s. It is a white-fleshed fruit, with small, black seeds interspersed throughout the flesh, and a bright red skin when mature.

'Thanh Long' (also known as 'Ruot Do' or red flesh)

This red-fleshed cultivar was imported from Thailand and Taiwan. The fruit is suitable for export.

'DF 16' (H. undatus)

This was selected by the researchers at SOFRI in 2016. It has medium-sized fruit (290 g), with a thick (2 mm), dark pink to red skin. The flesh colour is light pink and is very sweet (TSS

Fig. 5.19 (A) 'HU1' variety. (B) 'HU2' variety. Photographs courtesy of Ahmad Hafiz Baharom.

19°Brix), with an excellent flavour. It ripens from mid-May to the end of November in Long Dinh (Karp and Gasic, 2022).

*'DF 14' (*H. undatus*)*

This variety was selected by SOFRI in 2016. It has medium-sized fruit (225 g) and dark pink to red skin of average thickness (2 mm). The flesh is almost white with a pink stripe under the skin, and is medium sweet (TSS 16°Brix) with an excellent flavour. It ripens from mid-May to end November in Long Dinh (Karp and Gasic, 2022).

5.9.6 India

The most widely grown type of dragon fruit in India are the red-skinned, white-fleshed varieties (93%), followed by red-skinned, purple or red-fleshed varieties (6.5%) and yellow-skinned, white-fleshed varieties (<0.5%) (Wakchaure *et al.*, 2020). Punjab Agricultural University (Ludhiana, India) collected elite clones from Karnataka, Punjab, Maharashtra, Assam, West Bengal, and Andaman and Nicobar Islands, which were conserved and evaluated at the university research station. Two clones were selected, named 'Red Dragon 1' and 'White Dragon 1', which were released recently and have been recommended for cultivation for the growers of Punjab (Dr J.S. Brar, personal communication).

'Red Dragon 1'

The fruit of this variety is oblong with red-green bracts and an average weight of 325 g. The flesh is red and embedded with small, black seeds (Fig. 5.20). It has a TSS of 9.48°Bri × and TA of 0.41%. The fruit is rich in the antioxidant β-carotene, phenols and flavanols and in minerals such as potassium, calcium and magnesium. It bears fruit from July to November under Punjab conditions.

'White Dragon 1'

The fruit is oblong with reddish-green bracts and an average weight of 285 g. The flesh is white and embedded with small, black seeds (Fig. 5.21). The fruit has a TSS of 9.24°Brix, TA of 0.62%, and is rich in the micronutrients iron, zinc and manganese. It bears fruit from July to November.

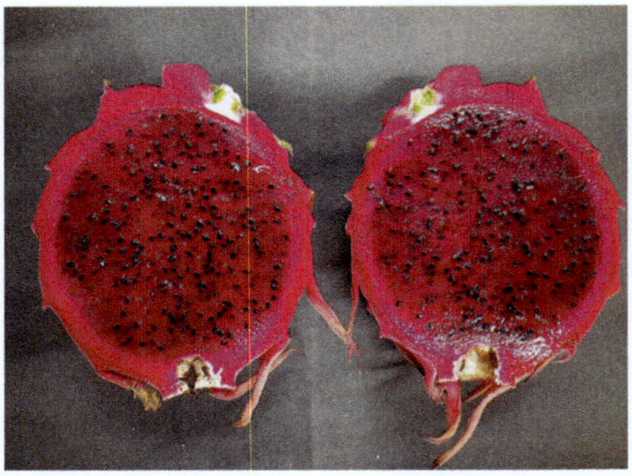

Fig. 5.20. 'Red Dragon 1' variety. Photograph courtesy of Dr J.S. Brar.

Fig. 5.21. Flesh colour of 'White Dragon 1' and 'Red Dragon 1'. Photograph courtesy of Dr J.S. Brar.

5.9.7 Ecuador

In Ecuador, red pitahaya and yellow pitahaya are cultivated; the latter is attractive for its appearance, comprising a yellow skin with thorns and aromatic white flesh with small, black seeds. There are two ecotypes of yellow pitahaya grown in Ecuador, 'Pichincha' also known as 'Nacional' (with fruit up to 150 g in weight), which is grown in the north-west of Pichincha, and the 'Palora' (with fruit up to 350 g in weight), which is grown in Morona Santiago (Trujillo, 2014).

5.9.8 Thailand

There are three species of dragon fruit grown in Thailand. More than 95% of the total production is red-skinned, white-fleshed dragon fruit (*H. undatus*), about 4% is a red-skinned, red-fleshed fruit (*H. costaricensis*/*H. polyrhizus*) and less than 1% is yellow-skinned, white-fleshed fruit (*S. megalanthus*). The red-skinned, red-fleshed fruit is gaining more popularity among dragon fruit lovers and is accepted for export to other countries.

5.9.9 Indonesia

In Indonesia, dragon fruit is called 'Buah Naga'. Before 2000, it was not an important crop and commercial planting was only started in the early 2000s. Commercial planting was established in the south of Yogyakarta, East Java, West Sumatra and Padang Pariaman district and in the Riau Islands region. However, it has spread to other places of Indonesia far from the coast, such as Kalimantan, Sulawesi and Bali (Muas *et al.*, 2021). Both red-fleshed and white-fleshed clones are grown in Indonesia.

5.9.10 Philippines

Dragon fruit was first introduced into the Philippines in 1992 from Vietnam and Taiwan. The white-fleshed variety was initially produced on a small farm in the Cavite region, which soon expanded to other farms in the region and then into other areas of the country. The Philippine Department of Agriculture has identified two species currently produced in this country: *H. monacanthus* and *H. undatus* (DAFF, 2022).

Breeders in the Philippines have developed the Saniata variety series: 'Saniata 1' ('Imee'), 'Saniata 2' ('Imelda'), 'Saniata 3' ('Miriam'), 'Saniata 4' ('Edit'), 'Saniata 5' ('Mimi') and 'Saniata 6' ('Elena') (Pascua *et al.*, 2015).

'Saniata 1'

The fruit of this variety is medium to large (337 g), with brick red-coloured flesh, a light red peel and an edible portion of 75%. The flesh has a TSS of 13.6°Brix and a grainy texture.

'Saniata 2'

The fruit is large (401 g), and the flesh and peel are light red. The edible portion is 75% and the TSS is 13.3°Brix. The flesh texture is grainy.

'Saniata 3'

The fruit is large (466 g) and white fleshed, with a dark pink peel. The edible portion is 81% with a TSS of 11.2°Brix and a grainy texture.

'Saniata 4'

The fruit is medium to large (374 g) with brick red-coloured flesh and a red peel. The edible portion is 73% and the TSS is 12.4°Brix. The flesh texture is grainy.

'Saniata 5'

The fruit is small (245 g) with brick red-coloured flesh and the peel has a sunset glow. The edible portion is 69% and the TSS is 13.5°Brix. The flesh texture is grainy.

'Saniata 6'

This variety has very small fruit (75 g) with white flesh and yellow peel. The edible portion is 41% and texture of the flesh is smooth.

5.9.11 Brazil

Dragon fruit has been cultivated in Brazil for over two decades. The external and internal appearance of the fruit, mainly the colouring, is what differentiates the types. Four species are of commercial importance in this country: *H. undatus*, *H. polyrhizus*, *H. setaceous* (syn. *S. setaceus*) and *H. megalanthus* (da Silva *et al.*, 2017). The Ministry of Agriculture, Livestock and Food Supply (MAPA, Brazil) recently registered five varieties: 'BRS Lua do Cerrado', 'BRS Luz do Cerrado', 'BRS Mini-pitaya do Cerrado', 'BRS Granada do Cerrado' and 'BRS Amber do Cerrado' (Faleiro and Junqueira, 2022).

*'BRS Lua do Cerrado' ('BRS LC') (*S. undatus*)*

The fruit is round in shape and red with white flesh. This variety is highly productive (Fig. 5.22) and self-compatible. The fruit is large (500–1000 g), with a high TSS and has good resistance to diseases.

*'BRS Luz do Cerrado' ('BRS LZC') (*S. undatus*)*

This variety has elongated, red fruit with white flesh and is self-compatible. The fruit is large (500–1000 g) and is highly productive (Fig. 5.23). The fruit has a high TSS content.

*'BRS Mini-pitaya do Cerrado' ('BRS MPC') (*S. setaceus*)*

The fruit is red with thorns and white flesh and is used in ornamental fruit growing and for the speciality fruit market. It is self-compatible, and the fruit is small (Fig. 5.24) but very sweet with a slight acidity, which gives a special flavour.

*'BRS Granada do Cerrado' ('BRS GC') (*S. undatus* × *S. costaricensis*)*

This is a purple-fleshed, red dragon fruit hybrid (Fig. 5.25) and is self-compatible. The fruit is medium in size (250 g), with a high TSS content, and is rich in antioxidants.

*'BRS Ambar do Cerrado' ('BRS AC') (*S. megalanthus*)*

The fruit is medium in size (Fig. 5.26) and very sweet with a special flavour. This variety is produced off-season under the conditions of Cerrado of the Central Plateau.

Fig. 5.22. 'BRS Lua do Cerrado' variety. Photograph courtesy of Embrapa Cerrados.

5.9.12 South Africa

It is estimated that less than 100 ha are under dragon fruit cultivation in South Africa. However, a number of varieties such as 'Azalea', 'Ballet', 'Blood', 'Fiery', 'Kiss', 'Flamingo', 'Fuchsia', 'India', 'Pink Moon', 'Snow' and 'Sweet Coral' (Fig. 5.27) are available in commercial nurseries in South Africa.

5.9.13 Israel

Ben-Gurion University of the Negev, Beer-Sheva, Israel, established a large live gene bank of *Hylocereus* genotypes and has selected several superior clones over the years. Some of the important clones and their characteristics are as follows.

'Yossi' (former name 'Omer') (04-25)

This is a summer-fruiting clone (Fig. 5.28), which ripens from August to December. It has a long shelf life (3 weeks at 10°C plus 5 days at 20°C). It is resistant to heat (>38°C) and cold temperatures (close to 0°C). The average fruit weight is 410 g. The taste score is 3.9 out of 5.

'Venus' (34-07)

This is a summer-fruiting clone (Fig. 5.29) and ripens from August to December. It has an air-freight shelf life (14 days at 10°C + 5 days at 20°C). It needs more pruning than 'Yossi' and is sensitive to soot in high-humidity zones. It has average tolerance to heat and cold. The average fruit weight is 380 g. The taste score is 3.5 out of 5.

'Golden' (98-338)

This is a summer-fruiting clone, ripening from August to December. It is a very strong plant, resistant to heat and cold temperatures. It has an air-freight shelf life (14 days at 10°C + 5 days at 20°C). The average fruit weight is 482 g (Fig. 5.30). The taste score is 3.1 out of 5.

Fig. 5.23. 'BRS Luz do Cerrado' variety. Photograph courtesy of Embrapa Cerrados.

Fig. 5.24. 'BRS Mini-pitaya do Cerrado' variety. Photograph courtesy of Embrapa Cerrados.

'Nilly' (S-107)

This is a summer-fruiting clone (Fig. 5.31), ripening from August to December. It is highly resistant to heat. It has an air-freight shelf life (14 days at 10°C + 5 days at 20°C). The average fruit weight is 280 g. It is the sweetest of all of the white-fleshed clones. The taste score is 3.9 out of 5.

'Mouyal' (S-91)

This is a summer-fruiting clone, ripening from August to December. It is resistant to moderate heat. It has an air-freight shelf life (14 days at 10°C + 5 days at 20°C). The average fruit weight is 320 g (Fig. 5.32). It has a very good taste score of 3.8 out of 5.

'Desert King' (S-75)

This is a winter-fruiting clone (Fig. 5.33), with the fruit harvested from late October to April. It cannot tolerate heat or cold temperatures. It performs well in warm climates. It has spines that are easy to remove upon ripening, and can be harvested using bare hands. It has a

Fig. 5.25. 'BRS Granada do Cerrado' variety. Photograph courtesy of Embrapa Cerrados.

long sea-freight shelf life (3 weeks at 10°C + 5 days at 20°C). The average fruit weight is 220 g. The fruit is very tasty, with a taste score of 4.3 out of 5.

'Julia' (J-21)

This is a summer-fruiting clone, ripening from August to December. It has a shorter air-freight shelf life (10 days at 10°C + 5 days at 20°C). The average fruit weight is 456 g (Fig. 5.34). The average taste score is 3.6 out of 5. The fruit is very much liked by European and North American visitors.

'Tantan' (clone 17)

This is a winter-fruiting clone, ripening from October to March. It is resistant to moderate heat. It has a sea-freight shelf life (21 days at 10°C + 5 days at 20°C). It is almost spineless, and the average fruit weight is 340 g (Fig. 5.35). The taste score is 4.2 out of 5.

5.10 Summary and Conclusion

The initial steps of dragon fruit improvement started by collecting elite lines in terms of yield,

Fig. 5.26. 'BRS Ambar do Cerrado' variety. Photograph courtesy of Embrapa Cerrados.

fruit quality, and abiotic and biotic stress tolerance. Collections were made from cactus collectors, botanical gardens, private gardens, wild habitation and countries of origin. Researchers in different growing countries collected and evaluated the genotypes, established a field gene bank and subsequently selected cultivars to grow in their countries. Today, the major commercial varieties of different countries are basically the selection from a large population or introduction of already established varieties from other countries.

Only about 10% of *Hylocereus* spp. are self-compatible. Cross-pollinated plants produce fruit that have a better size and weight than self-pollinated plants. The major barriers for intra- and interspecific hybridization of *Hylocereus* spp. are auto-infertility and asynchronicity in the flower aperture. However, fertility between species is evident, as cross-pollination among several species promotes 100% fruit set. The diploid *Hylocereus* spp. produce large, flavourless, red-purple fruit, whereas the tetraploid *H. megalanthus* bears sweeter but smaller yellow fruit. Crosses between *H. undatus* and *H. polyrhizus* bear fruit with a high sugar content. *H. megalanthus* × *H. undatus* crosses did not yield true hybrids or chimeras, but in reciprocal crosses, tetraploids, pentaploids, hexaploids and 6x aneuploids were produced indicating a unidirectional gene flow among these species. Interspecific homoploid and interploid crosses and embryo-rescue procedures produced improved hybrids. Molecular tools and morphological and phenological comparisons between the parental species and resulting hybrids provided valuable information on dominant/recessive traits and on genetic relationships that could be exploited for further hybridization. *In situ* chromosome doubling resulted in the production of autopolyploid lines, from which an understanding of the effect of increased ploidy on fruit traits was obtained. Gamete-derived lines were produced, which could be used for further hybridization programmes. Spontaneous chromosome doubling occurred in haploid gamete-derived *H. monacanthus* lines and in interspecific interploid *H. monacanthus* × *H. undatus* hybrids obtained from an embryo-rescue procedure, resulting in plants with double and expected ploidy.

The degree of heat tolerance has been investigated. Allotetraploid hybrids showed potential for cultivation in heat-challenging areas. Grafted plants were found to induce cross-tolerance.

Consumers are intrigued by the exotic nature of dragon fruit with their bright red skins, greenish scales, white or red flesh, nutritional value and antioxidant content. However, a common refrain from consumers is that, although it has a delicate flavour and antioxidant benefits, the taste is sometimes bland and not sweet. This has led to importation of the sweeter yellow dragon fruit from Ecuador and Colombia into North America and Europe. The current limited number of varietal selections used commercially has major long-term consumer appeal issues if the crop is moved more widely into global markets and beyond specialty stores and local markets. Future breeding should be targeted to produce hybrids with a delicate flavour and high sugar content, as well as higher productivity.

Fig. 5.27. Dragon fruit varieties of South Africa. Photograph courtesy of Amorentia Sweet Dragon Fruit.

Fig. 5.28. 'Yossi' clone. Photograph courtesy of Dr Y. Mizrahi.

Fig. 5.29. 'Venus' clone. Photograph courtesy of Dr Y. Mizrahi.

Fig. 5.30 'Golden' clone. Photograph courtesy of Dr Y. Mizrahi.

Fig. 5.31. 'Nilly' clone. Photograph courtesy of Dr Y. Mizrahi.

Fig. 5.32. 'Mouyal' clone. Photograph courtesy of Dr Y. Mizrahi.

Fig. 5.33. 'Desert King' clone. Photograph courtesy of Dr Y. Mizrahi.

Fig. 5.34. 'Julia' clone. Photograph courtesy of Dr Y. Mizrahi.

Fig. 5.35. 'Tantan' clone. Photograph courtesy of Dr Y. Mizrahi.

5.11 References

Abirami, K., Swain, S., Baskaran, V., Venkatesan, K., Sakthivel, K. and Bommayasamy, N. (2021) Distinguishing three dragon fruit (*Hylocereus* spp.) species grown in Andaman and Nicobar Islands of India using morphological, biochemical and molecular traits. *Scientific Reports* 11: 2894.

Barthlott, W. and Hunt, D.R. (1993) Cactaceae. In: Kubitzki, K., Rohwer, J.G. and Bittrich V. (eds) *The Families and Genera of Vascular Plants*. Springer, Berlin, pp. 161–197.

Bauer, R. (2003) A synopsis of the tribe *Hylocereus* F. Buxb. *Cactaceae Systematics Initiatives* 17, 1–63.

Benega Garcia, R., Cisneros, A., Schneider, B. and Tel-Zur, N. (2009a) Gynogenesis in the vine cacti *Hylocereus* and *Selenicereus* (Cactaceae). *Plant Cell Reports* 28, 719–726.

Benega Garcia, R., Schneider, B. and Tel-Zur, N. (2009b) Androgenesis in the vine cacti *Hylocereus* and *Selenicereus* (Cactaceae). *Plant Cell, Tissue and Organ Culture* 96, 191–199.

Calix de Dios, H. (2004) Geographical distribution of pithayas (*Hylocereus*) in Mexico. *Cactaceas y Suculentas Mexicanas* 49, 4–23.

Castillo, R., Livera, M. and Marquez, G.J. (2005) Caracterizacion morphologica y compatibilidad sexual de cinco genotipos de Pitahaya (*Hylocereus undatus*). *Agrociencia* 39, 183–194.

Chen, C., Wu, P., Xie, F., Sun, L., Xing, Y. *et al.* (2018) Breeding of 'Hongguan No.1' and Shuangse No.1' pitayas with superior quality. *HortScience* 53, 404–409.

Chu, Y.C. and Chang, J.C. (2020) High temperature suppresses fruit/seed set and weight, and cladode regreening in red-fleshed 'Da Hong' pitaya (*Hylocereus polyrhizus*) under controlled conditions. *HortScience* 55, 1259–1264.

Chu, Y.-C. and Chang, J.-C. (2022) Heat stress leads to poor fruiting mainly due to inferior pollen viability and reduces shoot photosystem II efficiency in "Da Hong" pitaya. *Agronomy* 12, 225.

Cisneros, A. and Tel-Zur, N. (2010) Embryo rescue and plant regeneration following interspecific crosses in the genus *Hylocereus* (Cactaceae). *Euphytica* 174, 73–82.

Cisneros, A. and Tel-Zur, N. (2010) Embryo rescues and plant regeneration following interspecific crosses in the genus *Hylocereus* (Cactaceae). *Euphytica* 174, 73–82.

Cisneros, A., Garcia, R. and Tel-Zur, N. (2011) Ovule morphology, embryology and seed development in three *Hylocereus* species (Cactaceae). *Flora* 206, 1076–1084.

Cisneros, A., Garcia, R. and Tel-Zur, N. (2013) Creation of novel interspecific-interploid *Hylocereus* hybrids (Cactaceae) via embryo rescue. *Euphytica* 189, 433–443.

Cohen, H. and Tel-Zur, N. (2012) Morphological changes and self-incompatibility breakdown associated with autopolyploidization in *Hylocereus* species (Cactaceae). *Euphytica* 184, 345–354.

Cohen, H., Fait, A. and Tel-Zur, N. (2013) Morphological, cytological and metabolic consequences of autopolyploidization in *Hylocereus* (Cactaceae) species. *BMC Plant Biology* 13: 173.

da Silva, A.C.C., Sabico, R., Chiamolera, F.M., Segantini, D.M. and Martins, A.B.G. (2017) Morphological traits as tool to verify genetic variability of interspecific dragon fruit hybrids. *Revista Brasileria Fruiticultura* 39, e-168.

DAFF (Department of Agriculture, Fisheries and Forestry) (2022) *Dragon Fruit from the Philippines: Biosecurity Import Requirements Draft Report*. DAFF, Australian Government, Canberra, Australia.
Dai, H.F., Li, J.C. and Sun, Q.M. (2022) Breeding report of a new pitaya cultivar Yuehong No. 5 *Journal of Fruit Science* 39, 2205–2208.
de Lima, C.A., Faleiro, F.G. and Juanqueira, N.T.V. (2013) Intra and interspecific genetic diversity of pitaya based on physical and chemical characteristics of fruits. *Revista Brasileira de Fruticultura* 35, 1066–1072.
Devaux, P. and Pickering, R. (2005) Haploids in the environment of Poaceae. In: Don Palmer, C.E., Keller, W. and Kasha, K. (eds) *Haploids in Crop Improvement II*. Springer, Berlin, pp. 215–242.
Dhooghe, E., Laere, K., Eeckhaut, T., Leus, L. and van Huylenbroeck, J. (2011) Mitotic chromosome doubling of plant tissue *in vitro*. *Plant Cell, Tissue and Organ Culture* 104, 359–373.
Dotor, M.Y., Gonzalez, L.A., Castro, M.A., Morillo, A.C. and Morillo, Y. (2016) Analysis de la diversidad genetic de la mora (*Rubus* spp.) en la departamento de Boyaca. *Bioteconologia en el Sector Agropecuario y Agroindustrial* 14, 10–17.
Faleiro, F.G. and Junqueira, N.T.V. (2021) *Pitayas: Atividades de Pesquisa, Desenvolvimento e Inovacao na*. Embrapa Cerrados, Planaltina, Brazil.
Faleiro, F.G. and Junqueira, N.T.V. (2022) Melhoramento genetic das pitaias na Embrapa e parceiros. In: V.R. Lacerda (ed.) *III Encontro Macional dos Produfores de Pitaia.* FEPAF, Botucatu, Brazil, pp. 171–183.
Fresh Plaza (2023) New dragon fruit varieties set to enter the global marketplace. Joint New Zealand–Vietnam fruit breeding programme. Available at: www.freshplaza.com/europe/article/9502657/new-dragon-fruit-varieties-set-to-enter-the-global-marketplace/ (accessed 9 November 2023).
Gasic, K., Preece, J.E. and Karp, D. (2016) Register of new fruit and nut cultivars list 48. *HortScience* 51, 620–652.
Goenaga, R., Marrero, A. and Perez, D. (2020) Yield and fruit quality traits of dragon fruit grown in Puerto Rico. *HortTechnology* 30, 803–808.
Gong, Y., Bi, X., Deng, L., Hu, J., Jiang, S. *et al.* (2019) Comparative study of cold resistance physiology of red pulp pitaya and white pulp pitaya. *E3S Web of Conference* 131: 01113.
Gonzalez, S. and Alvarado, J.Y. (2004) Utilizacion de caracteres culitativos y cuantitativos determinates en la variacon fenotopica de pitahaya (*Hylocereus undatus* Britt and Roose) que permiten propener una guia de descriptors. Thesis, Facultad de Agronomia, Universidad Nacional Agraria, Managua, Nicaragua.
Guha, S.I.P.R. and Maheshvari, S.C.D. (1964) *In vitro* production of embryos from anthers of *Datura*. *Nature* 204, 497.
Hang, N.T.M., Hooa, N., Chau, M. and Chang, W.N. (2014) Research strategies to increase sustainable production of dragon fruit and passion fruit. In: *Proceedings of the Workshop on Increasing Production and Marketing Access for Tropical Fruits in South East Asia, 13–17 October*. Southern Horticultural Research Institute, Vietnam, pp. 115–119.
Huang, L.F., Deng, H.Y., Wang, X.M., Ren, H., Liu, Y.Q. *et al* (2021) A new red-fleshed pitaya cultivar 'Mailong 1'. *Acta Horticulturae Sinica* 48, 1267–1268.
Hunt, D. (2006) *The New Cactus Lexicon*. DH Books, Milborne Port, UK.
Junqueira, K.P., Faleiro, F.G., Junqueira, N.T.V., Bellon, G., Lima, C.A.D. and Souza, L.S.D. (2010) Genetic diversity of native pitaya native from Brazilian savannas with basis on RAPD markers. *Revista Brasileira de Fruticultura* 32, 819–824.
Karp, J.E. and Gasic, K. (2022) Register of New Fruit and Nut Cultivars List 51. *HortScience* 57, 1174–1233.
Lee, W.-L. and Liu, P.-C. (2021) Shuttle breeding: a strategy for improving commercially grown pitaya. In: Fullerton, R.A. and Lee, W.L. (2021) *Dragon Fruit – From Production to Market*. Food and Fertilizer Technology Center for the Asian and Pacific Region, Taipei, Taiwan, pp. 161–168.
Li, D., Arroyave Martinez, M.F., Shaked, R. and Tel-Zur, N. (2018) Homozygote depression in gamete-derived dragon fruit (*Hylocereus*) lines. *Frontiers in Plant Science* 8, 1–13.
Li, J.C., Dai, H.F. and Sun, Q.M. (2022) Breeding report of a new pitaya cultivar Hongshuijing No. 6. *Journal of Fruit Science* 39, 1973–1976.
Lichtenzveig, J., Abbo, S., Nerd, A., Tel-Zur, N. and Mizrahi, Y. (2000) Cytology and mating systems in the climbing cacti *Hylocereus* and *Selenicereus*. *American Journal of Botany* 87, 1058–1065.
Luo, Z.W., Li, X.H., Peng, C., Han, B., He, F. *et al.* (2018) A new red pulp dragon fruit cultivar 'Dalong'. *Journal of Fruit Science* 35, 642–645.
Ma, M.B., Zhang, Z., Hu, G.B., Zhang, R., Yan, C.D. *et al.* (2023) A new pitaya cultivar Hongguan No. 2. *Journal of Fruit Science* 40, 1497–1502.
Miyajima, D. (2006) Ovules that failed to form seeds in zinnia (*Zinnia violacea* Cav.). *Scientia Horticulturae* 107, 176–182.
Mizrahi, Y. (2014) Vine-cacti pitayas: the new crop of the world. *Revista Brasilaria Fruticultura* 36, 124–138.

Mizrahi, Y. and Nerd, A. (1999) Climbing and columnar cacti: new and land fruit crops. In: Janick, J. (ed.) *Perspective on New Crops and New Uses*. American Society for Horticultural Science Press, Alexandria, Virginia, pp. 358–366.

Morillo, A.C., Mora, M.S. and Morillo, Y. (2022) Analysis of the genetic diversity of dragon fruit based in ISSR markers in Colombia. *Brazilian Journal of Biology* 82: e25465.

Morillo-Coronado, A.C., Hernandez, E.H.M. and Forero-Mancipe, L. (2021) Phenotypic diversity of morphological characteristics of pitahaya (*Selenicereus megalanthus* Haw.) germplasm in Colombia. *Plants* 10: 2255.

Muas, I., Mansyah, E. and Hendri, H. (2021) Production, quality and marketing of dragon fruit in Indonesia. In: Fullerton, R.A and Lee, W.L. (eds) *Dragon Fruit – From Production to Market*. Food and Fertilizer Technology Center for the Asian and Pacific Region, Taipei, Taiwan, pp. 107–124.

Nerd, A. and Mizrahi, Y. (1997) Reproductive biology of cactus fruit crops. *Horticulture Reviews* 18, 321–346.

Nerd, A., Sitrit, Y., Kaushik, R.A. and Mizrahi, Y. (2002) High summer temperatures inhibit flowering in vine pitaya crops (*Hylocereus* sp.). *Scientia Horticulturae* 96, 343–350.

Nobel, P.S. and de la Barrera, E. (2005) CO_2 uptake by the cultivated hemiphytic cactus *Hylocereus undatus*. *Annals of Applied Biology* 144, 1–8.

Ortiz-Hernandez, Y.D. (2000) *Hacia el Conocimiento y Conservacion de la Pitahaya (*Hylocereus *spp.)*. IPN-SIBEJ-CONACYTFMCN, Oaxaca, Mexico.

Ortiz-Hernandez, Y.D. and Carrillo-Salazar, J.A. (2012) Pitahaya (*Hylocereus* spp.): a short review. *Comunicata Scientiae* 3, 220–237.

Pagliaccia, D., Vidalakis, G., Douhan, G.W., Lobo, R. and Tarizaki, G. (2015) Genetic characterization of pitahaya accessions based on amplified fragment length polymorphism analysis. *HortScience* 50, 332–336.

Pascua, L.T., Pascua, M.E. and Gabriel, M.L.S. (2015) Dragon fruit production and marketing in the Philippines: its status, constraints and prospects. In: Jiang, Y.L. and Liu Huang, P.H. (eds). *Improving Pitaya Production and Marketing*. Food and Fertilizer Technology Center, Taipei, Taiwan, pp. 47–65.

Plume, O., Straub, S.C., Tel-Zur, N., Cisneros, A., Schneider, B. and Doyle, J.J. (2013) Testing a hypothesis of intergeneric allopolyploidy in vine cacti (Cactaceae: Hylocereeae). *Systematic Botany* 38, 737–751.

Qin, Y.H., Ye, Y.X., Hu, G.B., Li, J.Q., Wu, P.Y. *et al.* (2018) A new pitaya cultivar 'Shuangse 1'. *Acta Horticulturae Sinica* 45 (Suppl. 2), 2737–2738.

Ramirez, M.F.J. (1999) Caracterizacion compatibilidad en pithaya (*Hylocereus* spp.) MSc thesis, Universidad Autonoma Chapingo, Chapingo, Mexico.

Reed, S. (2004) Embryo rescue. In: Trigiano, R.N. and Gray, D.J. (eds) *Plant Development and Biotechnology*. CRC Press, Boca Raton, Florida, pp. 235–239.

Sage, T.L., Strumas, F., Cole, W.W. and Barret, S.C.H. (1999) Differential ovule development following self and cross pollination: the basis of selfsterility in *Narcissus triandrus* (Amaryllidaceae). *American Journal of Botany* 86, 855–870.

Sarip, J. (2020) Dragon fruit exportation practices. Challenges and country experiences: Malaysia. In: *Proceedings of the 3rd DFNet Workshop on Dragon Fruit Value Chain for Global Markets, 22–23 September, Taichung, Taiwan*, pp. 13–26.

Su, W., Wang, X., Wang, J., Ren, H., Wang, Y. *et al.* (2017) Breeding of new red-fleshed pitaya cultivar 'Jindu 1'. *Acta Horticulturae Sinica* 44 (Suppl. 2), 2665–2666.

Tel-Zur, N. (2013) Pitahayas introduction, agrotechniques and breeding. *Acta Horticulturae* 995, 109–115.

Tel-Zur, N. (2022) Breeding an underutilized fruit crop: a long term program for *Hylocereus*. *Horticultural Research* 9: uhac078.

Tel-Zur, N., Abbo, S., Bar-Zvi, D. and Mizrahi, Y. (2003) Chromosome doubling in vine cacti hybrids. *Journal of Heredity* 94, 329–333.

Tel-Zur, N., Abbo, S., Bar-Zvi, D. and Mizrahi, Y. (2004) Genetic relationship among *Hylocereus* and *Selenicereus* vine cacti (Cactaceae): evidence from hybridization and cytological studies. *Annals of Botany* 94, 527–534.

Tel-Zur, N., Abbo, S. and Mizrahi, Y. (2005) Cytogenetics of semi-fertile triploid and aneuploidy intergeneric vine cacti hybrids. *Journal of Heredity* 96, 124–131.

Tel-Zur, N., Dudai, M., Raveh, E. and Mizrahi, Y. (2011a) *In situ* induction of chromosome doubling in vine cacti (Cactaceae). *Scientia Horticulturae* 129, 570–576.

Tel-Zur, N., Mizrahi, Y., Cisneros, A., Mouyal, J., Schneider, B. and Doyle, J.J. (2011b) Phenotypic and genomic characterization of a vine cactus collection (Cactaceae). *Genetic Resources and Crop Evolution* 58, 1075–1085.

Tel-Zur, N., Dudai, M., Raveh, E. and Mizrahi, Y. (2012a) Selection of interspecific vine cacti hybrids (*Hylocereus* spp.) for self-compatibility. *Plant Breeding* 131, 681–685.
Tel-Zur, N., Mouyal, J., Cisneros, A. and Mizrahi, Y. (2012b) Intergeneric hybridization within the tribe *Hylocerecae*, subfamily Cactoideae (Cactaceae). *Israeli Journal of Plant Science* 60, 325–334.
Thomson, P. (2002) *Pitahaya (*Hylocereus *species), A Promising New Fruit Crop for Southern California*. Bonsall Publication, Bonsall, California.
Tomaz de Oliveira, M.M. and Tel-Zur, N. (2022) Cacti pear and pitaya: fruit production and orchard management. *Acta Horticulturae* 1343, 343–352.
Tomaz de Oliveira, M.M., Shuhua, L., Kumbha, D.S., Zurgil, U., Raveh, E. and Tel-Zur, N. (2020) Performance of *Hylocereus* (Cactaceae) species and interspecific hybrids under high- temperature stress. *Plant Physiology and Biochemistry* 153, 30–39.
Tomaz de Oliveira, M.M., Lu, S., Zurgil, U., Raveh, E. and Tel-Zur, N. (2021) Grafting in *Hylocereus* (Cactaceae) as a tool for strengthening tolerance to high-temperature stress. *Plant Physiology and Biochemistry* 160, 94–105.
Touraev, A., Pfosser, M. and Heberle-Bors, E. (2001) The microspore: a haploid multipurpose cell. *Advances in Botanical Research* 35, 53–109.
Trujillo, D. (2014) Microorganisms asociados a la pudrician blanda del tallo y manchado dd fruto en e cultivo de pitahaya amarilla en Ecuador. Thesis, Universidad Central del Ecuador, Quito.
Wakchaure, G.C., Kumar, S., Meena, K.K., Rane, J. and Pathak, H. (2020) *Dragon Fruit Cultivation in India: Scope, Marketing, Constraints and Policy Issues*. Technical Bulletin No. 46. ICAR-National Institute of Abiotic Stress Management, Baramati, India.
Wang, Y.L., Zheng, Y.F., Tan, B.S., Chen, H.P. and Sun, Q.M. (2022) A new pitaya cultivar 'Daquiu 4'. *Acta Horticulture Sinicia* 49 (Suppl. 1), 59–60.
Weiss, J., Nerd, A. and Mizrahi, Y. (1994) Flowering behaviour and pollination requirements in climbing cacti with a fruit crop potential. *HortScience* 29, 1487–1492.
Ye, Y.X., Hu, G.B., Li, J.Q., Qin, Y.H., Gu, W.Q. *et al*. (2018a) A new red-fleshed pitaya cultivar 'Guanhuahong'. *Acta Horticulturae Sinica* 45, 2267–2268.
Ye, Y.X., Hu, G.B., Li, J.Q., Qin, Y.H., Zhou, Y.W. *et al*. (2018b) A new white-fleshed pitaya cultivar 'Guanhuabai'. *Acta Horticulturae Sinica* 45(Suppl. 2), 2739–2740.
Zhang, Z., Ma, M.B., Hu, G.B., Yan, C.D., Ye, Y.X. *et al*. (2023) A new pitaya cultivar Fenhong No. 1. *Journal of Fruit Science* 40, 822–826.

6 Biotechnology

Leila Aparecida Salles Pio[1]* and Renato Paiva[2]
[1]*Department of Agriculture, Federal University of Lavras (UFLA), Brazil;*
[2]*Department of Biology, Federal University of Lavras (UFLA), Brazil*

6.1 Introduction

Pitaya is a climbing plant native to Central America, found from southern Mexico to northern South America. It is easy to grow and propagate and has a certain rusticity, as well as tolerance to drought and other adverse conditions such as salinity and cold. However, pitaya only reaches its maximum productive potential under ideal climatic, soil and humidity conditions.

Pitaya is a Crassulacean acid metabolism species with a higher water-use efficiency (about four to six times more) than other fruit trees grown under similar conditions. Therefore, pitaya cultivation is remunerative for farmers in dry regions or regions with increasing water scarcity and land degradation.

Research on pitaya has increased considerably in recent years as this crop has become more important. This chapter reviews the progress in pitaya biotechnology.

6.2 Taxonomy and Evolution

Pitaya, also known as dragon fruit, belongs to the family Cactaceae, which has approximately 100 genera and 1500 species native to the Americas. Although different pitaya species are cultivated (Santos *et al.*, 2022), the taxonomic classification is still controversial with revisions of the genera and species over time. Based on careful analysis, several researchers classify the most commercially important species as belonging to the genus *Selenicereus* (Arias *et al.*, 2005; Cruz *et al.*, 2016; Qin *et al.*, 2022).

The four species with greatest commercial potential differ in the size of their fruit, peel colour and pulp colour: *Selenicereus undatus* (Haw.) D.R. Hunt (large fruits, red peel and white pulp), *Selenicereus costaricensis* (F.A.C. Weber) S. Arias & N. Korotkova ex Hammel (medium fruits, red peel and red pulp), *Selenicereus megalanthus* (K. Schum. ex Vaupel) Moran (medium fruits, yellow peel with spines and white pulp) and *Selenicereus setaceus* (Salm-Dyck ex DC.) A. Berger ex Werderm (small fruits, red peel with spines and white pulp).

The basic chromosome number in Cactaceae is x = 11. *Hylocereus undatus*, *Hylocereus monacanthus*, *Hylocereus ocamponis*, *Hylocereus costaricensis* and *Selenicereus grandiflorus* are all diploid ($2n = 22$) (Tel-Zur *et al.*, 2004), while *Hylocereus megalanthus* is tetraploid ($2n = 44$) (Lichtenzveig *et al.*, 2000). Irregular chromosome disjunction at anaphase 1 in pollen mother cells of *H. megalanthus* is probably the major cause of its reduced pollen viability and may contribute to low seed set, low numbers of viable seeds and

* Email: leila.pio@ufla.br

 Dragon Fruit: Botany, Production and Uses (ed. S. Mitra)
DOI: 10.1079/9781800623156.0006

consequently low fruit mass (Lichtenzveig *et al.*, 2000). Allied with this, the process of endoreduplication occurs in pitaya. During the development of some plants, the normal cell cycle can be replaced by a partial cell cycle, in which mitosis does not occur (Chevalier *et al.*, 2011). This modified cycle is called endoreduplication and consists of one or several cycles of DNA synthesis in the absence of mitosis. Endoreduplication plays a role in the control of cell expansion, thus contributing to the growth of plant organs (Sliwinska and Lukaszewska, 2005). This phenomenon can occur in various cells, especially those undergoing differentiation (Sabelli and Larkins, 2009), which is the case of cladode cells. Menezes *et al.* (2012) confirmed the existence of endoreduplication in pitaya by flow cytometry analysis.

Species of Cactaceae are well known for being resistant to drought. To adapt to an arid environment, pitaya and other cacti have developed structures such as succulent stems and watery fruits. In fact, the family Cactaceae emerged approximately 28.8 million years ago (Arakaki *et al.*, 2011; Magallón *et al.*, 2015), when the Earth experienced a drop in atmospheric carbon dioxide concentration and a global expansion of aridity. As a result, plants of this family underwent rapid genome evolution including whole-genome duplication (endoreduplication) and gene expansion/contraction (Wang *et al.*, 2019; Kohler *et al.*, 2020), as well as drastic changes at the phenotypic level (Hernández-Hernández *et al.*, 2014).

6.3 Micropropagation

The propagation of pitaya by seeds gives rise to plants with a high degree of genetic diversity. This is interesting for breeding programmes but is not commercially viable. Furthermore, plants grown from seeds remain in the juvenile phase for several years before fruit production.

Although conventional vegetative propagation is carried out using cladode cuttings, biotechnological tools such as micropropagation can be employed as an alternative for obtaining rapid multiplication, high volume production and plant uniformity. *In vitro* production can be achieved directly by axillary bud proliferation or indirectly via callus formation (adventitious organogenesis) in a relatively short time.

Several studies have demonstrated the use of tissue culture for the *in vitro* propagation of *Hylocereus* spp. such as *H. undatus* (Menezes *et al.*, 2012; Hua *et al.*, 2015; Thiha, 2019), *H. costaricensis* (Viñas *et al.*, 2012; Kasim *et al.*, 2019), *H. megalanthus* (Mállap-Detquizán *et al.*, 2022) and *H. polyrhizus* (Román *et al.*, 2014; Ng *et al.*, 2021).

Bozkurt *et al.*, (2020, 2022) developed a micropropagation and rooting protocol for the varieties 'American Beauty', 'Halley's Comet', 'Bloody Mary' and 'Vietnam' cultured in Murashige and Skoog basal medium supplemented with 6-benzylaminopurine, gibberellic acid and indole-3-butyric acid (for more information, see Chapter 7, this volume).

Research using air-lift bioreactors has shown very promising results. These are pneumatically stirred bioreactors in which compressed air is used for aeration and agitation. Nutrient availability and the aerated microenvironment in the bioreactor system contribute to homogeneous culture conditions resulting in higher growth performance and sprouting compared with gel culture. Bello-Bello *et al.* (2021) reported that the multiplication of *H. undatus* cladodes increased twofold in a liquid temporary immersion system compared with gel culture. Dewir *et al.* (2023) used a similar micropropagation cultivation system for *H. polyrhizus* and added arbuscular mycorrhizal fungi (*Gigaspora margarita* and *G. albida*) during plantlet acclimation. This system led to high cladode multiplication, and cladodes with higher biomass and length. Inoculation with mycorrhizal fungi significantly increased vegetative growth during plantlet acclimation.

6.4 Flow Cytometry

Flow cytometry analyses verify genetic stability and the occurrence of endoreduplication. With this technique, it is possible to estimate the nuclear DNA index of plants in a fast, efficient and reliable way by comparing it with nuclei from a standard reference (Loureiro *et al.*, 2021).

The cytogenetics of pitaya is considered very complex due to the occurrence of diploid and tetraploid species, as well as endoreduplication. As the determination of the ploidy level by chromosome counting is technically challenging for this crop, flow cytometry is used as a reliable and rapid estimate of DNA content.

Tel-Zur *et al.* (1999) developed a simple protocol for DNA isolation from climbing cacti of the genera *Hylocereus* and *Selenicereus*. They suggested using roots as the source tissue due to the lower viscosity of the extracts relative to that of other tissues. The extraction and isolation procedure consists of: (i) three washes of ground tissue with extraction buffer to remove polysaccharides; (ii) extraction with high-salt (4 M NaCl) cetyltrimethylammonium bromide buffer to remove the remaining polysaccharides; (iii) removal of RNA by RNase; (iv) phenol:chloroform (1:1, v:v) extraction to remove proteins; and (v) chloroform extraction to remove remaining phenols. The yields ranged from 10 to 20 μg DNA g^{-1} of fresh roots. Tel-Zur *et al.* (2011) further evaluated the genomic and morphological characteristics of pitaya, including nuclear 2C DNA content by flow cytometry in 64 accessions of *Hylocereus*, *Selenicereus* and *Epiphyllum*. The DNA content ranged from 3.21 pg for *S. grandiflorus* to 8.77 pg for *H. megalanthus*. Except for the tetraploids *H. megalanthus* and *Selenicereus vagans*, all species were confirmed as diploid. Moreira *et al.* (2021) evaluated the viability of *S. megalanthus* micrografting on different rootstocks (*S. megalanthus*, *H. undatus*, *S. setaceus* and *H. polyrhizus*) based on DNA content analysis by flow cytometry and anatomy. The occurrence of endoreduplication was observed in all genotype combinations (rootstock × scion). Grafting *S. megalanthus* on to *H. polyrhizus* increased the amount of DNA and vigour. Based on DNA content and anatomical analyses, the micrografting of yellow pitaya was considered viable.

6.5 Molecular Markers

Molecular characterization is a very important tool and has numerous applications. For pitaya, it would be useful, as no certified pitaya plantlets are available for farmers in most producing regions. This is a serious problem, making farmers more vulnerable to different production and phytosanitary risks. Thus, genetic studies would help to identify elite materials that meet the needs of production chains. Garcia Aguilar *et al.* (2013) carried out a molecular characterization of three species of *Hylocereus* using the chloroplast regions *matk*, *rbcl* and *psbA* and the intergenic spacer *trnl-Fas* as a barcode. The results obtained allowed the identification of different levels of intra- and interspecific variation.

Genetic diversity studies are also useful to identify duplicate accessions within germplasm collections and establish genetic relationships between species (such as the ones from the genera *Hylocereus* and *Selenicereus*) (Martínez *et al.*, 2020; Oliveira *et al.*, 2020). In addition, they contribute to evaluating the fertility of triploid and aneuploid hybrids and demonstrate their allopolyploid origin (Oliveira *et al.*, 2020).

Several DNA markers have been used to assess the genetic diversity of pitaya varieties. Using random amplified polymorphic DNA (RAPD) analysis, Rifat *et al.* (2019) observed genetic variation and relatedness among different germplasms that was greater than expected. The Jaccard similarity coefficient (which ranged from 0.21 to 0.94) showed that 15 pitaya germplasms grouped into different clusters and that individuals within one cluster possessed similar morphological traits, although there were some discrepancies. Solano *et al.* (2005) evaluated the genetic diversity in natural populations of pitaya. Fifty pitaya collections from nine states of Mexico and one accession from Colombia were evaluated using RAPD markers. A high level of genetic polymorphism (92.5%) was detected between the populations, and fingerprinting allowed identification of individual accessions. The collection from Colombia was genetically similar to the majority of Mexican collections, suggesting a common origin. Additionally, a group of collections from Hidalgo, Mexico State and San Luis Potosi was detected, which were genetically different from the remaining collection, suggesting that in Mexico there is endemic genetic variability and this country is one of various possible centres of diversity of *H. undatus*. Analysis of amplified fragment length polymorphisms in a large collection of pitayas revealed that some varieties are identical, despite their different names (Pagliaccia *et al.*, 2015). These authors also showed that genotypic diversity occurs within varieties. Lack of reproductive barriers among most pitaya species leads to genetic mosaics. Nashima *et al.* (2021) developed simple sequence repeat (SSR) markers using the Roche 454 GS FLX+ sequencing platform, and out of 70 primer sets tested, 16 SSR markers successfully amplified fragments and detected polymorphisms among the accessions.

Among the microsatellite markers, inter simple sequence repeats (ISSRs) are characterized by their simple handling, low cost, high polymorphism and high stability. These markers are very useful for measuring genetic diversity in plants and identifying relationships among families and species and within species. Tao *et al.* (2014) first used the ISSR markers for unravelling genetic diversity as well as for fingerprinting the genotypes of pitaya in China. Fifteen morphological traits and 111 ISSR markers generated from 16 primers were first employed to discriminate 50 selected accessions. Taking 750 as a threshold, three major clusters were observed. The first included nine genotypes, which mainly belonged to wild accessions; the second cluster was of red-fleshed genotypes and the third consisted of white-fleshed accessions. Each ISSR primer generated four to ten obvious DNA bands ranging from 100 to 1500 bp, with an average of 6.9 per primer. The percentage of polymorphic bands varied from 25% to 100% with an average of 66.12%, and DNA markers generated from primers UBC 824, UBC 891 and UBC 900 efficiently fingerprinted 50 genotypes varying from 0.49 to 0.93 with an average of 0.85, suggesting a high genetic diversity among the tested genotypes. High genetic diversity was observed in *S. megalanthus* genotypes using ISSR markers (Morillo *et al.*, 2017, 2022), providing information to develop conservation strategies and breeding programmes for pitaya. Morillo *et al.* (2017) carried out the molecular characterization of 12 yellow pitaya samples using seven ISSR markers and observed genetic variability that could be exploited for the identification of elite materials. Working with 76 yellow pitaya genotypes and using eight ISSR markers, Morillo *et al.* (2022) further found genetic diversity with an average value of heterozygosity of 0.3% with a genetic differentiation coefficient of 0.26, indicating that there was wide genetic diversity. Zhijiang *et al.* (2021) evaluated 69 accessions with 13 ISSRs and detected a total of 247 loci, a percentage of polymorphic bands of 94% and expected heterozygosity of 0.27.

Abirami *et al.* (2021) distinguished three *Hylocereus* spp. based on morphological, biochemical and molecular characteristics. Morphological characterization revealed many genetic variations, such as the colour of fruit flesh/peel, number of spines and areole length. Higher phenol and flavonoid contents were observed in the fruit peel compared with the flesh, indicating higher antioxidant potential. Molecular characterization distinguished the three *Hylocereus* spp. based on their geographical origin and flesh colour.

6.6 Betalain Biosynthesis

Isobaric tags for relative and absolute quantitation (iTRAQ)-based proteomic analyses were used to reveal the molecular mechanism of betalain biosynthesis in *H. polyrhizus* fruits at the white- and red-fleshed stages, respectively (Hua *et al.*, 2016). RNA sequencing and iRTAQ analyses showed that some transcripts and proteins were positively correlated; they belonged to 'phenylpropanoid biosynthesis', 'tyrosine metabolism', 'flavonoid biosynthesis', 'ascorbate and aldarate metabolism', ' betalain biosynthesis' and 'anthocyanin biosynthesis'. In the betalain biosynthesis pathway, several proteins/enzymes such as polyphenol oxidase, CYP76AD3 and 4.5-dihydroxyphenylalanine (DOPA) dioygenase extradiol-like protein were identified. This study provides new insights into the molecular mechanism of betalain biosynthesis. Genome-wide characterization of R2R3-MYB transcription factors in pitaya revealed an R2R3-MYB repressor (*HuMYB1*) in fruit ripening through regulation of betalain biosynthesis by repressing betalain biosynthesis-related genes: *HuADH1*, *HuCYP76AD1-1* and *HuDODA1* (Xie *et al.*, 2021). These results provided the first genome-wide analyses of the R2R3-MYB subfamily involved in pitaya betalain biosynthesis.

6.7 Gene Mapping Based on RNA Sequencing

RNA sequencing technology has been used to verify the response of pitaya to abiotic stresses. Zhou *et al.* (2020a) investigated the pitaya transcriptome and proteome and found that chloroplast and mitochondria metabolism-related proteins may play an important role in pitaya plants submitted to cold stress. Transcriptome analysis was also employed to discover the genetic differences between white- and red-fleshed pitayas. Transcripts related to betalain and anthocyanin biosynthesis showed higher and lower expression levels in red-fleshed pitaya, respectively (Xi *et al.*, 2019). Two transcription factors (MYB and BvMYB1) were differentially expressed in both types of pitaya, and were suggested as putative candidates controlling betalain accumulation in this crop.

Xu *et al.* (2019) investigated the responses of *H. polyrhizus* varieties against *Neoscytalidium dimidiatum*, a fungal pathogen causing canker (the most serious disease of pitaya). Using Illumina RNA-Seq technology, the authors evaluated the expression of 33,584 unigenes, of which 6209 (18.49%) and 27,375 (81.51%) were contigs and singlets, respectively. These unigenes shared a similarity of 16.62% with *Vitis vinifera*, 7.48% with *Theobroma cacao*, 6.6% with *Nelumbo nucifera* and 5.35% with *Jatropha curcas*. In addition, the differential expression of 23 defence-related genes was confirmed by quantitative reverse transcriptase polymerase chain reaction (qRT-PCR). This study provides a basis for further research on the protein function of unigene assembly annotated with complementary DNA sequences.

Supplemental lighting at night can be used to induce pitaya flowering. Xiong *et al.* (2020) performed transcriptomic analysis to explore the molecular regulation mechanisms of floral induction in pitaya plants at different developmental stages exposed to supplemental lighting. Comparisons among samples revealed common and unique differentially expressed transcripts, including genes involved in energy metabolism, plant hormone signalling and direct regulation of flowering (e.g. many *CONSTANS-LIKE*, *FLOWERING LOCUS T*). Among these, 13 genes (key genes that control the floral induction network, flower induction and development) with different functions in the response to light-induced flowering were further verified by qRT-PCR. These findings enhance our understanding of the molecular mechanism involved in flowering regulation induced by supplemental lighting. Ye *et al.* (2021) made a comprehensive genome-wide analysis of aquaporins (*AQPS*) in *H. undatus* to understand the genes associated with flowering processes. A total of 33 *HuAQP* genes were identified from the *H. undatus* genome. The 33 *HuAQPs* were grouped into four subfamilies: ten plasma membrane intrinsic proteins, 13 tonoplast intrinsic proteins, eight nodulin 26-like intrinsic proteins and two small basic intrinsic proteins, which were distributed on nine out of 11 pitaya chromosomes (except for chromosomes 7 and 10). Results from expression profiles showed that *HuNIP6;1* may be involved in pitaya's floral opening. *HuNIP6;1* was localized exclusively in the cell membrane. Overexpression of *HuNIP6;1* in *Arabidopsis thaliana* significantly promoted early flowering by regulating negative flowering regulators of *MJM30*, *COL9* and *PRR5*, suggesting that *HuNIP6;1* plays key roles in regulating flowering time.

6.8 Whole-genome Duplication Events

Genome duplication is one of the main mechanisms by which additional numbers of gene copies are acquired, thereby introducing genetic and evolutionary novelty into organisms (Ohno, 1970). Genome doubling may have multi-level effects on the morphology, viability and physiology of polyploids compared with diploids. Cohen *et al.* (2013) studied the charges associated with autopolyploidization in two systems of somatic newly induced polyploids, diploid–autotetraploid and triploid–autohexaploid, belonging to the genus *Hylocereus*. They observed that induced autotetraploid and autohexaploid lines exhibited morphological and cytological characteristics that differed from those of their donor plants and that were accompanied by significant metabolic alterations. It was suggested that a developmental arrest occurs in the fruits of the autohexaploid lines, as their pericarp shows a genetic abundance of acids and reduced sugars. Cohen *et al.* (2013) concluded that genome doubling does not necessarily confer a fitness advantage and that the extent of alterations induced by autopolyploidization depends on the genetic background of the donor genotypes. In *H. monacanthus*, autotetraploid lines break down the self-incompatibility system (Cohen and Tel-Zur, 2012).

The basic platform for molecular and genomic studies aimed at closing the gaps between phenotypes and DNA sequences was established at the South China Agricultural University in Guangzhou, China. In the first stage, the expression stability of different potential pitaya reference genes for qRT-PCR (including different tissues, growing conditions and developmental stages) was investigated and validated, providing the essential reference genes needed to calibrate the experimental conditions (Chen *et al.*, 2019). In this study, six candidate reference genes (*Actin(1)*, *GAPDH*, *UBC(1)*, *UBC(2)*, *EF-a(1)* and *Histone(1)*) were selected from 39 typical candidate reference genes to determine the most stable genes for qRT-PCR normalization in different tissues, temperature stresses and fruit developmental stages of pitaya. *Actin(1)* was found to be the most stably-expressed gene in different tissues and fruit developmental stages of pitaya (Chen *et al.*, 2019).

Zheng *et al.* (2021) pioneered sequencing of the *H. undatus* genome at the chromosomal level to study the duplication of the entire pitaya genome. Compared with the five previously published Cactaceae genomes (*Stenocereus thurberi*, *Lophocereus schottii*, *Carnegiea gigantea*, *Pachycereus pringlei* and *Pereskia humboldtii*), *H. undatus* has the longest N50 value (1700-fold longer than *C. gigantea*).

The chromosome-level genome assembly contains the 11 longest scaffolds corresponding to the 11 chromosomes of *H. undatus*. Genome annotation of *H. undatus* found approximately 29,000 protein-coding genes, similar to *C. gigantea*. Whole-genome duplication analysis revealed a whole-genome duplication event in the last common ancestor of the Cactaceae followed by extensive genome rearrangements (e.g. chromosome breakage and reorganization). Chromosome 11 (Scaffold 2055) shares large syntenic gene blocks with chromosome 8 (Scaffold 33675) and chromosome 3 (Scaffold 33676), while chromosome 10 (Scaffold 3410) shares large syntenic gene blocks with three chromosomes: 2, 8 and 9 (Fig. 6.1). It also appears that the genomic synteny is fairly conserved between the Cactaceae and Chenopodiaceae genomes according to the intergenome *H. undatus–Beta vulgaris* and *H. undatus–Spinacia oleracea* synteny alignment. The divergence time between *H. undatus* and *C. gigantea* was estimated to be 9.18 million years ago. Functional enrichment analysis of orthologous gene clusters in six Cactaceae plants found significantly enriched orthologous gene clusters in drought resistance. Fruit flavour-related functions were expanded in *H. undatus*. The *H. undatus* draft genome also enabled the discovery of carbohydrate and plant cell wall-related functional enrichment in pitaya treated with typsin for a longer storage time. Finally, genes of the betacyanin (red and violet betalain) biosynthetic pathway were found to be co-localized on a 12 Mb region of the chromosome, which may imply a more efficient form of betacyanin biosynthesis in this species (Zheng *et al.*, 2021).

6.9 PacBio-SMRT Sequencing

Pitaya is the only commercial edible fruit that contains high levels of betalains, which have potent antioxidant activity. To date, three main enzymes are known to participate in betalain biosynthesis: (i) arogenate dehydrogenase (ADH), which catalyses tyrosine formation (Lopez-Nieves *et al.*, 2018); (ii) 4,5-DOPA extradiol dioxygenase (DODA), which is responsible for betalamic acid biosynthesis (Gandía-Herrero and García-Carmona, 2012); and (iii) the P450 enzyme CYP76AD1, which catalyses cyclo-DOPA synthesis (Hatlestad *et al.*, 2012). In addition, transcription factors are reported to play a critical role in betalain biosynthesis. For example, the MYB anthocyanin-like protein BvMYB1 directly regulates the transcription of CYP76AD1 and DODA1, thereby activating the betalain red pigment pathway (Hatlestad *et al.*, 2015).

To date, several betalain biosynthetic genes, such as TYR, GT-like, DOD-like and CytP450-like, have been discovered in pitaya via transcriptomic analysis (Xie *et al.*, 2020). The combination of various sequencing approaches (PacBio-SMRT, Illumina HiSeq paired-end, 10x Genomics, and high-throughput chromosome conformation capture (Hi-C)) have provided a 'GHB' chromosome-level genome assembly of *H. undatus* ($2n = 2x = 22$ chromosomes) (Chen *et al.*, 2021). The size of the assembled genome was 1.41 Gb, with 27,753 protein-coding genes and 896.31 Mb of repetitive sequences.

A total of 29,328 intact long terminal repeat retrotransposons (~696.45 Mb) were obtained in *H. undatus*, of which two significantly expanded lineages, Ty1/copy and Ty3/gypsy, were the main drivers of the genomic expansion. A high-density genetic map of F_1 hybrid populations of 'GHB' × 'Dahong' (*H. monacanthus*) pitaya and their parents were constructed and a total of 20,872 bin markers were identified (56,380 single-nucleotide polymorphisms) for

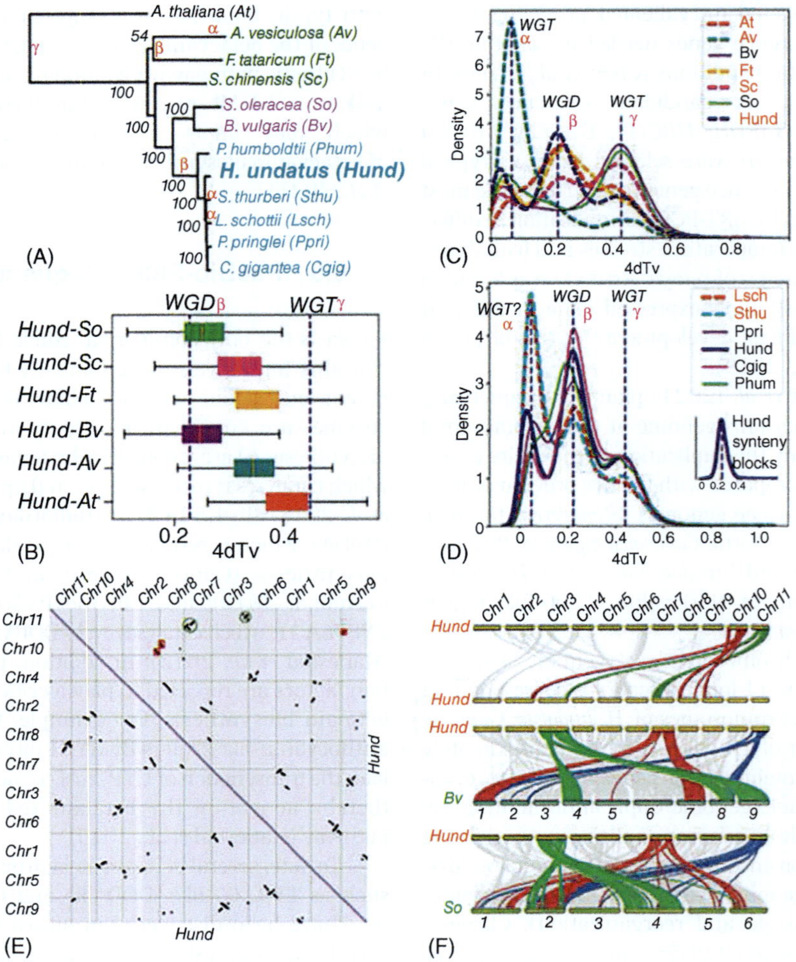

Fig. 6.1. Whole-genome duplication (WGD) analysis of *Hylocereus undatus* (Hund) and 11 other species. The full names and acronyms of the 11 plants are: *Stenocereus thurberi* (Sthu), *Lophocereus schottii* (Lsch), *Carnegiea gigantea* (Cgig), *Pachycereus pringlei* (Ppri) and *Pereskia humboldtii* (Phum) of the family Cactaceae, *Beta vulgaris* (Bv) and *Spinacia oleracea* (So) of the subfamily Chenopodiaceae, *Aldrovanda vesiculosa* (Av) of the family Droseraceae, *Simmondsia chinensis* (Sc) of the family Simmondsiaceae, and *Fagopyrum tataricum* (Ft) of the family Polygonaceae, and *Arabidopsis thaliana* (At). (A) A phylogeny is built with single-copy orthologous genes (identified by OrthoFinder) in the 12 genomes (concatenated alignment with MAFFT and tree built with RAxML). The WGD events characterized in (C) and (D) are indicated beside the branches. (B) Orthologous genes between Hund and other species were identified using the reciprocal best hit (RBH) approach, and fourfold degenerate synonymous sites (4dTv) values were calculated and plotted as box plots. The dashed lines correspond to the WGD events characterized in (C). (C) Intragenome 4dTv distribution of paralogous genes (identified by the RBH approach) in Hund and six other genomes. (D) Intragenome 4dTv distribution of paralogous genes (identified by the RBH approach) in the six Cactaceae genomes. The inset figure shows the 4dTv distribution of only paralogous genes present in the synteny blocks (identified by MCScanX) of Hund. (E) Whole-genome self-alignment and syntenic blocks within the 11 largest Hund scaffolds/pseudochromosomes are presented as a dot plot. Strings of dots (paralogous genes) that correspond to duplicated regions in chromosome 11 (scaffold 2055) and chromosome 10 (scaffold 3410) are highlighted using the same circles or boxes in the same colour. (F) A linear representation of the synteny analysis results is shown for intragenome Hund–Hund, as well as inter-genome Hund–Bv and Hund–So genome pairs. Synteny blocks of selected chromosomes are coloured in red, green and blue as examples. The chromosome sizes are drawn proportional to the actual genome size within each species but not between different species (total genome sizes: Hund, 1.33 Gb; Bv, 544 Mb; So, 969 Mb). Adapted from Zheng *et al.* (2021) with permission.

11 linkage groups. Through transcriptomics and weighted gene co-expression network analysis, Chen *et al.* (2021) presented an overview of the gene regulatory network including structural genes and transcription factors involved in betalain biosynthesis.

6.10 Plastome Sequencing

As already mentioned, the specific definitions of *Hylocereus* and *Selenicereus* have always been controversial. Britton and Rose (1920) presented initial definitions of the two genera based on morphology. Decades later, analyses of a large number of nuclear and plastidial DNA sequences, morphology and anatomy proved that these two genera that were treated as separate were not distinct, with *Hylocereus* now encompassed in *Selenicereus* (Arias *et al.*, 2005; Cruz *et al.*, 2016; Qin *et al.*, 2022).

Genome sequencing of the genetic information included in chloroplasts (the plastome) is an important tool to resolve phylogenetic relationships of species in addition to taxonomic and evolutionary studies (Daniell *et al.*, 2016; Ivanova *et al.*, 2017) due to their small size, simple structure and maternal inheritance (Maliga, 2002).

Qin *et al.* (2022) sequenced, assembled and analysed the plastid genomes of red-fleshed pitaya (*S. monacanthus*, previously classified as *H. polyrhizus*) and three *Selenicereus* spp. (*S. anthonyanus*, *S. grandiflora* and *S. validus*). The main steps of this work were as follows: (i) provision of four high-quality *Selenicereus* reference plastomes; (ii) analysis of the characteristic structure and sequence divergence of the plastomes; (iii) identification of simple repeat sequence loci and repeated sequences for further studies on the genetic structure of the population; (iv) conference of the phylogenetic relationships of *Selenicereus* and *Hylocereus* in Cactaceae based on the complete plastome sequence; and (v) identification of hypervariable regions that could be used as DNA barcodes for commercial identification of pitaya varieties. Phylogenomic analysis revealed a close relationship between *Selenicereus* spp. and that these four *Selenicereus* spp. are monophyletic. However, it is worth noting that *S. monacanthus* is most closely related to *S. anthonyanus*, the traditional species of *Selenicereus*.

Although plastid genomes are generally conserved in terms of sequence differences and structural organization, some non-coding regions may experience an unexpectedly high frequency of nucleotide substitutions, and these hypervariable regions can be used as DNA barcodes for species identification (Qin *et al.*, 2022).

6.11 Gamete-derived Progeny

Effective breeding programmes are needed to develop improved hybrids with superior fruit quality, good yields and greater tolerance to extreme temperatures. However, several difficulties make pitaya breeding challenging, such as limited genomic information, the large genome size and high heterozygosity. In addition, six to ten generations of self-pollination are required to obtain phenotypically stable homozygous lines. As pitaya is a perennial plant with a prolonged juvenile stage, many varieties have self-incompatibility (Cisneros and Tel-Zur, 2013; Plume *et al.*, 2013; Pagliaccia *et al.*, 2015).

Gamete-derived progeny have, theoretically, a lower level of heterozygosity than that of the parental lines (Verdoot *et al.*, 1998). In both *H. megalanthus* and *H. monacanthus*, successful haploid induction via androgenesis or gynogenesis (Garcia *et al.*, 2009a,b) have paved the way for breeding programmes and genetic studies. To accelerate breeding programmes, Li *et al.* (2018) studied the gamete-derived progeny of two *Hylocereus* spp., the diploid *H. monacanthus* and the tetraploid *H. megalanthus*. The objectives were to confirm their gamete origin, evaluate their potential use as genetic resources and determine the origin (allotetraploid versus autotetraploid) of *H. megalanthus* by exploring morphological variations in the diploid (2x) *H. megalanthus* progeny. The authors concluded that chromosome duplication, as occurred spontaneously in the double-haploid *H. monacanthus* and the double dihaploid *H. megalanthus*, was not sufficient to restore fertility in *Hylocereus*. Very low gametoclonal variation was also observed among the dihaploid (2x) lineages of *H. megalanthus*, which supported an autotetraploid rather than an allotetraploid origin of this species.

6.12 Abiotic Stress Tolerance

Proteogenomic analysis of pitaya has revealed a cold stress-related molecular signature. Zhou *et al.* (2020b) identified 116 differentially abundant proteins (DAPs) associated with cold stress in pitaya, of which 18 proteins were upregulated and 98 proteins were downregulated. Kyoto Encyclopedia of Genes and Genomes (KEGG) analysis and other results showed that these DAPs mainly related to chloroplast and mitochondrial metabolism. It was concluded that the chloroplast and mitochondrial metabolism-related proteins may play an important role in the response to cold stress in pitaya.

Transcriptomic and metabolomics analysis of red pitaya (*H. polyrhizus*) seedlings in response to heat stress revealed numerous genes and metabolites involved in the response to heat stress suggesting a complex stress-response mechanism. Jiao *et al.* (2021) stated that overexpression of one of the upregulated genes (*contig10820*) in *Arabidopsis*, which is a homologue of *PR-1* and is named *HuPR-1*, significantly increased tolerance to heat stress. This was the first study showing that *HuPR-1* plays a role in the response of pitaya to abiotic stress.

6.13 Near-infrared Spectroscopy

Near-infrared (NIR) spectroscopy uses the near-infrared region of the electromagnetic spectrum (~700–2500 nm). By measuring the light scattered in and through a sample, NIR reflectance spectra can be used to quickly determine the properties of a material without altering the sample.

This technique is based on the vibration of organic chemical molecules and their interaction with radiation from the infrared spectrum (Pasquini, 2003). In combination with chemometrics, NIR spectroscopy has shown great potential for determining physical and chemical parameters in agricultural products, such as total soluble solids (TSS), total titratable acidity and moisture (de Moraes *et al.*, 2022), phenolic compounds (Tian *et al.*, 2021) and biochemical changes such as fruit ripening (Loutfi *et al.*, 2015).

Ferreira *et al.* (2023) performed a rapid and non-destructive assessment of pitaya quality using NIR and obtained predictions of TSS, pH, titratable acidity, moisture and phenolic compounds, as well as classified pitaya according to their shelf-life stages.

6.14 Conclusions

Pitayas are increasingly being consumed worldwide, quickly gaining popularity and attracting the attention of both farmers and consumers. Pitaya is easy to cultivate and has good profitability. Furthermore, it has a pleasant taste and is rich in nutritional substances, especially betalain, a powerful antioxidant capable of preventing diseases.

Research on pitaya has also rapidly advanced, and several breeding studies have aimed to contribute to creating new varieties adapted to different growing regions and with different flesh and peel coloration, among other characteristics. In parallel, a series of biotechnological tools can be used and are helping to solve the controversial botanical classification of the species. However, the genetic origin of many varieties is still uncertain. Thus, molecular markers are crucial for the genetic characterization of pitaya species and varieties. Advanced genetic mapping techniques have been helping to understand the sequenced pitaya genome. With an assembly at the chromosome level and high completeness of the genome, this will also be a great reference for the study of various cacti. Recent research has provided initial insights into tolerance to abiotic stresses and into the processes involved in flowering and fruit ripening. The recently published *H. undatus* draft genome has paved the way for further molecular breeding programmes.

In vitro cultivation has major applications in the large-scale production of plantlets and crop improvement through micropropagation. Flow cytometry and other refined techniques will provide new insights into this important crop.

Acknowledgements

The authors recognize support from the Brazilian Nacional Council for Scientific and Technological Development (CNPq) and the Minas Gerais State Foundation for Research Support (FAPEMIG).

6.15 References

Abirami, K., Swain, S., Baskaran, V., Venkatesan, K., Sakthivel, K. *et al.* (2021) Distinguishing three dragon fruit (*Hylocereus* spp.) species grown in Andaman and Nicobar Islands of India using morphological, biochemical and molecular traits. *Scientific Reports* 11: 2894.

Arakaki, M., Christin, P.-A., Nyffeler, R., Lendel, A., Eggli, U. *et al.* (2011) Contemporaneous and recent radiations of the world's major succulent plant lineages. *Proceedings of the National Academy of Sciences USA* 108, 8379–8384.

Arias, S., Terrazas, T., Arreola-Nava, H.J., Vázquez-Sánchez, M. and Cameron, K.M. (2005) Phylogenetic relationships in *Peniocereus* (Cactaceae) inferred from plastid DNA sequence data. *Journal of Plant Research* 118, 317–328.

Bello-Bello, J.J., Schettino-Salomón, S., Ortega-Espinoza, and Spinoso-Castillo, J.L. (2021) A temporary immersion system for mass micropropagation of pitahaya (*Hylocereus undatus*). *3 Biotechnology* 11: 437.

Bozkurt, T., İnan, S. and Dündar, İ. (2020) Micropropagation of different pitaya varieties. *International Journal of Agricultural and Natural Sciences* 13, 39–46.

Bozkurt, T., İnan, S., Dündar, İ. and Özdemir, S.K. (2022) Effect of different plant growth regulators on micropropagation of some pitaya varieties. *Journal of Tropical Life Science* 12, 183–190.

Britton, N.L. and Rose, J.N. (1920) *The Cactaceae: Descriptions and Illustrations of Plants of the Cactus Family*, Vol. 2. Carnegie Institute, Washington, DC.

Chen, C., Wu, J., Hua, Q., Tel-Zur, N., Xie, F. *et al.* (2019) Identification of reliable reference genes for quantitative real-time PCR normalization in pitaya. *Plant Methods* 15: 70.

Chen, J.-Y., Xie, F.-F., Cui, Y.-Z., Chen, C.-B., Lu, W.-J. *et al.* (2021) A chromosome-scale genome sequence of pitaya (*Hylocereus undatus*) provides novel insights into the genome evolution and regulation of betalain biosynthesis. *Horticulture Research* 8: 164.

Chevalier, C., Nafati, M., Mathieu-Rivet, E., Bourdon, M., Frangne, N., *et al.* (2011) Elucidating the functional role of endoreduplication in tomato fruit development. *Annals of Botany* 107, 1159–1169.

Cisneros, A. and Tel-Zur, N. (2013) Genomic analysis in three *Hylocereus* species and their progeny: evidence for introgressive hybridization and gene flow. *Euphytica* 194, 109–124.

Cohen, H. and Tel-Zur, N. (2012) Morphological changes and self-incompatibility breakdown associated with autopolyploidization in *Hylocereus* species (Cactaceae), *Euphytica* 184, 345–354.

Cohen, H., Fait, A. and Tel-Zur, N. (2013) Morphological, cytological and metabolic consequences of autopolyploidization in *Hylocereus* (Cactaceae) species. *BMC Plant Biology* 13: 17.

Cruz, M.A., Salvador, A. and Teresa, T. (2016) Molecular phylogeny and taxonomy of the *genus Disocactus* (Cactaceae), based on the DNA sequences of six chloroplast markers. *Willdenowia* 46, 145–164.

Daniell, H., Lin, C.S., Yu, M. and Chang, W.J. (2016) Chloroplast genomes: diversity, evolution, and applications in genetic engineering. *Genome Biology* 17: 134.

de Moraes, I.A., Cruz-Tirado, L.J.P. and Barbin, D.F. (2022) Online measurement of carambola (*Averrhoa carambola* L.) physicochemical properties and estimation of maturity stages using a portable NIR spectrometer. *Scientia Horticulturae* 304: 111263.

Dewir, Y.H., Habib, M.M., Alaizari, A.A., Malik, J.A., Al-Ali, A.M. *et al.* (2023) Promising application of automated liquid culture system and arbuscular mycorrhizal fungi for large-scale micropropagation of red dragon fruit. *Plants* 12: 1037.

Ferreira, M.V.S., Moraes, I.A., Passos, R.V.L., Barbin, D.F. and Barbosa, J.L. Jr (2023) Determination of pitaya quality using portable NIR spectroscopy and innovative low-cost electronic nose. *Scientia Horticulturae* 310: 111784.

Gandía-Herrero, F. and García-Carmona, F. (2012) Characterization of recombinant *Beta vulgaris* 4,5-DOPA-extradiol-dioxygenase active in the biosynthesis of betalains. *Planta* 236, 91–100

Garcia Aguilar, M.A., Terrazas, T., Leon, S.O., Arias, S., Vibrans, H. and López-Mata, L. (2013) Caraterization molecular de tres species de *Hylocereus* (Cactaceae) presents en Mexico. *Revista Eitotecnia Mexicana* 36, 13–22.

Garcia, R.B., Cisneros, A., Schneider, B. and Tel-Zur, N. (2009a) Gynogenesis in the vine cacti *Hylocereus* and *Selenicereus* (Cactaceae). *Plant Cell Reports* 28, 719–726.

Garcia, R.B., Schneider, B. and Tel-Zur, N. (2009b) Androgenesis in two vine cacti, *Selenicereus* and *Hylocereus* (Cactaceae). *Plant Cell Tissue and Organ Culture* 96, 191–199.

Hatlestad, G.J., Sunnadeniya, R.M., Akhavan, N.A., Gonzalez, A., Goldman, I.L. *et al.* (2012) The beet *R* locus encodes a new cytochrome P450 required for red betalain production. *Nature Genetics* 44, 816–820.

Hatlestad, G.J., Akhavan, N.A., Sunnadeniya, R.M., Elam, L., Cargile, S. *et al.* (2015) The beetroot *Y* locus encodes a MYB-like anthocyanin protein that activates the betalain red pigment pathway. *Nature Genetics* 47, 92–96.

Hernández-Hernández, T., Brown, J.W., Schlumpberger, B.O., Eguiarte, L.E. and Magallón, S. (2014) Beyond aridification: multiple explanations for the elevated diversification of cacti in the New World Succulent Biome. *New Phytologist* 202, 1382–1397.

Hua, Q., Chen, P., Liu, W., Ma, Y., Liang, R. *et al.* (2015) A protocol for rapid *in vitro* propagation of genetically diverse pitaya. *Plant Cell, Tissue and Organ Culture* 120, 741–745.

Hua, Q., Zhou, Q., Gan, S., Wu, J., Chen, C. *et al.* (2016) Proteomic analysis of *Hylocereus polyrhizus* reveals metabolic pathway changes. *International Journal of Molecular Sciences* 17: 1606.

Ivanova, Z., Sablok, G., Daskalova, E., Zahmanova, G., Apostolova, E. *et al.* (2017) Chloroplast genome analysis of resurrection tertiary relict *Haberlea rhodopensis* highlights genes important for desiccation stress response. *Frontiers in Plant Science* 8: 204.

Jiao, Z., Xu, W., Nong, Q., Zhang, M., Jian, S. *et al.* (2021) An integrative transcriptomic and metabolomics analysis of red pitaya (*Hylocerus polyrhizus*) seedlings in response to heat stress. *Genes* 12: 1714.

Kasim, D.P., Kishore, N.S., Suneetha, P., Rao, K.B., Kumar, M.N. *et al.* (2019) Multiple shoot regeneration in seed-derived immature leaflet explants of red dragon fruit (*Hylocereus costaricensis*). *Research Journal of Pharmacy and Technology* 12, 1491–1494.

Kohler, M., Reginato, M., Souza-Chies, T.T. and Majure, L.C. (2020) Insights into chloroplast genome evolution across *Opuntioideae* (Cactaceae) reveals robust yet sometimes conflicting phylogenetic topologies. *Frontiers in Plant Science* 11: 729.

Li, D., Arroyave Martinez, M.F., Shaked, R. and Tel-Zur, N. (2018) Homozygote depression in gamete-derived dragon-fruit (*Hylocereus*) lines. *Frontiers in Plant Science* 8: 2142.

Lichtenzveig, J., Abbo, S., Nerd, A., Tel-Zur, N., and Mizrahi, Y. (2000) Cytology and mating systems in the climbing cacti *Hylocereus* and *Selenicereus*. *American Journal of Botany* 87, 1058–1065.

Lopez-Nieves, S., Yang, Y., Timoneda, A., Wang, M., Feng, T. *et al.* (2018) Relaxation of tyrosine pathway regulation underlies the evolution of betalain pigmentation in Caryophyllales. *New Phytologist* 217, 896–908.

Loureiro, J., Kron, P., Temsch, E.M., Koutecký, P., Lopes, S. *et al.* (2021) Isolation of plant nuclei for estimation of nuclear DNA content: overview and best practices. *Cytometry Part A* 99, 318–327.

Loutfi, A., Coradeschi, S., Mani, G.K., Shankar, P. and Rayappan, J.B.B. (2015) Electronic noses for food quality: a review. *Journal of Food Engineering* 144, 103–111.

Magallón, S., Gómez-Acevedo, S., Sánchez-Reyes, L.L. and Hernández-Hernández, T. (2015) A metacalibrated time-tree documents the early rise of flowering plant phylogenetic diversity. *New Phytologist* 207, 437–453.

Maliga P. (2002) Engineering the plastid genome of higher plants. *Current Opinion in Plant Biology* 5, 164–172.

Mállap-Detquizán, G., Vilca-Valqui, N.C., Meléndez-Mori, J.B., Huaman-Huaman, E. and Oliva, M. (2022) *In vitro* multiplication of yellow dragon fruit (*Hylocereus megalanthus*) from seedlings obtained *in vitro*. *Agronomia Mesoamericana* 33: 45472.

Martínez, M.A., Morillo, A.C. and Reyes-Ardila, L. (2020) Characterization of the genetic diversity in *Passiflora* spp. in the Boyacá Department, Colombia. *Chilean Journal of Agricultural Research* 80, 342–351.

Menezes, T.P., Gomes, W.A., Pio, L.A.S., Pasqual, M. and Ramos, J.D. (2012) Micropropagation and endoreduplication in red pitaya, *Hylocereus undatus* Haw. *Bioscience Journal* 28, 868–876.

Moreira, R.A., Rodrigues, M.A., Magalhães, D.S., Pio, L.A.S., Santos, D.N.D. *et al.* (2021) DNA index and anatomical aspects of the micrografting of dragon fruit on different rootstocks. *Pesquisa Agropecuária Brasileira* 56: e01867.

Morillo, A.C., Tovar, Y.P. and Morillo, Y. (2017) Caracterización molecular de pitahaya amarilla (*Selenicereus megalanthus* Haw.) en la provincia de Lengupá, Boyacá-Colombia. *Biotecnologia en el Sector Agropecuario y Agroindustrial* 15, 11–18.

Morillo, A.C., Mora, M.S. and Morillo, Y. (2022) Analysis of the genetic diversity of dragon fruit based on ISSR markers in Colombia. *Brazilian Journal of Biology* 82: e256451.

Nashima, K., Hosaka, F., Shimajiri, Y., Matsumura, M., Tarora, K. *et al.* (2021) SSR marker development and genetic identification of pitaya (*Hylocereus* spp.) collected in Okinawa prefecture, Japan. *Horticulture Journal* 90, 23–30.

Ng, Z.C., Shiekh Mahmud, S.H. and Hian, T.S. (2021) Effects of cytokinin in enhancing the multiplication of vegetative *Hylocereus polyrhizus*. *Materials Science Forum* 1025, 97–103.

Ohno, S. (1970) *Evolution by Gene Duplication*. Springer, New York.

Oliveira, M.M., Shuhua, L., Kumbha, D.S., Zurgil, U., Raveh, E. and Tel-Zur, N. (2020) Performance of *Hylocereus* (Cactaceae) species and interspecific hybrids under high-temperatures stress. *Plant Physiology and Biochemistry* 153, 30–39.

Pagliaccia, D., Vidalakis, G., Douhan, G.W., Lobo, R. and Tanizaki, G. (2015) Genetic characterization of pitahaya accessions based on amplified fragment length polymorphism analysis, *HortScience* 50, 332–336.

Pasquini, C. (2003) Near infrared spectroscopy: fundamentals, practical aspects and analytical applications. *Journal of the Brazilian Chemical Society* 14, 198–219.

Plume, O., Straub, S.C., Tel-Zur, N., Cisneros, A., Schneider, B. *et al.* (2013) Testing a hypothesis of intergeneric allopolyploidy in vine cacti (Cactaceae: Hylocereeae). *Systematic Botany* 38, 737–751.

Qin, Q., Li, J., Zeng, S., Xu, Y., Han, F. and Yu, J. (2022) The complete plastomes of red fleshed pitaya (*Selenicereus monacanthus*) and three related *Selenicereus* species: insights into gene losses, inverted repeat expansions and phylogenomic implications. *Physiology and Molecular Biology of Plants* 28, 123–137.

Rifat, T., Khan, K. and Islam, M.S. (2019) Genetic diversity in dragon fruit (*Hylocereus* sp.) germplasms revealed by RAPD marker. *Journal of Animal and Plant Sciences* 29, 809–818.

Román, R.S.S., Caetano, C.M. and Ramírez, H. (2014) Multiplication of *Selenicereus megalanthus* (yellow pitahaya) and *Hylocereus polyrhizus* (red dragon fruit) by somatic organogenesis view. *Acta Agronómica* 63, 272–281.

Sabelli, P.A. and Larkins, B.A. (2009) The contribution of cell cycle regulation to endosperm development. *Sexual Plant Reproduction* 22, 207–219.

Santos, D.N., Pio, L. and Faleiro, F. (2022) *Pitaya: Uma Alternativa Frutífera*. ProImpress, Brasília.

Sliwinska, E. and Lukaszewska, E. (2005) Polysomaty in growing *in vitro* sugar-beet (*Beta vulgaris* L.) seedlings of different ploidy level. *Plant Science* 168, 1067–1074.

Solano, L., Porfirio, J., Cano, A., Elisa, M. and Hernandez, R.G. (2005) Diversidad genetic en pithaya (*Hylocereus undatus* Haworth, Britton y Rose). *Revista Fitotecnia Mexicana* 28, 179–185.

Tao, J., Qiao, G., Wen, X.P., Gao, G.L., Liu, T. *et al.* (2014) Characterization of genetic relationship of dragon fruit accessions (*Hylocereus* spp.) by morphological traits and ISSR markers. *Scientia Horticulturae* 170, 82–88.

Tel-Zur, N., Abbo, S., Myslabodski, D. and Mizrahi, Y. (1999) Modified CTAB procedure for DNA isolation from epiphytic cacti of the genera *Hylocereus* and *Selenicereus* (Cactaceae). *Plant Molecular Biology Reporter* 17, 249–254.

Tel-Zur, N., Abbo, S., Bar-Zvi, D. and Mizrahi, Y. (2004) Genetic relationships among *Hylocereus* and *Selenicereus* vine cacti (Cactaceae): evidence from hybridization and cytological studies. *Annals of Botany* 94, 527–534

Tel-Zur, N., Mizrahi, Y., Cisneros, A., Mouyal, J., Schneider, B. and Doyle, J.J. (2011) Phenotypic and genomic characterization of vine cactus collection (Cactaceae). *Genetic Resources and Crop Evolution* 58, 1075–1085.

Thiha, S. (2019) Effects of explants and growth regulators on *in vitro* regeneration of dragon fruit (*Hylocereus undatus* Haworth). PhD thesis, Yezin Agricultural University, Zeyathiri, Myanmar.

Tian, W., Chen, G., Zhang, G., Wang, D., Tilley, M. and Li, Y. (2021) Rapid determination of total phenolic content of whole wheat flour using near-infrared spectroscopy and chemometrics. *Food Chemistry* 344: 128633

Verdoot, L., van Haute, A., Goderis, I., de Witte, K., Keulemans, J. and Broothaerts, W. (1998) Use of the multi-allelic self-incompatibility gene in apple to assess homozygosity in shoots obtained through haploid induction. *Theoretical and Applied Genetics* 96, 294–300.

Viñas, M., Fernández-Brenes, M., Azofeifa, A. and Jiménez, V.M. (2012) *In vitro* propagation of purple pitahaya (*Hylocereus costaricensis* [F.A.C. Weber] Britton & Rose) cv. Cebra. *In Vitro Cellular & Developmental Biology – Plant* 48, 469–477.

Wang, N., Yang, Y., Moore, M.J., Brockington, S.F., Walker, J.F. *et al.* (2019) Evolution of Portulacaceae marked by gene tree conflict and gene family expansion associated with adaptation to harsh environments. *Molecular Biology and Evolution* 36, 112–126.

Xi, X., Zong, Y., Li, S., Cao, D., Sun, X. *et al.* (2019) Transcriptome analysis clarified genes involved in betalain biosynthesis in the fruit of red pitayas (*Hylocereus costaricensis*). *Molecules* 24: 445.

Xie, F., Hua, Q., Chen, C., Zhang, L., Zhang, Z. *et al.* (2020) Transcriptomics-based identification and characterization of glucosyltransferases involved in betalain biosynthesis in *Hylocereus megalanthus*. *Plant Physiology and Biochemistry* 152, 112–124.

Xie, F., Hua, Q., Chen, C., Zhang, Z., Zhang, R. *et al.* (2021) Genome-wide characterization of R2R3-MYB transcription factors in pitaya reveals R2R3-MYB repressor *HuMYB1* involved in fruit ripening through regulation of betalain biosynthesis by repressing betalain biosynthesis-related genes. *Cells* 10: 1949.
Xiong, R., Liu, C., Xu, M., Wei, S.S., Huang, J.Q. *et al.* (2020) Transcriptomic analysis of flower induction for long-day pitaya by supplementary lighting in short-day winter season. *BMC Genomics* 21, 1–17.
Xu, M., Liu, C.L., Luo, J., Qi, Z., Yan, Z. *et al.* (2019) Transcriptomic de novo analysis of pitaya (*Hylocereus polyrhizus*) canker disease caused by *Neoscytalidium dimidiatum*. *BMC Genomics* 20: 329.
Ye, X., Gao, Y., Chen, C., Xie, F., Hua, Q. *et al.* (2021) Genome-wide identification of *Aquaporin* gene family in pitaya reveals an *HuNIP6;1* involved in flowering process. *International Journal of Molecular Science* 22: 7689.
Zheng, J., Meinhardt, L.W., Goenaga, R., Zhang, D. and Yin, Y. (2021) The chromosome-level genome of dragon fruit reveals whole-genome duplication and chromosomal co-localization of betacyanin biosynthetic genes. *Horticulture Research* 8: 63.
Zhijiang, W., Fengzhu, H., Lifang, H., Guifeng, L., Guidong, L. *et al.* (2021) Genetic characterization of pitaya accessions based on ISSR analysis. *Redai Zuowu Xuebao* 42, 47–53.
Zhou, Z., Gao, H., Ming, J., Ding, Z., Lin, X.E. *et al.* (2020a) Combined transcriptome and metabolome analysis of pitaya fruit unveiled the mechanisms underlying peel and pulp colour formation. *BMC Genomics* 21: 734.
Zhou, J., Wang, Z., Mao, Y., Wang, L., Xiao, T. *et al.* (2020b) Proteogenomic analysis of pitaya reveals cold stress-related molecular signature. *PeerJ* 8: e8540.

7 Propagation

Taner Bozkurt[1]* and Özhan Şimşek[2]
[1]*Tekfen Agricultural Research Production and Marketing Inc., Adana, Türkiye;*
[2]*Horticulture Department, Agriculture Faculty, Erciyes University, Kayseri, Türkiye*

7.1 Introduction

The homeland of pitaya (also known as dragon fruit; *Hylocereus* spp.) is tropical America, southern Mexico, Costa Rica, El Salvador, Venezuela, Colombia, Ecuador, Panama, Brazil and Uruguay. Pitaya is widely cultivated in tropical and subtropical America, southern Florida, the Caribbean, Hawaii, Australia, Taiwan, Vietnam, Malaysia and Israel. This plant species, which can grow in limited climatic conditions, has become widespread in countries with suitable microclimates (Bozkurt *et al.*, 2022). Accurate and efficient propagation of pitaya is of critical importance in all countries that are cultivating or will cultivate pitaya. Pitaya can be propagated by plant tissue culture, seeds and cuttings. While tissue culture and seed propagation are used in breeding studies, propagation with cuttings is preferred for planting and developing orchards.

7.2 Seeds

Pitaya seeds are black, very small and numerous (Fig. 7.1), and are embedded and scattered in the fruit flesh. The seeds are about 3 mm in diameter and are shiny and smooth (Ortiz *et al.*, 2018). After collecting the seeds from ripe fruit, the seeds should be washed thoroughly and then planted in a medium of sand and clay (1:1, v:v) or on moist blotting paper for germination (Fig. 7.2) (Gunasena *et al.*, 2007). A single fruit can contain over 7000 seeds (Tel-Zur *et al.*, 2011). Seedcoat colour is linked to seed germination potential; seeds with hard black coats have greater germination rates than seeds with soft brown coats, which are mostly aborted (Cisneros *et al.*, 2011). However, propagation by seed is expected to cause genetic variations among the seedlings, which directly affects production and fruit quality.

One of the most important disadvantages of propagation by seed is juvenility. It has been reported that the fruiting of pitaya plants raised from seeds may take approximately 4–5 years after planting due to juvenility (Chhetri *et al.*, 2021). Different factors such as temperature, light and growing medium affect the germination of pitaya seeds. It has been reported that seeds gave the best germination percentage when germinated at 20–24°C. Alves *et al.* (2011) studied the effects of environment and temperature on seed germination and found increased seed germination on filter paper at a constant temperature of 25°C. Ahmed (2006) investigated the effects of four different temperatures (16°C, 20°C, 24°C and 28°C) and four different white light intensities (0, 500, 1000 and 2000 lux) on the germination of pitaya

* Email: taner.bozkurt@tekfen.com.tr

DOI: 10.1079/9781800623156.0007

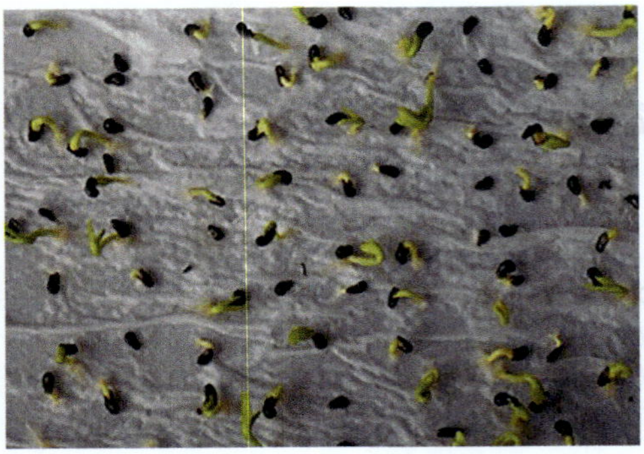

Fig. 7.1. Pitaya seeds. Photograph courtesy of Dr Hugo Viallachica Leon.

Fig. 7.2. Pitaya seedlings. Photograph courtesy of Dr Hugo Viallachica Leon.

seeds. They noted that the germination rate of seeds varied between 71% and 83%. Germination was higher in the dark and in 500 lux light intensity. Lone *et al.* (2014) investigated the effects of different light colours and wavelengths on germination of seeds of five different pitaya species (*Hylocereus undatus*, *H. polyrhizus*, *Selenicereus megalanthus*, *H. undatus* × *H. costaricensis* and *H. costaricensis* × *H. undatus*). They reported that the most suitable light intensities for germination of *H. undatus*, *H. polyrhizus* and *H. undatus* × *H. costaricensis* seeds were 10.1 and 17.9 μmol m^{-2} s^{-1}. The most suitable light intensity for *S. megalanthus* was 17.9 μmol m^{-2} s^{-1}, and light intensities of 3.4, 10.1 and 17.9 μmol m^{-2} s^{-1} were suitable for *H. costaricensis* × *H. undatus*. They emphasized that *H. undatus*, like *H. costaricensis* × *H. undatus*, showed germination at all tested light wavelengths. These studies show that the germination of seeds of different pitaya species and varieties is affected by light intensity.

Pitaya seed germination is affected by drought and salinity conditions. Ortiz *et al.* (2014) studied pitaya seed germination and survival abilities under natural stress conditions. The effects on seeds of three different pitaya species of six salt concentrations (0.0, –0.2, –0.4, –0.6, –0.8 and –1.0 MPa) and four different salt sources (PEG 6000 polymer, potassium chloride (KCl), sodium chloride (NaCl) and magnesium chloride ($MgCl_2$)) in terms of germination percentage, germination rate and average germination time were investigated. They found that salt doses had a more significant effect on germination than salt sources, and that application of 6000 PEG at –0.2 MPa affected seed viability. Seeds of the pitaya hybrid *H. undatus* × *H. costaricensis* were the most resistant to stress conditions and gave better results in terms of germination percentage and germination rate. A water potential of –0.2 MPa is limiting for germination and initial growth of *H. costaricensis* and *H. undatus*, the latter being more tolerant to water stress in the germination phase (Carvalho *et al.*, 2023). Pre-germination treatments with hydropriming, gibberellic acid (150 mg l^{-1}), salicylic acid (1 μM) and thiamethoxam (1 ml kg^{-1} of seed) promoted the physiological performance of *H. costaricensis* seeds, with gibberellic acid being the best water stress attenuator. Gibberellic acid treatment also improved the physiological performance of *H. undatus* seeds under water-deficit conditions (Carvalho *et al.*, 2023). The germinating medium pH level is another criterion that can affect the germination of seeds. Ortiz *et al.* (2015) observed that when the pH of the germinating medium was between 6.0 and 6.5, it decreased the average germination time of the seeds and increased the germination rate. Seed propagation of pitaya is not preferred in commercial cultivation due to the long juvenile period and genetic variability among seedlings; it is mostly used to develop new hybrids.

7.3 Vegetative Production

7.3.1 Cuttings

Using cuttings is one of the most practical and the preferred method of propagating pitaya.

Propagation with cuttings involves removing a small piece from a mother plant for rooting. Studies have been carried out to understand the effect of different factors such as cutting type, cutting length, rooting medium, planting depth of cuttings and application of root-inducing hormones on the rooting of cuttings. Cuttings are taken from mature shoots in 15–20 cm lengths (Fig. 7.3). The cuttings are planted directly or transferred to their final places after rooting. For rooting, a growing medium with good drainage (e.g. sand, perlite, vermiculite) should be used (Zee *et al.*, 2004). Bastos *et al.* (2006) reported that the treatment of cuttings with a length of 15–25 cm with indole-3-butyric acid (IBA) at 3000 ppm for 20 s increased the rooting rate. Marques *et al.* (2011) used five different cutting sizes (5, 10, 15, 20 and 25 cm) to determine the effects of cutting size on the production of pitaya saplings. They suggested using cuttings of 15 and 25 cm for production of good-quality planting materials. Drawing attention to the planting depth, different researchers (Marques *et al.*, 2012) emphasized that planting red pitaya (*H. undatus*) cuttings at a planting depth of 1 cm and with apical dominance gave more successful results. Additionally, it has been emphasized that apical dominance has positive effects on shoot elongation. Rahad *et al.* (2016) suggested the use of soil and/or soil and sawdust (1:1, v:v) as a rooting medium for planting of cuttings, while Sunita *et al.* (2022) used sand:soil:farmyard manure (2:1:1, v:v:v) for rooting of cuttings. Ayesha and Thippesha (2018) studied the effect of growth regulators for rooting pitaya cuttings. An IBA application of 7000 ppm was found to give satisfactory results in terms of root development and was recommended for commercial production, while Sunita *et al.* (2022) found the highest level of rooting using 3000 ppm IBA plus 100 ppm α-naphthaleneacetic acid (NAA). Although IBA is mostly used as a root-promoting hormone, it was reported that an algal extract (SprintAlga TS) is a promising alternative or complement to IBA (de Freitas *et al.*, 2021). Studies on the effect of 6-benzyloaminopurine (BAP) showed that the survival rate of cuttings was markedly increased by BAP treatment (100%) (Cruvinel *et al.*, 2019).

Fig. 7.3. The ideal size of a cutting.

Another factor that influences the rooting of pitaya cuttings is the period in which the cuttings are taken. The correct selection of this period is important, as propagation at the right time significantly affects success. It was reported that the most suitable seasons are between May and September (Wasielewski and Balerdi, 2019). Therefore, as spring and summer may be the best seasons for propagation with cuttings, it is relatively common to propagate with this method in greenhouse conditions. For rootstock production, the cuttings should be older than 1 year (Merten, 2003; Zee *et al.*, 2004). If the cutting sizes are 50–70 cm long, they can be propagated and planted directly in the field or in pots (Moreira *et al.*, 2017). It is also imperative that the cuttings should be healthy and the state of health should be at a level that can be seen with the naked eye (Fig. 7.4).

7.4 Micropropagation

One of the most critical biotechnological methods that can be used effectively in pitaya production is the *in vitro* tissue-culture method. *In vitro* tissue culture is a collection of techniques used to preserve or grow plant cells, tissues or organs under aseptic conditions in an artificial culture medium of known composition (Damankeshan and Panahi, 2013). These methods include the propagation of endemic plants, virus-free production of plants, genome transformation, production of secondary plant metabolites of high commercial value and protection of plant genetic resources (Espinosa-Leal *et al.*, 2018).

Fig. 7.4. Rooted cuttings are ready for planting in the field in 4–6 months. Photograph courtesy of Dr Hugo Viallachica Leon.

In plant tissue-culture studies, the plants must be kept in a sterile environment and transferred to artificial medium. For this, laminar flow cabinets with HEPA filters are used (Twaij *et al.*, 2020). Studies on surface sterilization, one of the first and most essential steps of tissue-culture studies, should be conducted in this environment. Tissue or organ explants taken from selected living plants are subjected to surface sterilization with appropriate chemicals because they are contaminated with organic or inorganic contaminants on or inside the tissue (Yeole *et al.*, 2016). Chemicals such as ethanol, sodium hypochlorite and mercury chloride are widely used for this purpose. Ethanol is a widely used and powerful sterilizing agent but is reportedly phytotoxic to plant cells (Singh, 2018).

After healthy surface sterilization, the plant materials are transferred to the nutrient medium, where they will be cultured. The most widely used and well-known medium used in plant tissue-culture studies is MS medium developed by Murashige and Skoog (1962). This nutrient medium consists mainly of organic and inorganic salts, carbon sources, vitamins and plant growth regulators (PGRs) in different concentrations. Although plants naturally synthesize the growth hormones they need, they also need to be supplemented externally to provide better growth and development. These hormones are cytokinins, auxins and gibberellins (Gupta *et al.*, 2020).

With the plant tissue-culture methods, it is possible to produce pitaya species quickly, with high quality and efficiency. This method is a crucial step for the commercial production of pitaya by tissue culture. First, the plant variety desired to be propagated clonally should be selected, and materials should be healthy explant sources. For surface sterilization of the selected plant materials to be brought to the laboratory, they should be washed with detergent. The detergent and other impurities should be removed under running water for about 20–30 min. After this process, the explants should be treated with 70% ethyl alcohol for 20–40 s and 1–2% (v:v) sodium hypochlorite for 20–30 m. Surface sterilization is completed by washing three to four times with distilled water (Fig. 7.5) (Bozkurt *et al.*, 2022).

After surface sterilization, the plant materials are transferred to the nutrient medium. Commonly, MS medium is prepared by supplementing with an average of 20–30 g l^{-1} of sucrose, various PGRs and agar, which is mainly

Fig. 7.5. Surface sterilization.

used as a thickener. The medium should be sterilized in an autoclave at 121°C and 1.2 PSI pressure for 15–20 min, and different concentrations of plant hormones added such as BAP and NAA for micropropagation. After the sterile nutrient medium has cooled in suitable culture containers, the explants are cut to the desired sizes in a sterile cabinet and transferred to the medium. Plant material transferred to the medium should be cultured for 4–6 weeks in culture rooms at an average temperature of 25±2°C and a photoperiod of 16 h light:8 h dark. At the end of this culture period, the plantlets are transferred back to new medium, called a subculture (Fig. 7.6). In this way, the desired quality and number of micropropagated pitaya plants can be produced.

Pitayas propagated in micropropagation medium should be rooted *in vitro* by transferring to MS medium supplemented with root-promoting hormones and cultured for the same photoperiod and duration specified above. The plantlets are then transferred to the soil in a healthy way to increase the success of adaptation to the soil.

The most critical step in the micropropagation of pitaya plants grown *in vitro* is the transfer of these plants under *ex vitro* conditions, as is the case with many plants. The fact that plantlets propagated *in vitro* have weak root systems, non-functional stomas and underdeveloped cuticles causes a high rate of death in the transfer of plants to the *ex vitro* environment (Mathur *et al.*, 2008). For this reason, plantlets should be gradually acclimatized when transferred from the field or greenhouse to the outside environment (Hazarika, 2003; Chandra *et al.*, 2010).

An additional consideration in micropropagation of pitaya is that different varieties or clones respond differently to the same culture medium (Drew and Azimi-Tabrizi, 2002; Mohamed-Yasseen, 2002; Hua *et al.*, 2015; Bozkurt *et al.*, 2020, 2022).

The yellow pitaya has been regenerated from epicotyls of seedlings via somatic embryogenesis (Infante, 1992). Drew and Azimi-Tabrizi (2002) described a protocol for multiplication of *H. undatus*. Shoot apices imposed a strong dominance and limited the growth of shoots from axillary areoles. Shoot and root initiation from stem sections with the apices removed and from intact areoles was achieved on PGR-free MS medium, although addition of 5 μM 2-isopentenyl purine produced 100% shoot initiation and reduced the time to shoot initiation. Shoot production was affected by the size of stem explants, and shoots could be produced from stem explants as small as one whorl of stem.

Fig. 7.6. Subculture.

A large number of shoots (containing seven areoles) increased shoot production (8.2 shoots per 250 ml flask), thus economizing the cost of production.

In another study for the micropropagation of pitaya, explants extracted from young shoots of the mature plant were cultured in MS medium containing 0.5 µM NAA and 0.5 µM thidiazuron (TDZ) (Mohamed-Yasseen, 2002). The shoots produced from these first explants were transferred to MS medium containing PGRs at different concentrations by cutting the head and dividing it into three longitudinal parts to produce secondary explants. These media were created by adding 0.5 µM NAA and 0.01, 0.09, 0.5 and 0.9 µM TDZ. After 4 weeks, it was observed that the decapitated explants produced 11 more shoots than the longitudinal explants. In this medium, 3.2 shoot formations were recorded in the longitudinally cut explants. Shoots produced from secondary explants were rooted in MS medium and transferred to the soil.

A protocol for rapid *in vitro* propagation of genetically diverse pitaya was developed by Hua *et al.* (2015) (Fig. 7.7). Various types of PGRs were used to determine the most effective hormone combination. Zeatin (ZT), TDZ, 2,4-dichlorophenoxyacetic acid (2,4-D) and 6-benzylaminopurine (6-BA) alone could each induce multiple shoots. The best PGR combination was MS medium with 3.0 µM ZT and 0.5 µM IBA, producing the most shoots per explant and the most vigorous shoots. The results showed that MS medium with 13.68 µM ZT and 2.46 µM IBA was suitable for shoot propagation of diverse pitaya varieties and selection. This was the first report on the effect of ZT and 2,4-D in pitaya propagation. The protocol developed by Hua *et al.* (2015) avoids shoot production from callus and production of multiple shoots directly from areoles. Sabha and Saad (2020), however, obtained the best result by supplementing MS medium with 4.55 µM TDZ for shoot development of explants collected from stem nodal segments. The best rooting was obtained with 2.69 µM NAA. High-efficiency micropropagation of dormant buds in the spine base of red pitaya for industrial breeding was developed by Qin *et al.* (2017). The dormant buds in the spine base should be surface sterilized in 0.2% mercuric chloride for 7 min. The medium formulation for adventitious shoot induction and proliferation was suggested as MS medium supplemented with 5.5 mg l^{-1} 6-benzyladenine (BA) and 0.1 mg l^{-1} NAA. Half-strength MS medium supplemented with 0.5 mg l^{-1} NAA and 0.3 mg l^{-1} IBA was suggested for the maximum rooting response of proliferated shoots. The micropropagation protocol for *H. costaricensis* was standardized by Ismail *et al.* (2023) in Egypt. Surface-sterilized areole sections successfully induced multiple shoots on MS basal medium supplemented with BA, kinetin and TDZ at 1.5 mg l^{-1}. The maximum response was, however, recorded with BA (13.67 shoots). The maximum number of rootlets was obtained by using a combination of IBA and NAA at 0.25 mg l^{-1}. The rooted regenerants were transferred to pots containing a mixture of peat moss:sand:clay at a ratio of 11:1:1 (v:v) with a 100% survival rate.

Fig. 7.7. A schematic representation of the *in vitro* propagation procedure of pitaya. MS, Murashige and Skoog. Adapted from Hua *et al.*, 2015, with permission.

Pal *et al.* (2022) suggested a pot mixture of soil:sand:vermicompost:cocopeat (1:1:1:1, v:v) for hardening of *H. undatus* plantlets. Bozkurt *et al.* (2022) evaluated the success of different varieties of pitaya in micropropagation. The highest micropropagation coefficient of the varieties 'Physical Graffiti' (4.27) and 'Siam Red' (3.25) was obtained using MS medium supplemented with BAP at 2 mg l^{-1}, while for 'Royal Red' (3.25) and 'Seoul Kitchen' (2.87), the highest coefficients were noted with MS supplemented with BAP at 4 mg l^{-1}. However, the best rooting response was seen with IBA at 1 mg l^{-1} in 'Royal Red' (54.47%) followed by 'Siam Red' (50.33%), 'Physical Graffiti' (47.75%) and 'Seoul Kitchen' (44.82%).

7.5 Conclusion

Pitaya has become a widespread tropical fruit species with importance in human nutrition. It is crucial to propagate pitaya economically and efficiently in different areas of the world. Pitaya can be propagated by seeds, cuttings and *in vitro* techniques. The propagation of pitaya by seed is essential for breeding programmes. The hybrid seeds that result from cross-pollination provide a valuable source of new varieties. Seedlings may take between 3 and 7 years to bear fruit. Asexual reproduction by cuttings is quite common, and cuttings may produce fruit in as little as 10–15 months. *In vitro* micropropagation of members of the genus *Selenicereus* has been successful and a number of protocols have been developed by researchers from different countries for mass multiplication. The genotype effect is one of the most crucial factors for propagation by cuttings and plant tissue-culture techniques. Therefore, preparing propagation strategies specific for the studied species and variety is necessary. In addition, the length of the cutting, the period it was taken and the use of root promoters for rooting are significant in propagation. This chapter has illustrated the various propagation methods used for different pitaya species and varieties.

7.6 References

Ahmed, A.E. (2006) Mass propagation of pitaya (dragon fruit). *Fruits* 61, 313–319.

Alves, C.Z., Godoy, A.R. and Correa, L.S. (2011) Methodology adjustment for the germination test of red pitaya seeds. *Ciencia Rural* 41, 779–784.

Ayesha, S. and Thippesha, D. (2018) Influence of plant growth regulators on rooting of stem cuttings in dragon fruit [*Hylocereus undatus* (Haworth) Britton & Rose]. *International Journal of Chemical Studies* 6, 1834–1839.

Bastos, D.C., Pio, R., Scarpare, J.A., Libardi, M.N., Almeida, L.F.P. *et al.* (2006) Propagation of red pitaya (*Hylocereus undatus*) by cuttings. *Ciencia e Agrotecnologia* 30, 1106–1109.

Bozkurt, T., İnan, S. and Dündar, İ. (2020) Micropropagation of different pitaya varieties. *International Journal of Agricultural and Natural Sciences* 13, 39–46.

Bozkurt, T., İnan, S., Dundar, I. and Ozdemir, S.K. (2022) Effect of different plant growth regulators on micropropagation of some pitaya varieties. *Journal of Tropical Life Science* 12, 183–190.

Carvalho, S.M.C., de Paiva, E.P., Torres, S.V., Neta, M.L.S., Leite, M.S. *et al.* (2023) Pre-germination treatments in pitaya (*Hylocereus* spp.) seeds for water stress mitigation. *Revista Caatinga* 36, 80–86.

Chandra, S., Bandopadhyay, R., Kumar, V. and Chandra, R. (2010) Acclimatization of tissue cultured plantlets from laboratory to land. *Biotechnology Letters* 32, 1199–1205.

Chhetri, S., Hasan, M.A. and Tamang, A. (2021) Influence of varying length of stem cutting and IBA concentrations on root and shoot growth in dragon fruit cv Giant White (*Hylocereus undatus*). *Environment and Ecology* 39, 1114–1118.

Cisneros, A., Garcia, R.B. and Tel-Zur, N. (2011) Ovule morphology, embryogenesis and seed development in three *Hylocerrus* species (Cactaceae). *Flora* 206, 1076–1084.

Cruvinel, F.F., Vasconcellos, M.D.S. and Martelleto, L.A.P. (2019) Effects of cytokinin benzylaminopurine in pitaya cuttings. *Nativa: Pesquisas Agrárias e Ambientais* 7, 43–49.

Damankeshan, B. and Panahi, B. (2013) A comparative study on the growth characteristics of offshoot and tissue culture propagated palm trees in orchards. *International Journal of Agriculture and Crop Sciences* 5: 2221.

de Freitas, F.R., Brighenti, A., Coutinho, M., Voltolini, J., Malohlava, I. *et al.* (2021) Effect of algae extract on the rooting of pitaya cutting. *Agropecuária Catarinense* 34, 34–36.

Drew, R. and Azimi-Tabrizi, M. (2002) Micropropagation of red pitaya (*Hylocereus undatus*). *Acta Horticulturae* 575, 93–98.

Espinosa-Leal, C.A., Puente-Garza, C.A. and García-Lara, S. (2018) *In vitro* plant tissue culture: means for production of biological active compounds. *Planta* 248, 1–18.

Gunasena, H.P.M., Pushpakumara, D.K.N.G. and Kariyawasam, M. (2007) Dragon fruit *Hylocereus undatus* (Haw.) Britton and Rose. *Underutilized Fruit Trees in Sri Lanka* 1, 110–141.

Gupta, N., Jain, V., Joseph, M.R. and Devi, S. (2020) A review on micropropagation culture method. *Asian Journal of Pharmaceutical Research and Development* 8, 86–93.

Hazarika, B.N. (2003) Acclimatization of tissue-cultured plants. *Current Science* 85, 1704–1712.

Hua, Q., Chen, P., Liu, W., Ma, Y., Liang, R. *et al.* (2015) A protocol for rapid *in vitro* propagation of genetically diverse pitaya. *Plant Cell Tissue and Organ Culture* 120, 741–745.

Infante, R. (1992) *In vitro* axillary shoot proliferation and somatic embryogenesis of yellow pitaya *Mediocactus coccineus* (Salm-Dyck). *Plant Cell, Tissue and Organ Culture* 31, 155–159.

Ismail, A.M., El-Kader, E.M.A., Eid, A.M. and Hussein, E.A. (2023) Clonal micropropagation of purplish-red dragon fruit (*Hylocereus costaricensis*) newly introduced to Egypt. *Alfarama Journal of Basic and Applied Science* 4, 235–254.

Lone, A.B., Unemoto, L.K., Ferrari, E.A.P., Takahashi, L.S.A. and Faria, R.T. (2014) The effects of light wavelength and intensity on the germination of pitaya seed genotypes. *Australian Journal of Crop Science* 8, 1475–1480.

Marques, V.B., Moreira, R.A., Ramos, J.D., Araujo, N.A. and Cruz, M.D.M. (2011) Cladode size in the production of red pitaya seedlings. *Revista Caatinga* 24, 50–54.

Marques, V.B., Moreira, R.A., Ramos, J.D., Araujo, N.A. and Cruz, M.D.M. (2012) Profundidade de plantio e dominância apical naestaquia de pitaia vermelha. *Semina Ciencias Agrarias* 33, 2091–2097.

Mathur, A., Mathur, A.K., Verma, P., Yadav, S., Gupta, M.L. and Darokar, M.P. (2008) Biological hardening and genetic fidelity testing of micro-cloned progeny of *Chlorophytum borivilianum* Sant. et Fernand. *African Journal of Biotechnology* 7, 1046–1053.

Merten, S. (2003) A review of *Hylocereus* production in the United States. *Journal of the Professional Association for Cactus Development* 5, 98–105.

Mohamed-Yasseen, Y. (2002) Micropropagation of pitaya (*Hylocereus undatus* Britton et Rose). *In Vitro Cellular & Developmental Biology – Plant* 34, 427–442.

Moreira, A.R., de Souza, F.L.B., da Silva, R.T.L., de Oliveira, R.L.L., dos Santos Alonço, A. *et al.* (2017) Determinação do comprimento da estaca para a produção de mudas de pitaia (*Hylocereus costaricensis*) emambiente protegido. *Tecno-Lógica* 21, 41–45.

Murashige, T. and Skoog, F. (1962) A revised medium for rapid growth and bio assays with tobacco tissue cultures. *Physiologia Plantarum* 15, 473–497.

Ortiz, T.A., Gomes, G.R., Takahashi, L.S.A., Urbano, M.R. and Strapasson, E. (2014) Water and salt stress in germinating seeds of pitaya genotypes (*Hylocereus* spp.). *African Journal of Agricultural Research* 9, 3610–3619.

Ortiz, T.A., Mortiz, A., Oliveira, M.A. and Takahashi, L.S.A. (2015) Effects of the hydrogen potential and fungicide treatment on pitaya seed germination. *Acta Scientiarum Agronomy* 37, 69–74.

Ortiz, T.A., Becker, G.F. and Takahashi, L.S.A. (2018) Pitaya genotypes (*Hylocereus* spp.) seed germination at different pH levels based on statistical models. *Australian Journal of Crop Science* 12, 1200–1204.

Pal, A., Kumar, A., Rajbhar, Y.P., Prakash, S., Kumar, V. *et al.* ((2022) Standardization of acclimatization and field establishment of *in vitro* regenerated dragon fruit (*Hylocereus undatus*). *Pharma Innovation Journal* 11, 1849–1852.

Qin, J., Wang, Y., He, G., Chen, L., He, H. *et al.* (2017) High-efficiency micropropagation of dormant buds in spine base of red pitaya (*Hylocereus polyrhizus*) for industrial breeding. *International Journal of Agriculture and Biotechnology* 19, 193–198.

Rahad, M.A.B.K., Islam, M.A., Rahim, M.A. and Monira, S. (2016) Effect of rooting media and varieties on rooting performance of dragon fruit cuttings (*Hylocereus undatus* Haw.) *Research in Agriculture, Livestock and Fisheries* 3, 67–77.

Sabha, S.M. and Saad, M.M.A. (2020) Propagation and preservation of *Hylocereus undatus* via tissue culture technique. *Journal of Basic and Applied Scientific Research* 10, 7–12.

Singh, C.R. (2018) Review on problems and its remedy in plant tissue culture. *Asian Journal of Biological Sciences* 11, 165–172.

Sunita, M.M.L., Choudhary, A., Nagori, A. and Nishad, U. (2022) Impact of plant growth regulators on development of dragon fruit cutting [*Hylocerus costaricensis* (Web.) Britton and Rose]. *Annals of Horticulture* 15, 162–167.

Tel-Zur, N., Mizrahi, Y., Cisneros, A., Mouyal, J., Sehneider, B. and Doyle, J.J. (2011) Phenotypic and genomic characterization of the vine cactus collection (Cactaceae). *Genetic Resource and Crop Evolution* 58, 1076–1084.

Twaij, B.M., Jazar, Z.H., and Hasan, M. (2020) Trends in the use of tissue culture, applications and future aspects. *International Journal of Plant Biology* 11: 8385.

Wasielewski, J. and Balerdi, C.F. (2019) Tropical and subtropical fruit propagation. *EDIS* 2019: 7.

Yeole, M.P., Gholse, Y.N., Gurunani, S.G. and Dhole, S.M. (2016) Plant tissue culture techniques: a review for future view. *Critical Review in Pharmaceutical Sciences* 5, 16–25.

Zee, F., Yen, C.R. and Nishina, M. (2004) Pitaya (dragon fruit, strawberry pear). University of Hawaii, Honolulu, Hawaii. Available at: https://scholarspace.manoa.hawaii.edu/items/c675febb-db04-4236-8739-cbfba6a6e68b (accessed 9 November 2023).

8 Orchard Management

Mai Van Tri[1,2*], Dinh Thi Yen Phuong[3] and Sisir Mitra[4]
[1]*Former Deputy Director, Southern Horticultural Research Institute (SOFRI), Tien Giang, Vietnam;* [2]*Director, Southeast Horticultural Research Center, Ba Ria-Vung Tau, Vietnam;* [3]*Applied Biotechnology Institute, Tan Phu, Ho Chi Minh City, Vietnam;* [4]*Former Professor, Faculty of Horticulture, Bidhan Chandra Krishi Viswavidyalaya (State Agriculture University), Mohnapur, West Bengal, India*

8.1 Introduction

Dragon fruit cultivation has increased rapidly during the past 10 years. It is a new crop for most countries that grow it, and standardized countrywide growth practices are not available. Dragon fruit is a shallow-rooted plant and needs extra care in terms of management of soil, nutrition, irrigation and weed management. It is a vine-cactus plant and needs support for growth. The use of posts to support the plant as well as different trellis systems depends on the growing conditions and situation, plant density, variety and other factors. During the initial years, the interspaces between the rows and plants can be utilized for growing intercrops, which not only provides additional income but also improves the physicochemical properties of the soil and checks weed growth. With advancements in technology, it is possible to harvest dragon fruit out of season. Establishment and management of a dragon fruit orchard is a long-term investment and requires critical planning. Orchard management covers a broad range of practices designed to maximize production of high-quality dragon fruit and increase profitability for growers. To maintain the health of the orchard and achieve good fruiting every year, it is necessary to consider and make appropriate decisions on the variety and planting materials, site location and preparation, water availability, type of trellis, density and spacings, soil and nutrient management, irrigation management, pruning and training systems, orchard floor management, crop load and pest management, flowering induction and crop regulation, and harvest and postharvest practices. In this chapter, these aspects of orchard management are described.

8.2 Site Selection

Although dragon fruit can be grown on a wide range of soils, survive temperature variations, and requires less water, care and maintenance than some crops (Lim, 2012), a number of factors need to be considered before planting: the availability of an irrigation water resource and its quality, surface water drainage conditions in the wet season, the type and texture of the soil, the water table, soil pH, the soil slope and topography, rainfall and rainfall patterns, sunshine duration, temperatures and relative humidity. Some other factors should also be considered while selecting the site for establishment of a commercial orchard including availability of employees, transportation and conditions for orchard mechanization.

* Email: maivantri2017@gmail.com

DOI: 10.1079/9781800623156.0008

Dragon fruit has the potential to grow in tropical and subtropical climatic regions with average temperatures of 21–29°C but can stand temperatures up to 38–40°C and as low as 0°C for a short time (McMahon, 2003); however, the site should be free of frost and freezes. In Vietnam, sunburn usually occurs when the temperature is higher than 38°C. In growing areas at more than 35°N, dragon fruit is often grown in greenhouses, with the inside temperature maintained at 5°C or above during winter. According to Karunakaran *et al.* (2019), the rainfall requirement is 1145–2540 mm year^{-1}. High-rainfall areas are not suitable for the crop, and excessive rain can cause flower and fruit drop (Karunakaran and Arivalagan, 2019), waterlogging, soil erosion, and increased fungal and bacterial diseases (Mai *et al.*, 2020). In many growing areas in Vietnam, dragon fruit grows well in regions with rainfall of 1200–1800 mm year^{-1}. The plant is also successfully grown commercially in areas with rainfall of only 800–1000 mm year^{-1} (north-eastern region of Binh Thuan province, Vietnam). The ideal rainfall for dragon fruit in the monsoon tropical climate is 1000–1400 mm year^{-1}. In areas with low rainfall, effective irrigation is necessary for successful growth. Irrigation water needs will be higher in growing areas with lower rainfall. Excessive rain promotes stem canker disease caused by the fungus *Neoscytalidium dimidiatum* and other important diseases in dragon fruit. Dragon fruit is vulnerable to strong winds and tropical storms. Reinforced concrete posts should be used instead of bamboo or wooden posts to grow dragon fruit in areas at risk of strong winds.

Dragon fruit has a shallow root system, with most of the roots concentrated in the top 15–35 cm of soil. The plants can be grown on a wide range of soils and tolerate poor soil conditions. The ideal soil type for commercial production is clay loam that is well drained, rich in organic matter, neutral or slightly acidic (pH 5.5–7.0) and not waterlogged (Mai *et al.*, 2020).

8.3 Spacing and Density

As a sun-loving tree, dragon fruit needs to be planted at an appropriate planting distance for good growth and fruit production. The optimal spacing for dragon fruit depends on a number of factors, including the environment, climate, variety, soil type, fertility status, soil slope and topography, type of trellis system to be used, using light for flower induction in the off-season or not, mechanization level, pest status and other factors (Mai *et al.*, 2021). Optimum spacing improves air circulation around the vines and allows more light penetration, which reduces the incidence of many pests and diseases (Mai *et al.*, 2021). In warm and humid areas, when designing the orchard, rows should be oriented parallel to the prevailing wind direction to assist ventilation and help reduce the duration of plant wetness.

The planting spacing and density of dragon fruit also depends on the type of trellis system used. Generally, higher densities are found on fence trellis systems compared with the single-post system. The single-row system, square system, rectangular system and triangular system can all be used for planting dragon fruit. The single-row system is usually applied using the fence trellis system. Square and rectangular systems are usually applied using a single post, while the triangular system is not common.

Various planting distances of 1.5–3.7 m between plants in a row and 1.5–5.0 m between rows have been used in commercial production on a single post (Thomson, 2002; Zee *et al.*, 2004; Ben-Asher *et al.*, 2006; Gunasena *et al.*, 2007). Planting distances of 3 × 5 m have been recommended in Nicaragua (Jacobs, 1999), 2.0 × 1.7 m in the USA (Thomson, 2002), 1.5 × 1.5 m in Israel (Weiss *et al.*, 1994) and 3–3.7 × 3–4.9 m in Sri Lanka (Gunasena *et al.*, 2007), as well as 1.5 × 2.5 m (Nerd *et al.*, 1999) and 1.5 × 3 m (Raveh *et al.*, 1998). In Malaysia, a distance of 2.0–3.0 m is recommended between two single posts (Le Bellec *et al.*, 2006). Plant spacing of 3.0 × 2.4 m with 1361 posts ha^{-1} is the most common for cultivation of red dragon fruit (*Hylocereus polyrhizus*) in Malaysia (Kek Hoe, 2017). In Vietnam, planting densities are usually between 950 posts ha^{-1} (3.0 × 3.5 m) and 1600 posts ha^{-1} (2.5 × 2.5 m). The densities widely applied by local growers are 1275 posts ha^{-1} (2.8 × 2.8 m) and 1371 posts ha^{-1} (2.7 × 2.7 m), while 1111 posts ha^{-1} (3 × 3 m) and 1190 posts ha^{-1} (3 × 2.8 m) are less common. In a field experiment with a high-density planting of red pitaya in Malaysia, three planting densities –1815 trellis plants ha^{-1} (3.0 × 1.8 m), 1556 trellis plants ha^{-1} (3.0 × 2.1 m) and 1361 trellis plants ha^{-1}

(3.0 × 2.4 m) – were compared. Although the higher density of 1815 trellis plants ha^{-1} caused overcrowding of the canopy and a shading effect from the dense branches among the trellis plants, the cumulative fruit numbers and fruit yield were significantly higher compared with those planted at 1556 trellis plants ha^{-1} (Kek Hoe, 2017)

In the fence trellis system, the distance between rows is commonly 3.0 m and between plants in the row is 0.6–0.8 m (5550–4166 cuttings ha^{-1}) (Son *et al.*, 2019; Mai *et al.*, 2020). The distance between two cement posts in a row is 3.0–4.2 m. After planting, the cuttings are supported to grow up the fence by wood posts (e.g. using bamboo, acacia (*Acacia auriculiformis*), paperbark (*Melaleuca alternifolia*) or tall-stilt mangrove (*Rhizophora apiculata*)).

The distance between rows should be wider if mechanical equipment is to be used in the orchard. The distance between the rows is determined primarily by the machinery width and the size of the plant canopy. The distance should be wide enough to provide the minimum space necessary if machinery or vehicles are needed to access or move through the rows.

8.4 Land Preparation

Land preparation should be done 6–12 months before planting. The following aspects should be considered: type of soil texture and composition (clay, loam or sand); soil pH and fertility; cleaning the selected site, including removal of weeds, grasses and other undesirable materials (should not be removed by burning); consideration of the drainage pattern and the slope of the selected area (this helps to determine the orientation of the rows and prevent soil erosion); improving the selected area to ensure adequate drainage, aeration and space for root growth; determining the trellis type to be used (single-post or fence trellis system) for proper preparation of the planting area; and making ditches/drains to allow excess water runoff.

Soil analysis should be done before cleaning of the land and ploughing to measure the levels of nutrients and estimate the nutrient requirements. Samples should be taken at depths of 15 and 30 cm of the soil profile. It is also important to measure soil pH to estimate the need for liming or gypsum application to adjust the pH to a suitable level (pH 5.5–7.0) for dragon fruit. A hand trowel can be used to remove soil from a depth of 10–20 cm in each spot. Gypsum or sulfur is usually used for correcting soil with a pH of over 7, while calcium carbonate or calcium phosphate is usually used to adjust low-pH soils. Soil organic matter is a major contributor in enhancing soil fertility and quality. Adding organic matter helps improve soils that are high in clay or sand. Soil amendments should be added to obtain the desired level of soil organic matter of at least 2%.

In well-drained soils, clearing of the land is required to remove the native cover including trees, bushes and shrubs from the surface. This should be done when the soil is dry, ideally late in the dry season. On clay-loam soils, ploughing and harrowing may be required to achieve a fine tilth, removing weeds, roots, stumps, wood, rocks and other undesirable materials. The field should be levelled by passing a blade harrow or laser leveller where necessary. The orientation of the rows is then determined as part of the choice of the orchard layout system, which includes building the trellis system, planting holes and basal pit fertilization before planting. In sandy soils, ploughing and harrowing are not necessary. These soils are light, with quick water drainage and low in nutrients. Most sandy soils are low in organic matter, and adding organic matter including organic manures improves the soil structure, enhancing biological function in the soil, increasing water-holding capacity and supplying nutrients (Bruand *et al.*, 2005). For sandy soils or lighter loam soils with good drainage, level ground planting can be appropriate. A mound will be made later and gradually built up in the years after planting (Fig. 8.1) (Mai *et al.*, 2020). On sloping soils, several ways of planting dragon fruit are recommended, depending on the severity of the slope. On gentle slopes, a series of terraces or flat areas with the same width are built on the slope for planting (Fig. 8.1). The posts should be located on the terrace close to the top of the slope. Growing dragon fruit on stepped slopes is a component of contour farming, aiming to protect the soil and conserve water. A series of contour lines on the slope are created. The posts should be located on the contour line close to the top of the slope.

Fig. 8.1. Dragon fruit orchards in the uplands with low mounds established on sandy soil (A) and on sloping soil (B).

The obstacles to dragon fruit cultivation in the lowlands are a high water table, poor drainage and being prone to flooding in the rainy/wet season. Therefore, the type of soil, texture, soil pH and fertility, depth of the water table and drainage pattern in the wet season are important considerations for site selection and land preparation (Mai *et al.*, 2020). The depth of the water table can change depending on the season of the year. In areas with monsoons, the water table usually rises during the wet/rainy season and will fall during the dry season. Good drainage is therefore vital for growing dragon fruit in the lowland regions.

Most of the planting area of dragon fruit in the Mekong Delta (Vietnam) is located in the lowland regions with alluvial soils and about 1.0–2.5 m above sea level (Torell and Salamanca, 2003). Dragon fruit orchards have been established on rice fields. The clay soils are predominantly flat, poorly permeable, with poor natural drainage and easily become waterlogged in the wet season. Thus, successful adoption of dragon fruit into rice fields under lowland conditions requires proper drainage to remove excess water. The trench-bed system is constructed to drain away the excess water accumulated in the rainy season and to improve soil conditions. Trenches are made by removing the topsoil from the inter-row, and the topsoil is used to make the planting beds/mounds. The width and depth of these trenches depends on the terrain elevation and the thickness of the topsoil above the sulfuric horizon (Fig 8.2). In lower areas, wider trenches are required to have enough topsoil to raise the planting bed to a desired height and also to help with better drainage in the wet season. These beds/mounds need to be high enough to accommodate sufficient soil for the plant roots and to avoid waterlogged conditions in the rainy season. The height of the raised bed must be at least 40 cm above the water level in the trench.

In the lowland Mekong Delta, trenches of 30–60 cm wide × 30–45 cm deep are usually made between two rows in the lower areas. Larger trench sizes are also used in areas with poor drainage and/or a higher water table. In higher areas of the lowlands, ditches can be made between three or four rows depending on the drainage conditions in the orchard (Fig. 8.2). The trenches between rows should be connected with deeper trenches or canals that connect to rivers to drain away excess water in the orchard.

Fig. 8.2. Dragon fruit orchard in the lower area of the Mekong Delta with trenches and mounds. (A) A larger trench. (B) A smaller trench. (C) An orchard in a higher area with raised beds for improved drainage.

The mounds are built from the topsoil around the mound or from soil from the trenches or both. They may be small and low at first, 15–25 cm in height and 60–90 cm in diameter, and will gradually be built up to reach 100–120 cm in diameter and 30–40 cm in height (Fig. 8.2). The soil of the bed/mound is a blend of topsoil, compost, organic manures, rice straw and other organic matter such as mulch. The trench–raised bed/mound system should be built during the dry season. A bed-on-bed system with a small temporary bed on a larger raised bed system is also practised (Fig. 8.3B). In Florida, a planting mound is recommended for areas with a high water table that readily experience flooding in the wet season (Crane *et al.*, 2020). The size of the mound is 60–90 cm in height and 120–310 cm in diameter, made from a native soil. The planting hole is prepared for planting at the required distance.

8.4.1 Planting hole

The planting hole should be prepared before planting. Making a planting hole allows the soil to be broken up, creating abundant cracks between clods, and helping oxygen and water to infiltrate and move through the soil to allow new roots to grow. This helps the plant to grow more roots and ultimately more shoots. The root system of dragon fruit plants is concentrated in the top 15–35 cm of the soil. The size of the planting hole varies depending on soil type and topography, the depth of the water table and the planting bed/mound conditions. It may be larger on heavier, well-drained soils and smaller in lowland poorly-drained soils. The hole is about 30 cm deep and 20 cm wide for each cutting. A planting trench can be made around the planting post for all cuttings to be planted next to the post. This is about 20–30 cm deep and 20–30 cm wide. In the alluvial soils of the Mekong Delta, due to the high water table, planting holes are small with an average recommended size of 20–30 cm wide and 10–20 cm deep. In hilly regions and in sandy soils, a planting hole of 20–40 cm wide and 30–50 cm deep or a planting trench of the same dimensions should be made near or around the planting post (10–20 cm away from the post). If the soil is poor in organic matter, larger planting holes should be made to accommodate organic manures as well as chemical fertilizers to improve the soil condition and increase the amount of organic matter and nutrients in the soil. In the uplands of Brazil, planting holes with a size of 0.5 × 0.5 × 0.5 m are suggested (Fernandes *et al.*, 2018; Alves *et al.*, 2021).

The plants can also be planted on a mound or bed without making a planting hole. In lowlands with poor drainage, a mound or raised bed is made from the topsoil, and top-dressed with organic manure, topsoil and mulch. A raised bed can also be made using a home-made planting ring with potting mixture added to it (Fig. 8.3A). In sandy soils, plants can be planted on a mound

Fig. 8.3. A low planting mound (A) and the bed-on-bed system with a home-made planting ring (B) in the lowlands of the Mekong Delta.

but are commonly planted on a level with the ground. Mounds or beds are made taller and larger according to the diameter of the canopy. Annual topdressing with a few centimetres of compost, organic manure, quality topsoil and mulch around the post is required to build up these mounds or beds.

8.5 Planting

8.5.1 Planting materials

Dragon fruit can be propagated by seeds, micropropagation (Mohamed-Yassen, 2002; Trivellini *et al.*, 2020) or vegetative propagation (Ortiz-Hernández, 1999). Seed propagation is an easy method and can create a large number of seedlings from a small amount of material. However, the variation in the plants produced by seed makes it difficult to create uniform commercial farms or orchards, and it also takes longer to produce fruit from plants grown from seed – the time from planting to fruit production may be up to 6 years (Trivellini *et al.*, 2020). Seed-propagated plants may have variations in growth and may mature at different times. Additionally, plants grown from seed are slow growing and genetically different from the parent (Le Bellec *et al.*, 2006). This method is used mainly in breeding programmes.

Grafting is also used for some weaker varieties or varieties with a small quantity of material that can be grafted on to a stronger rootstock such as 'Vietnam White'. Another benefit of grafting is to speed up the growth of seedlings that have been raised from breeding programmes. *In vitro* propagation producing disease-free and true-to-type planting materials is the key method for the management of viral diseases of dragon fruit. This method is widely used in breeding programmes and for research purposes. Tissue-cultured plantlets have not yet been sold widely for local growers. They require more care at the early stage after planting compared with planting from cuttings. Generally, taking cuttings is the most commonly used method for propagation of dragon fruit in commercial production.

8.5.2 Preparation of cuttings

The size of the cutting has a significant effect on overall rooting performance. Trivellini *et al.* (2020)

reported that cuttings of 20 cm developed a better radical apparatus in terms of rooting frequency and number and length of roots compared with cuttings of 10 cm. This may be because longer dragon fruit cuttings have higher carbohydrate reserves available to initiate and maintain root growth and development. Kakade *et al.* (2019) also observed that larger cuttings (35–40 and 40–45 cm) produced greater shoot and root growth compared with intermediate (20–25, 25–30 and 30–35 cm) and smaller (10–15 and 15–20 cm) cuttings. Chhetri *et al.* (2021) reported that cuttings of 30 cm showed significantly better results than cuttings of 15 or 20 cm with respect to various rooting and shooting parameters.

In commercial production, larger cuttings are recommended to maximize the survival rate, and these also result in earlier flowering and fruiting. In contrast, smaller cuttings should be used when propagation material is limited, or can be used in breeding programmes.

Cuttings should not be taken from branches that show symptoms of mineral nutrient deficiency or sunburn, or that are in bloom. It is better to select vigorous branches growing in full sunlight that are 8–24 months old for cuttings. At this age, they have grown sufficiently to have the strength to grow and propagate on their own. Stem cuttings must be taken from disease-free plants. The risk of cuttings carrying diseases, especially viral diseases, should be taken into consideration (Li *et al.*, 2015; Mai *et al.*, 2020), and disease-free planting material is essential for management of targeted pathogens. Early morning is the best time to take cuttings.

There are four ways to make a cut for the cuttings: segment cut, flat cut, semi-tapered cut and tapered cut (Fig. 8.4). The tapered-cut method is popular in Vietnam, while the segment cut is an easier and quicker method. A sharp knife is used to make a clean cut. The wounds can be treated with a copper-based fungicide if they are to be planted in wet conditions. The cuttings should be dried vertically in a shaded area to help heal the wounds and reduce the risk of infection. It takes 10–20 days for the wounds to dry up and heal. Once the wounds are sealed and the areoles turn white, they are ready for planting. These cuttings can be planted directly into the ground or into plastic pots/polythene bags of potting mixture before being planted into the orchards. The cuttings used in this method are usually small-sized cuttings (no more than 30 cm in length). Rooting hormone should be applied to encourage root and shoot development.

The recommendations for dragon fruit cuttings in Vietnam are as follows (Suu *et al.*, 2017; Mai *et al.*, 2020). Cuttings must be taken from 3–4-year-old, healthy, disease-free plants and from branches that are 1–2 years old, dark green, stout and plump with thorns or nodes that look healthy and strong. The cuttings should be 50–70 cm in length with tapered cuts.

8.5.2 Rooting hormones

Plant growth regulators (PGRs) are used to improve the rooting ability of dragon fruit cuttings. Indole-3-acetic acid (IAA), indole-3-butyric acid (IBA) and naphthalene acetic acid (NAA) have been tested and their effects on rooting of stem cuttings of dragon fruit have been reported.

Fig. 8.4. Preparation of cuttings for planting. (A) A segment cut (1), a flat cut (2), a semi-tapered cut (3) and a tapered cut (4). (B) After 10 days, the cut wounds are dry and ready for planting. (C) After preparation, the cuttings are kept in a dry, shaded place before planting.

Rahad *et al.* (2016) suggested treatment of dragon fruit cuttings with a solution of IAA at 200–300 ppm and planting in a medium containing soil (100%), 50% soil:50% sawdust or 50% soil:50% sand for better rooting. Ravikant *et al.* (2022) noted that cuttings of 10–15 cm treated with NAA at 200 ppm gave good results with respect to rooting and shooting parameters, followed by NAA at 175 ppm, while Ahmad *et al.* (2016) recorded better rooting with IBA at 100 ppm. Raut *et al.* (2022) treated cuttings by dipping them for 10 min into different concentrations of IBA solution (and then allowing them to dry for 15 min before planting) and obtained the best results with IBA at 2000 ppm. Other methods of propagation of dragon fruit are discussed in detail in Chapter 5 (this volume).

8.6 Trellises

Dragon fruit plants can be grown in the ground or in a pot without any support, but this is not a common practice in commercial production. As a climbing cactus, dragon fruit plants may become large and spread out, and therefore strong and durable support structures such as trellises should be established for commercial orchards to allow the stems to climb up and then hang out, which stimulates budding and fruiting while allowing for ease of care and harvesting for a period of 10 years or more.

There are several types of trellis system used for supporting dragon fruit plants, including single posts and wires stretched across a row supported by evenly spaced posts (also called a fence trellis). Fences made of stones or bricks are also used as a trellis. In Taiwan, single wooden or cement posts are usually used in commercial orchards. Other trellis systems include horizontal trellises, galvanized fence trellises and T-bar trellises (Zee *et al.*, 2004). Single posts are typical for dragon fruit trellis systems in commercial orchards in South Asian countries. The vertical posts (made of wood or concrete) are 15–20 × 15–20 cm in diameter and up to a height of 1.5–2.2 m. The plants can branch and hang down, forming an 'umbrella canopy' around the post. These poles can fit up to four plants grown from cuttings planted on the four sides of the post (Mai *et al.*, 2020). In some countries, standard grapevine trellis has also been used for growing dragon fruit with central posts with or without cross-arms. These are known as fence trellises. In Israel and Taiwan, plants are also grown on galvanized steel fence trellises but without any cross-arms (Merten, 2003; Zee *et al.*, 2004). The T-bar trellis is a newly developed support system. The main vines are grown to the top of the wire with all side branches removed and are then allowed to branch and hang down from the top on two sides. The two systems of trellis for dragon fruit commonly used in commercial production in Vietnam are the single trellis or single posts, and the fence trellis (T-bar trellis).

8.6.1 Single trellis

There are many types of materials that can be used to make a single-post trellis support with or without a top support. The post needs to be strong and durable, and is usually made from steel, concrete or wood, with or without some top support; car tyres or bike wheel rims fixed to a protruding rebar have even been used, with wire mesh fixed to the bar. The most commonly used single trellis has the rebar protruding from the top of the pole (Suu *et al.*, 2017). This is the typical trellis system seen in dragon fruit orchards across regions in Vietnam and also in many other countries, and allows the formation of the 'umbrella canopy' once the plants become established. The post can be living or non-living.

Living posts

Living posts are when a live plant is used as a supporting structure for the dragon fruit vines (Fig. 8.5C). Living posts were common in Vietnam before 2000 but are gradually being replaced by concrete posts. The plants commonly used for living poles are *Erythrina variegata*, *Samanea saman*, *Tamarindus indica*, *Cassia siamea* and *Wrightia annamensis*. In Mexico, dragon fruit are planted on *Pithecellobium dulce* (monkeypod trees), *Bursera simaruba* and other plants (Sosa *et al.*, 2020). In the Philippines, *Gliricidia sepium*, locally known as Kakawate, is used as a living post for dragon fruit (Eusebio and Alaban, 2018). White mangrove (*Avicennia marina*) is also used as a support for dragon fruit in the mangrove forests of Ca Mau, Vietnam (Luu *et al.*, 2021); in adverse conditions, its aerial root

Fig. 8.5. The trellis system used for supporting dragon fruit plants. (A) A single cement post. (B) A single wooden post. (C) A living post. (D) A fence trellis.

system acts like the root system of an epiphytic orchid plant, with water and nutrients for the plant coming mainly from rain water, dew and fog, mineral dust, organic detritus, animal droppings and other substances collected on the surfaces of the living post. Cuttings should be grown in plastic bags or in planting rings with organic potting mix.

The yield in dragon fruit orchards that use living posts is often lower than in orchards using wooden/concrete posts. This may be due to lower planting densities and competition for light and nutrients between the post plants and the dragon fruit. Living posts can reduce hot temperatures in orchards during the summer period and are recommended as a method for management of sunburn.

Non-living posts

The most commonly used materials for non-living posts are wood, concrete and brick (Fig. 8.5A, B). The economic lifespan of a dragon fruit orchard is 15 years or more. If wood is used to make the trellis, the type of wood should be strong and durable. Wood holds moisture so the aerial roots of the plant can easily latch on. However, in tropical conditions, wood may decay and collapse after some years and need to be replaced.

The height of a single concrete post varies: it is 2.0–2.5 m in Indonesia (Mansyah *et al.*, 2019), 1.5–2.0 m in India (Karunakaran *et al.*, 2019), 1.5–1.7 m in Hawaii (Taitano *et al.*, 2021) and 1.2–1.3 m in Vietnam (Thoai *et al.*, 2021).

A higher canopy can utilize maximum light and avoid the build-up of a microclimate that may lead to disease and infection. However, a higher canopy is less convenient for carrying out care and protection, and also has increased risk of falling due to strong winds. Concrete poles are durable and are widely used for posts. The top support is created using a rebar from the top of the post. A typical concrete pole is 160–210 cm long and 12–15 cm wide, with four horizontal, symmetrical arms of the rebar at the top to support the vines. The poles are buried 40–50 cm deep so that the height of the pole from the ground is 120–150 cm (Mai *et al.*, 2020). Single wooden or cement posts about 15 × 15 × 200 cm set at spacings of 2.75 × 4.60 m are used for commercial cultivation in Taiwan (Zee *et al.*, 2004).

8.6.2 Fence trellis systems

Fence trellis

Using a fence trellis for planting dragon fruit is perhaps the easiest and most effective way of growing them (Fig. 8.5D). Many newly established farms use this method in Vietnam. It is also commonly used in Taiwan and other growing regions.

A fence trellis is a continuous connection of single trellis posts in a row, connected using various materials such as steel bars or wire, steel pipes, wood or bamboo bars; the methods of connection vary depending on the cultivation conditions and available materials.

The closest spacing used in California with a three-wire trellis is 2.5 m between the rows. If the spacing is closer, it is difficult to maintain the plants and harvest the fruit. Spacing between the plants within the row varies. If plants are not in short supply, 1.5 m between plants will give quicker production than a larger spacing. High-density plantings will produce returns earlier, but the plants will begin to crowd each other sooner, so they are usually planted 3 m or more apart.

T-bar trellis

The T-bar fence trellis system was developed and introduced into dragon fruit production in Vietnam (Son *et al.*, 2019). Using the T-bar system combined with a suitable training and pruning practice helps improve ventilation and sunlight transmission in the canopy, reducing the occurrence of diseases, and cultivating healthy fruiting branches to increase fruit yield and quality. Growing dragon fruit using a fence trellis also makes it easier in terms of orchard management and maintenance compared with the conventional use of single poles. The marketable yield of dragon fruit was found to be 1.5 to three times higher compared with the conventional system using single reinforced cement concrete posts over a 3-year experiment (Son *et al.*, 2019). Although the set-up cost of the T-bar system is higher compared with the single concrete post system, the higher fruit yield makes it more efficient.

The poles in the T-bar system can be made of wood, steel or concrete. Concrete poles are more popular because they are more durable than wood and can be made on site. They are widely used in coastal areas due to their resistance to rust and corrosion caused by saltwater and salty air.

A T-bar generally looks like a power pole, having a vertical pole with two steel crossbars and wires connecting the pole and crossbars

(Fig. 8.6A). The poles at the end of the rows are called termination poles. They are usually larger and stronger than the other poles and have guy-wires to transfer some of the weight of the structures to the ground (Fig. 8.6B). The guy-wires are attached to an anchor and are concreted into the ground to hold the guy-wires steady. The anchor must be enough strong to resist the maximum tensile load of the guy-wires – both the dead load of the tension of the wire (with the canopy weight) and the maximum possible live load due to strong winds.

Concrete poles can be purchased from a professional manufacturer or can be cast on site to save on costs. The posts have a square cross-sectional area of around 12 × 12 cm and are about 180 ± 10 cm in height (with the foot part buried in the ground up to around 50 cm). Four hot-rolled deformed bars of 8 mm in diameter are used as steel reinforcement in ordinary concrete posts, while hot-rolled deformed bars of 8–10 mm are used in termination posts to provide additional strength. These bars protrude from the top of the post by 5 cm to serve as a place to fix the steel wire connecting the posts.

The posts have two round holes that are 10–15 and 70–75 cm from the top of the post for attachment of the two crossbars. The upper crossbar is 60 cm long and the lower one is 80 cm long. In Vietnam, galvanized steel tubes of 27 mm in external diameter and 2.1–3.25 mm wall thickness are recommended for the crossbars.

After burying the posts in a row, steel wires are stretched to connect the posts through the post top and crossbars. Two steel wires of 4 mm diameter are used to connect through the post tops (bracing and other posts), while two steel wires of 3 mm diameter are used to connect the upper crossbar and the lower bars at a position 1 cm from the top of the crossbar. A turnbuckle is used to adjust the tension and reduce slack in the wire.

8.7 Planting

Planting time varies from country to country. It is based mainly on climatic conditions, especially rainfall pattern, and the availability of planting

Fig 8.6. The T-bar trellis systems. (A) Using cement posts, galvanized steel pipes and galvanized steel wires. (B) The post at the end of row showing the guy-wires attached to an anchor.

materials. In irrigated areas of the tropics, dragon fruit can be planted throughout the year except in severely dry months and during heavy rains when the soil remains very wet. In subtropical regions, the plants can be planted in warmer months. In the monsoon tropical conditions of southern Vietnam (rainy season May–October), there are two main recommended planting seasons: at the end of the rainy season (October and November) and at the beginning of the rainy season (May–June). The advantage of planting in October and November is to avoid the risk of waterlogging from heavy rains in lowland areas, as well as the local availability of cuttings. However, after planting, these young plants should be mulched with rice straw (or other local suitable materials) and watered regularly to help keep the roots cool and also to help them retain moisture during the hot dry season (Mai *et al.*, 2021).

The recommended number of planted cuttings for each post can be from one to four (Luders and McMahon, 2006; Gunasena *et al.*, 2007; Pascua *et al.*, 2015; Nangare *et al.*, 2020; Mai *et al.*, 2021; Thoai *et al.*, 2021). In general, four cuttings per post are recommended, which will probably be beneficial to the overall health/growth of the plants (Nangare *et al.*, 2020; Mai *et al.*, 2021). In organized commercial orchards in Vietnam, four cuttings per post is the most common. The cuttings are planted on the four sides of the square post. As the vines grow up the post, they are tied in to create an erect and strong trunk to help support their extensive weight until they overhang the trellis. The aerial roots of vines can easily attach to the post, which helps to secure the vines to the post and encourage the vines to grow (Fig. 8.7).

8.8 Mulching

Mulching refers to the covering of bare soil with a layer of material to enrich the soil and benefit the plants. Mulching is one of the most important practices in horticultural crops. A wide range of materials can be used as mulches, which are broadly classified as organic and inorganic. Chopra and Koul (2020) compared various organic and inorganic mulches. Different mulch types are applied depending on the plant and soil type, the climate conditions, crop management practices and the purpose of mulching (Kader *et al.*, 2017; Mai *et al.*, 2021). Mulches can help to retain soil moisture content, control weed growth and enhance the yield of the crop (Chopra and Koul, 2020; Luo *et al.*, 2021). However, they can also affect the pH of the soil, and may increase its acidity (Luo *et al.*, 2021). Selection of a suitable mulch is important to achieve the goal of improving the soil and increasing the crop yield.

8.8.1 Organic mulch

Organic mulch is any organic material that covers the soil's surface to enrich the soil or improve plant growth. Mulching of dragon fruit plants

Fig. 8.7. Planting with two (A), four (B) and eight (C) cuttings per post.

helps to reduce soil water loss by reducing evaporation, and contributes to overcoming the lack of irrigation water in the dry season in some growing areas. It also helps reduce the impact of rain and soil erosion and weed problems around the plants. After application, organic mulches will break down gradually over time and add organic matter and plant nutrients to the soil. However, the mulch must be reapplied periodically. By use of suitable combinations of organic mulches and mineral fertilizers, the fruit yield can be increased (Mai *et al.*, 2021). Mulching can be carried out in combination with organic manures to enhance soil health and increase crop yield (Lasmini *et al.*, 2019). The application of rice straw and cow manure mulch is a popular choice for dragon fruit cultivation in tropical Vietnam (Fig. 8.8).

Organic materials used for mulching are usually natural materials from leaves, trees, grasses and other plant material. Rice straw, burnt rice husk, coconut fibre/coco peat, hay, bark, wood chips, dried sugarcane leaves and tops, and sugarcane bagasse are materials that are widely used for mulching in tropical regions (Fig. 8.9A, B). Sugarcane mulch and rice straw mulch are increasingly used and are now processed and baled into compressed, wrapped packages for easy transportation.

Fig. 8.8. Rice straw used for mulching.

The layer of mulch should be 5–12 cm deep and should be placed at least 10–20 cm away from the base of the plants. However, these requirements may vary depending on the type of mulch material used, the age of the plant, climate/weather conditions and soil conditions (Singh *et al.*, 1991; Luo *et al.*, 2019; Mai *et al.*, 2020). There should never be too thick a mulch around the base of the plant over the root zone, and the mulch should not touch the trunk as this can cause excess moisture in the root zone and reduced soil oxygen to the roots, as well as creating the perfect conditions for disease development causing root rot. The roots may also end up growing up through the mulch in search of air, causing stem girdling of the roots. A deep mulch can hide rot-damaged spots on the trunk and major roots, and provides the perfect place for slugs, snails, rodents, termites and other pests to reside (Mai *et al.*, 2020).

In general, the mulch is likely to be thicker and placed closer to the base of the plant in regions of sandy soils and a dry climate, while in tropical regions with heavy clay soils and a humid climate, the mulch should be thinner and further away from the base of plant. In the tropical conditions of Vietnam, a rice straw mulch of 3–5 cm thick is used in areas of heavy clay soils and of 6–10 cm thick in areas of sandy soils. In conditions of heavy rainfall on heavy clay lowlands, the mulch can be removed during the rainy season so that the soil and mulch are not too wet. Periodically throughout the dry season, compost may need to be added to maintain the benefits of the mulch.

8.8.2 Mulching with compost

Compost can be used for mulching. It is also a way of applying organic manure to the plant (Fig. 8.9C). Compost can be an ideal material for mulching when supplies are plentiful. The soil environment below the cover becomes more favourable for earthworms, thereby helping to increase aeration of the soil. Organic matter and plant nutrients are gradually added to the soil as the mulch decomposes. The compost mulch can be 5–15 cm thick (McLaurin and Gary, 1991;

Fig. 8.9. Mulching in the tree basin with coconut fibre/coco coir (A), rice straw (B) or organic manure (C).

Mai *et al*., 2020). When it is used as mulch, complete decomposition of the compost is not necessary (McLaurin and Gary, 1991).

8.8.3 Inorganic mulch

Inorganic materials can also be used for mulching in dragon fruit orchards. They consist of synthetic materials (e.g. plastic sheets, landscape fabrics (geotextiles), plastic shading nets, rubber mulch) and also natural materials (e.g. stones, rocks, gravel). While organic mulch is the preferred choice due to its soil-improving capacity, inorganic options still have a place because they do not need to be reapplied or topped off as frequently as organic mulches.

Plastic mulch can help minimize temperature fluctuations in the soil, reduce soil moisture loss during crop growth, increase radiation and water-use efficiency, and prevent weeds from growing. Plastic is the most commonly used material as a mulch in agriculture because of its high flexibility, durability and chemical resistance. The plastic mulch acts as an efficient barrier as it is completely impermeable to water, prevents the direct evaporation of moisture and is useful for soil solarization. The thickness of the mulch may have an effect on soil solarization and hence a thin film of 15 μm of plastic mulch is used nowadays. Different-coloured plastic mulches are available such as white film, transparent film, black film and silver film, and the colour of the plastic mulch may affect soil salinity, soil temperature and air temperature. Black plastic mulch has become popular and has given good results in arid and semi-arid regions (Bhardwaj *et al*., 2011).

Plastic film mulches are the most widely used inorganic mulch material in agricultural production in general and for dragon fruit production in particular. Different materials have been used for producing mulch films, but polyethylene plastic is widely used due to its excellent mechanical strength and lower cost. It can be used to cover the soil between rows as a mulch and also enables workers to move easily between the rows. Plastic silver-black mulch may be more suitable for dragon fruit in cooler regions as it reflects light to the plants and fruit, which helps improve colour development in the fruit. The disadvantages are that most inorganic mulches are made using fossil fuels in a process that is believed to increase greenhouse-gas emissions, and they can remain in the environment for many years and potentially enter the food chain.

Rubber can also be used as an inorganic mulch. Rubber mulch is made from recycled and pulverized old tyres but is not the preferred choice as toxins in the rubber may cause harm to the soil and may not be suitable for organic production standards.

8.9 Weed Management

Weed management is an important practice in dragon fruit production. Weeds compete with the dragon fruit plants for water, nutrients, sunlight and space and hence affect growth of the plant. Weed management in dragon fruit orchards involves eradication, prevention and other appropriate methods.

8.9.1 Herbicide application

The use of herbicides is one of the most popular and widely used control methods for weed control in dragon fruit orchards. It requires less cost and human effort than other methds, and is the most effective and time-efficient method of managing weeds.

The use of herbicides should comply with the regulations of the importing countries and also of the growing countries. Depending on the types of weed in the orchard, an appropriately registered herbicide can be chosen. Choosing the appropriate herbicide can be a difficult task when faced with the sheer volume of products available. Post-emergent herbicides are widely used to kill weeds in dragon fruit orchards. They are applied mainly by spot spraying (targeting individual areas of weeds) to avoid spraying on the plants (Mai *et al.*, 2020). Herbicides are usually sprayed every 45–60 days in the rainy season and every 80–120 days in the dry season for dragon fruit orchards in southern Vietnam (Mai *et al.*, 2020).

Chemical herbicides should be combined with other methods such as mulching or mowing and cutting to reduce the amount of chemicals used. Growers should use a proper management strategy including a combination of effective and more eco-friendly measures for sustainable management of weeds in dragon fruit orchards (Fig. 8.10).

Glufosinate is a post-emergence, non-selective herbicide for broadcast use on some fruit crops including dragon fruit. In Vietnam, glufosinate ammonium is commonly used for controlling weeds in dragon fruit orchards (Mai *et al.*, 2020). Glufosinate ammonium (200 g l^{-1}) is usually applied at rate of 3 l ha^{-1} (in 400–600 l of water). It should be applied when the weeds are young and succulent. It is sprayed evenly on the weeds, avoiding contact with the green parts of the plant. Indaziflam is a pre-emergence herbicide for control of annual grasses and broadleaf weeds, and is used for controlling weeds in dragon fruit orchards. Indaziflam (500 g l^{-1}) is usually applied at rates of 0.1 l ha^{-1} (in 400 l of water) (Mai *et al.*, 2020). It is sprayed evenly over the ground area, avoiding contact with the green parts of the plant (Fig. 8.10). Instructions for application on the labels of the products should be followed.

8.9.2 Hand weeding

Manual weeding is the most common practice in developing countries, including Vietnam (Fig. 8.11A) (Mai *et al.*, 2020). All weeds under the canopy including around the roots and rhizome are removed. The use of herbicides or hand

Fig. 8.10. Weed management in dragon fruit orchards. (A) Using chemical herbicides. (B) Using rice straw mulching. (C) Using rice straw mulching combining with a herbicide spray.

Fig. 8.11. Weed management in orchard. (A) Hand weeding. (B) The mowing and cutting method.

weeding is difficult and potentially risky, especially in the early stages and in the first year after planting. The weeds should be removed before they get too big or start setting seed. Hoes and knives are usually used, combined with pulling weeds by hand. Hand weeding is done every 30–45 days in the wet conditions of the rainy season and every 45–60 days in the dry season (Mai *et al.*, 2020).

8.9.3 Mowing and cutting

Mowing and cutting are more common for weed management in dragon fruit orchards in Vietnam (Fig. 8.11B) (Mai *et al.*, 2020). The aim is to slow the growth, reduce biomass and prevent seed production. Keeping the weeds actively growing and regularly cutting them short is a recommended practice in dragon fruit orchards (Mai *et al.*, 2020). Keeping the weeds helps increase organic carbon in the soil. The best time for mowing and cutting is when the weeds are growing and their flowers are about to bloom. Doing it at this stage helps prevent the spread of seeds and also eventually weakens their root system. More frequent mowing is required to achieve a higher degree of weed control, especially in dragon fruit orchards where sprinklers or overhead irrigation are used.

Depending on the type of weed, soil and weather conditions, the mowing cycle can change. During the monsoon season, mowing should be more frequent, and less frequent during the dry season. Growth of weeds is typically less vigorous during the dry season so they will only need to be cut every 60–90 days (Mai *et al.*, 2020).

8.9.4 Mulching for weed management

Mulching the soil with other materials to suppress weeds, as discussed above, is one of the preferred management methods for dragon fruit orchards.

A ground-cover crop can be used as a living mulch (Fig. 8.12). This is known as cover cropping and aims to manage soil fertility, soil quality, water, weeds, pests, biodiversity and wildlife in an agroecosystem (Lu *et al.*, 2000). Cover cropping is a practice widely adopted in tropical fruit crops for many purposes. Cover crops are cultivated in the orchard to prevent the growth of weeds, resulting in reduced use of chemical herbicides; they mitigate runoff of irrigation and rainwater, improve the stored soil moisture and

Fig. 8.12. Examples of living mulch (cover crops) in dragon fruit orchards. (A) Pinto groundnuts. (B) Natural grasses. Photographs courtesy of R. Martins.

organic matter, and reduce mud and dust in the orchards. Other value benefits of cover crops in dragon fruit orchards are mitigation of orchard temperatures on hot days, and reduced heat stress and sunburn of the plants during summer. Luo *et al.* (2019) reported that mulching with self-seeding cover crops in orchards helps harbour useful fungal species, such as *Trichoderma* and *Chaetomium* spp. However, there are only a few studies documenting the use and benefits of cover crops in dragon fruit orchards.

Different species of ground-cover crops are used in dragon fruit orchards. Legumes that can fix nitrogen from the air are often recommended for use in tropical orchards, including dragon fruit orchards (Thoai *et al.*, 2021). Pinto groundnut (*Arachis pintoi*) is a perennial tropical legume commonly used in dragon fruit cultivation (Fig. 8.12A). It is grown in the wet tropics and subtropics, and is widely distributed in tropical and subtropical regions. This nitrogen-fixing legume species spreads by rhizomes and creeping stolons that root at nodes to form a dense and leafy mat, forming low (15–30 cm high) ground cover, and is well adapted to dragon fruit orchards. Pinto groundnut was recommended for use on sloping soils due to its ability to limit soil erosion, and in sandy and dry regions for reducing temperature and soil moisture fluctuations (Tuan *et al.*, 2007). It can be used as a cover crop for dragon fruit to offset some of the required nitrogen input. Pinto groundnut can be planted using both stolons and seeds before or after the dragon fruit is planted. Planting using stolons is more common in Vietnam. The stolons or runner sections from the mother plant are removed and replanted 2–3 cm deep in soil at a spacing of 30–50 cm × 30–50 cm. Growth of the ground cover suppresses natural weeds. However, in dragon fruit orchards infected with root mealybugs, these pests are also found abundantly on the roots and basal stems of pinto groundnuts, so this plant can be a host of certain species of root mealybugs.

Wedelia (*Sphagneticola trilobata*/*Wedelia trilobata*) is often planted as an ornamental ground cover and spreads quickly to form a dense cover, preventing the regeneration of weeds in orchards. This species provides rapid establishment and coverage on sloping soils. Once established, wedelia is difficult to eradicate, if herbicides are not used, and this species is listed as an invasive alien species in many countries.

Low-growing natural grasses are used as ground cover in many dragon fruit orchards. This is one of the most common living ground covers used in Vietnam (Fig. 8.12B). Growing grass in orchards is an easy practice with lower costs compared with other ground-cover plants. Grass in orchards helps prevent soil erosion and hardening, increases soil fertility, improves soil structure, increases soil water-holding capacity, promotes soil biodiversity, and controls weeds and pests (Hartwig and Ammon, 2002; Ingels *et al.*, 2005; Tang *et al.*, 2022).

Ground-cover plants need water and nutrients for growth. Therefore, additional water and fertilizers are necessary. The amount of nutrients required varies according to the species of cover crop. Some ground-cover plants may compete with dragon fruit for nutrients and water and need to be managed by cutting or spraying with herbicides (at 40–50% lower than the recommended doses for reducing growth; Mai Van Tri, personal communication) to reduce competition with the crop.

8.10 Training and Pruning

Dragon fruit is a fast-growing plant, forming a dense mass of stems and branches. Training and pruning are therefore important to: (i) control the size, shape and direction of plant growth so that the plants can get the maximum amount of light falling on the branches, making the plants easier to care for and manage; (ii) decrease the risk of pests by removing diseased and pest-damaged plant parts; (iii) increase ventilation and sunlight distribution, thus regulating the fruit crop and improving fruit size and quality due to better distribution of light, nutrients and energy; and (iv) decrease the damage caused by sunburn and wind damage.

The plants should be carefully pruned and trained from the start of their orchard life. Pruning should start soon after planting. Pruning of dragon fruit can be divided into formative pruning and production/maintenance pruning, as well as renewal pruning for old plants or orchards that have been abandoned for months. Formative pruning is usually carried out in the first 2–3 years on a young plant to create an appropriate framework of main branches and a good canopy shape. Maintenance pruning is carried out on established trees and usually starts from year 4 after planting. Maintenance/production pruning is mainly to maintain or improve plant health, structure, shape and size once orchards have become established.

8.10.1 Formative pruning

The first year

In the first year, formative pruning or training should be practised shortly after planting. At 2–4 weeks after planting, shoots will emerge from the planted cuttings. All shoots should be eliminated except for one or two strong shoots on each cutting (at least 10 cm apart) until they reach the end of the support. Soft ropes (plastic/polypropylene ropes/cloth strips) can be used to tie the stem to the post. All lateral shoots emerging from the selected shoots below the support top must be removed. Once the top is reached and the shoots are about 30–40 cm higher than the post top, they are bent to grow vertically on the surface of the post top in opposite directions (Figs 8.13 and 8.14), allowing the development of lateral shoots from these selected shoots. These lateral shoots will become the main branches of the canopy and will produce more branches.

For orchards using a fence trellis, a temporary post (e.g. a wooden, bamboo or metal post) is required to support the stem climbing up to the trellis (Fig. 8.15A). Aerial roots formed later will provide further support for the stems by attaching to the post. As with the single posts, once the shoots reached the top of the post, they should be trained to have a drooping posture to obtain an umbrella canopy. When the stems have grown above the post top, they are bent and secured to the rebar protruding from the top of the post, parallel to each other but in opposite directions. This should be done in the afternoon, when the branches are softer because of the heat of the day. Stopping watering in the morning also helps the trunk to become softer and easier to bend without breaking. Shoots will emerge from the stems on the top, and the branches that develop from these shoots will form the structure of the umbrella-shaped canopy (Fig. 8.15B). Initially, the branches need to be left because they

Fig. 8.13. Pruning and training of dragon fruit vines along the top of the post at 3–4 months (A), at 6–8 months (B) and 12 months (C) after planting.

Fig. 8.14. Canopy pruning and training. The canopy shape and structure on a single post is shown at 3 months (A), 6 months (B), 9 months (C), 12 months (D) and 3 years (E) after planting.

Fig. 8.15. Dragon fruit grown on a T-bar trellis system. (A) After planting. (B) After branches have been trained vertically on the wires.

will provide food to help the plant develop. They can then gradually be removed, several times a year, as the tree grows more branches and the canopy expands. These branches and sub-branches should be arranged evenly, distributed in different directions around the planting post. This process is continued to create an umbrella-shaped plant in the second and following years.

The second and third year

Formative pruning is continued to create a canopy with the desired structure and shape. First, the whole canopy is examined to estimate and plan how to prune it. Any sprouts from the stems below the top of the post are removed. Then dead and diseased branches, broken or drooping branches, and overhanging branches are removed, followed by old or weak branches, crowded branches and crossing branches. Healthy branches with good flowering potential are left. There can be 40–80 branches on the post in the second year (Fig. 8.16). When removing any diseased parts of the plant, the pruning tools must be cleaned after each use to prevent spreading diseases to other plants. By the end of the third year, the number of branches per post will have increased and the canopy will be larger and thicker (Fig. 8.16B). An unbalanced canopy with branches dominating on one side has a risk of falling. Therefore, pruning, training and arranging the branches should be done monthly so that they are evenly distributed in all directions, creating a balanced canopy. Depending on the planting density, soil, climate, variety, and pest and disease severity, about 80–100 branches should be maintained on the canopy. The tops of the long branches (1.2–1.5 m) and those that are close to the ground will be cut or removed to enable better air ventilation and spray coverage, and to allow more sunlight to penetrate the canopy.

8.10.2 Production pruning

Production pruning is applied in the fourth and following years. The plant can now grow strongly and produces an extensive amount of vegetation if not pruned. The canopy needs to be trimmed to maintain the umbrella shape of the correct size and height. During this period, the aim of pruning is to maintain the desired canopy structure, shape and size, and to get the right balance between vegetative growth, flower production and fruiting. According to Hieu *et al.* (2018), pruning unproductive or poorly productive and pest-infected branches by 60% did not increase dragon fruit yield but increased fruit weight and reduced canker disease infection caused by *Neoscytalidium dimidiatum* during off-season production.

For this purpose, all dead or dying branches, damaged, sunburnt or rotting branches, and severely diseased branches should be removed. Suckers and branches on the plant trunk below the top of the post/trellis should also be removed. Branches that cause canopy imbalance or structural instability, branches that are covered within the canopy and not exposed to the

Fig. 8.16. Pruning and training on a fence trellis system. (A) Year 2 after planting. (B) Years 3 and 4 after planting.

sunlight, branches that are small, thin and weak, and branches that are touching or close to the ground or that obstruct the walking or working space in the orchard should also be removed (Suu *et al.*, 2017; Mai *et al.*, 2020; Thoai *et al.*, 2021). The tips of long branches (≥1.2 m) should be cut short. When dragon fruit plants grow too large, canopy reductions should be applied to recreate the shape and structure by shortening the overgrown branch lengths. The distance between two canopies in the row should be 50–80 cm. Mature plants should be skirted to a height of at least 30–50 cm from the ground (Mai *et al.*, 2021).

Arredondo *et al.* (2022) compared three different methods for pruning pitaya (*Hylocereus undatus*) including spur, cane and combined pruning under greenhouse conditions using the fence trellis system. Spur pruning consists of eliminating approximately two-thirds of each cladode, leaving only the four or five basal areoles or compound buds (in each cladode edge). Cane pruning consists of just tipping all cladodes, cutting back only the last apical buds. Combined pruning is a mixture of both types of pruning, with half of the cladodes spur pruned and the other half cane pruned. The control plants underwent only sanitary pruning, leaving the productive cladodes intact and removing only decayed cladodes. The result showed that cane pruning was the best pruning option under the conditions of this trial (i.e. for plants in full production), as it allowed a high-intensity flowering, leading to higher fruit production.

In Vietnam, dragon fruit plants may need to be pruned four to six times a year when they are mature and productive. The main pruning is performed after harvest or before starting a new season crop (there are three to four crops a year). Pruning after harvest encourages the growth of new young shoots and new branches that will bear flowers the following year (Le Bellec *et al.*, 2006). The number of branches maintained on the canopy per post depends on factors such as plant spacings, the trellis system, variety, weather conditions, the type of soil and pests. The number of branches that need to be maintained after pruning on each post are shown in Table 8.1.

For the main pruning, about 30–50% of the branches are removed. If the number of branches on the canopy is less than required (Table 8.1), the best branches are chosen and

Table 8.1. Number of branches left on post after pruning in Vietnam (Mai *et al.*, 2020).

Age of plant (years)	Number of branches per post
1	20–30
2	60–80
3	90–110
4–8	120–180
>8	>150

their ends cut to reduce their length to about 25–35 cm from the planting post. One or two strong shoots, at least 10 cm apart, are selected on the branch and allowed to develop into sub-branches (Suu *et al.*, 2017; Mai *et al.*, 2020).

After the main pruning, the orchard needs to be monitored weekly to remove diseased, weak and inappropriate shoots or branches to create the canopy structure and shape. Each month, the branches should be pruned and rearranged to avoid an imbalance or structural instability of the canopy. The tips of long branches and any branches growing near the ground should be cut.

Pruning of dragon fruit on a fence trellis system is basically the same as for the single-post system. Depending on the type of fence trellis, it may be more difficult to tie branches up without a supporting stake. It is therefore recommended that one stake is planted for each cutting because the branches will grow faster when they are supported. Stakes can be made from wood, bamboo, cement, or plastic or steel pipes. This also helps to reduce wind damage, especially in the early weeks (Figs 8.15 and 8.16).

8.10.3 Thinning of flower buds and young fruit

Thinning of flower buds and young fruit is widely carried out in commercial dragon fruit production in Vietnam (Suu *et al.*, 2017; Mai *et al.*, 2021; Thoai *et al.*, 2021). Dragon fruit plants tend to produce a large number of flowers in summer, especially in subtropical conditions. Extra flower buds should be thinned to improve the balance between the canopy and the fruit, which allows the remaining buds to develop in a more favourable way. This may also allow the remaining flower buds on the branches and canopy to be distributed in a more efficient way. However, removal of areoles on branches to reduce the number of flowers on the branch is not common because it is time consuming and labour intensive (Mai *et al.*, 2020).

At 5–7 days after the appearance of the flower buds, they should be checked, and any deformed, abnormal, pest-infected, small or poorly positioned buds removed. Flower buds on the top of the canopy (30–50 cm away from the post) and buds near the branch tips (two-thirds of the way along the branch length) should be removed. Only healthy flower buds at should be left in favourable positions. Depending on the total number of flower buds on branches and canopy, two to four flower buds should be left on each stem, with at least 15 cm between buds (Suu *et al.*, 2017; Mai *et al.*, 2020).

Thinning of young fruit is done at 5–7 days after flowering and follows a similar method to flower bud thinning. Any damaged, diseased, ugly or dying fruit are removed, leaving healthy fruit. In Vietnam, for light-induced flowering crops, the fruit should be thinned to no more than two fruit per branch, with 15–30 branches bearing fruit per post, to produce an expected yield of 9–18 t ha^{-1} for average spacings of 2.8–3.0 × 3.0–3.5 m (Mai *et al.*, 2021; Thoai *et al.*, 2021).

8.10.4 Removing petals after flowering

Removal of flowers after anthesis is an important practice to reduce fruit rot under high disease pressure conditions in warm and humid regions (Fig. 8.17). Parts of the flower are highly susceptible to some diseases of dragon fruit (see Chapter 11, this volume). Their removal helps the young fruit stay free of fungal and bacterial diseases. Under wet conditions, some pathogens can infect flowers and will spread directly from the petals to the young fruit through the lesions on the petals and then cause fruit rot. Removal also helps to reduce sticky sap from the petals getting on to the peel of the fruit, thus keeping the fruit clean and more attractive at harvest. Removal of flowers should be done 3–4 days after anthesis (Fig. 8.18). A stainless Chinese soup spoon can be used to cut and remove the petals. After removal, the diseased flowers should be destroyed by burning or composting appropriately.

Fig 8.17. (A) After flowering (4–6 days), the petals may become wet and get infected. (B) The water-soaked area quickly expands towards young fruit. Removal of the petals after flowering helps prevent bacterial and fungal rots.

Fig. 8.18. (A) Petals at 3–4 days after anthesis. (B) Removal of the petals after flowering. (C) Young fruit after removal of the petals; the style is not removed.

8.10.5 Pruning tools

The tools used for pruning are mainly shears and scissors, long-handle sickles and knives (Fig. 8.19) (Mai *et al.*, 2021). In areas where diseases of virus, bacteria and vascular fungi are established, the tools must be disinfected before pruning a plant; soil, sap, debris and other material on the tool surface should be removed before disinfecting. There are a number of products used for disinfecting pruning tools. Several are available in shops, including chlorine bleach and trisodium phosphate. Chlorine bleach is inexpensive and widely

Fig. 8.19. Tools used for pruning dragon fruit in Vietnam. (A) A short-handled sickle. (B) A long-handled sickle. (C) Pruners.

used but can be corrosive to metal. Pruning tools need to be soaked in 10% bleach solution (one part bleach:nine parts water) for about 30 min. They are then rinsed with clean water and allowed to dry before use. The bleach solution should be used within 2 h of preparation, after which it becomes less effective.

Pruning tools can also be soaked or sprayed for at least 1 min in a 1:5 dilution of Clorox or Lysol; spray cleaning is as effective as soaking. Clorox is less expensive than Lysol but Lysol causes less damage to metals (Teviotdale *et al.*, 1991). The labels should be read carefully and precautions followed when using these products. Note that none of these disinfectants is 100% effective for all pathogens. To avoid the risk of spreading disease through tools, cuts should not be made into diseased areas of active lesions or oozing cankers (Kleinhempel *et al.*, 1987).

Only sharp tools should be used for pruning in order to make clean cuts. Infection may occur on cuts that are crushed or ragged from dull tools. Sharp tools cut with less effort and make pruning much easier, and clean cuts heal quicker. Sharp pruners are more efficient and safer for pruning (Fig. 8.19) (Mai *et al.*, 2021).

Gloves are important to protect the hands during plant care, including pruning. Cloth gloves help protecting the hands from thorns, blisters, splinters and sharp blades, and enable a better grip when necessary.

8.11 Fertilization

Most orchard soils growing dragon fruit do not have enough of the right nutrients for the plants to reach their full potential. Therefore, applying fertilizers is important to meet the nutritional requirements of the plants and to promote the growth, development, yield and fruit quality of dragon fruit. It is necessary to perform soil analysis to determine the nutrients that need to be added by fertilization to improve the soil and meet the requirements of the plant. Common fertilizers used in commercial production of dragon fruit include organic manures, microbial organic fertilizers and chemical fertilizers. The recommended quantity and time of application of fertilizers vary widely and depend on a variety of factors including soil type, planting density, trellis system, variety, climate and weather conditions and season. Generally, the higher the yield in the previous season, the more nutrients are required to help the plants recover faster for the next season.

8.11.1 Organic manure

Applying organic manure helps improve the soil structure and nutrients, as well as the ability of the soil to take up and retain water. Organic manure also serves as a reservoir of nutrients, releasing significant amounts of nitrogen, phosphate and sulfur as it decomposes. Organic manure can introduce microorganisms, including beneficial ones, into the soil. Several organisms found in compost are important for mineralization of organic nitrogen, converting organic nitrogen into inorganic forms. Additionally, organic manure used as a compost mulch can help reduce erosion and increase the water-holding capacity of the soil. Application of organic manure also reduces the need to use chemical fertilizers for dragon fruit. Integrated use of organic manure along with chemical fertilizers may be a sustainable approach to improving soil health and increasing the efficiency of fertilizers, while also reducing both nutrient losses and environmental impact.

In Malaysia, Then (2014b) reported that the total number of fruit, average fruit weight and total yield were increased with a yearly application of 6–12 kg of compost per post per year alone, or incorporated with 1.2 kg fertilizer mixture of nitrogen, phosphorus, potassium and magnesium (N:P:K:Mg at a ratio of 9.6:4.8:17.6:2.4) per post. In Indonesia, four organic manures were tested: cow manure, goat manure, chicken manure and rice husk compost used at 5 kg per post (Herawati *et al.*, 2021). The results showed that cow and goat manure application had the greatest effect on shoot formation, while a higher number of fruit was obtained using goat and chicken manure. Verma *et al.* (2019) observed that application of farmyard manure enhanced plant height, the number of branches per plant and the stem diameter of young dragon fruit plants. In Florida, Crane and Balerdi (2019) recommended application of manures or compost at 1.2 kg per post in the first year, increasing to 2.7 kg per post in the second and 4.4 kg per post in the fourth and following years.

In Vietnam, N.V. Son *et al.* (2020) and N.H. Son *et al.* (2020) suggested application of 10 kg of chicken manure or 15 kg of vermicompost combined with inorganic fertilizer comprising $N:P_2O_5:K_2O$ (500:500:500 g per post), which increased fruit weight and yield compared with the control. Field surveys in Vietnam showed that most growers use organic manures on dragon fruit orchards, with average applied rates being cow manure at 10–30 kg per post, chicken manure at 5–10 kg per post and microbial organic fertilizers at 2–3 kg per post (Mai *et al.*, 2020). Cow manure is the most widely used of all organic manures, with other commonly used types being chicken, goat, pig and duck manures (Mai *et al.*, 2020). N.V. Son *et al.* (2020) reported that yearly application of 15 kg per tree of worm castings manure or vermicompost, or 10 kg poultry manure per post combined with chemical fertilizers of $N:P_2O_5:K_2O$ (500:500:500 g per post) increased fruit weight, yield and colour. The general recommended rates for application of organic/farmyard manure are 10–15 kg at planting time and 15–40 kg for plants that are 3 or more years old (Ke, 1997; Suu *et al.*, 2017; Mai *et al.*, 2020). When the price of dragon fruit is low, the application rate can be reduced to save on input costs (Table 8.2).

A comparison of biogas residue from organic fertilizers with cattle manure showed that application of biogas residue increased fruit yield as much as the use of cattle manure fertilizer, but the fruit sugar:acid ratio was higher (Huang *et al.*, 2019). Microbial organic fertilizers are also used as an alternative to organic/farmyard manures if there is an insufficient supply or unsuitable conditions for application of the latter. Yearly application rates are 1–3 kg per plant for plants less than 3 years old and 2–10 kg per plant for plants of 3 or more years old. N.V. Son *et al.* (2020) studied the pink-fleshed

Table 8.2. Recommended rates of organic manure/compost and microbial organic manure/vermicompost for dragon fruit in Vietnam (Hang and Chau, 2001; Hoa *et al.*, 2015; Suu *et al.*, 2017; N.H. Son *et al.*, 2020; Mai *et al.*, 2021; Thoai *et al.*, 2021).

Plant age (years)	Organic manure (kg per post)	Microbial organic fertilizer/vermicompost (kg per post)	Application
At planting	10–15	3–5	Applied in planting hole
1	10–15	5–7	Two applications in the early and late rainy season
2	15–20	8–10	Two applications in the early and late rainy season
3	15–25	10–13	Two applications in the early and late rainy season
≥4	15–30	10–15	Two applications in the early and the late rainy season

dragon fruit variety 'LD5', and found that yearly application of $N:P_2O_5:K_2O$ (500:500:500 g per post) did not significantly change the yield of dragon fruit compared with yearly application of 10–15 kg per post of microbial organic fertilizer A ($N:P_2O_5:K_2O:CaO:MgO$:organic materials at a ratio of 1:2:0.5:1:0.5:24 and microorganisms) or 10–15 kg of microbial organic fertilizer B ($N:P_2O_5:K_2O:CaO:MgO$:organic materials at a ratio of 2:1:0.5:4:0.5:22 and microorganisms) or 10 kg of chicken manure.

Organic manure is usually applied in the early rainy season and/or late rainy season. The most common method of application in established orchards is broadcasting on the soil (top-dressing) under the canopy without touching the stems, and then hoeing or raking it into the top 5–10 cm of soil or covering it with mulch or other organic matter. If the soil is dry, irrigation is carried out after fertilization. At planting, use of organic manure mixed with topsoil and chemical fertilizers (when necessary) to refill the planting hole is suggested.

8.11.2 Chemical fertilizers

Application of chemical fertilizers at the required rate enhances the growth and increases the yield of dragon fruit. The growth and yield were found to be significantly affected by increasing N and P doses, and influenced the number of shoots and flowers and fruit set (Hariyanto *et al.*, 2023). Other researchers have also reported increased plant growth, fruit yield and quality of dragon fruit following application of N (Muchjajib and Muchjajib, 2012; Chakma *et al.*, 2014). The yield of dragon fruit increased significantly after application of NPK fertilizer mixture (Tri *et al.*, 2000; Chakma *et al.*, 2014). Chemical fertilizers had an effect on the number of fruit per trellis, fruit weight and the content of total soluble solids (TSS) but did not affect the vitamin C content (Muchjajib and Muchjajib, 2012).

The application rates of chemical fertilizers vary widely depending on the type of soil, planting density, application method, weather and seasonal conditions, irrigation system, number of fruit crops in the year and growing conditions. In Indonesia, Hariyanto *et al.* (2023) tested application rates of N and P fertilizers and suggested yearly applications per post of 50 g of N fertilizer and 75 g of P fertilizer combined with 100 g of KCl and 10 kg of cow manure, which produced the maximum yield of *H. polyrhizus*. In Vietnam, Tri *et al.* (2000) recommended a yearly application for plants of 3 or more years old of 540 g N, 720 g P_2O_5 and 300 g K_2O combined with 20 kg organic manure per post. In India, Karunakaran *et al.* (2014) recommended that fertilizer mixture of NPK 19-19-19 (19 g of each per 100 g) per post combined with organic manure should be applied every 3–4 months. Parween and Hasan (2019) tested yearly application rates of N, P_2O_5 and K_2O and concluded that rates of 50 g N, 350 g P_2O_5 and 300 g K_2O combined with 20 kg organic manure per post produced the maximum individual fruit weight, number of fruit and yield. Nangare *et al.* (2020) recommended application of 500 g N, 500 g P and 300 g K per post four times a year at 3-month intervals during the first 2 years and 800 g N, 900 g P and 550 g K, which should be well distributed, applied six times a year during the third and following years. According to Gunasena *et al.* (2007), the recommended rates in Taiwan are 4 kg of organic manure and 100 g of commercial NPK 13-13-13 fertilizer per post every 4 months, while rates of 180–230 g per post of NPK 16-16-16 fertilizer were applied at 4–6 monthly intervals in Hawaii. In Florida, Crane and Balerdi (2019) suggested application of NPK fertilizer at frequent intervals (every 2 months) using 118 g per post of NPK fertilizer (6-6-6, 8-3-9 or 8-4-12) with 2–3% Mg, or a similar formula, in the first year. In the second and third years, the amount is gradually increased to 136–182 g per post every 2 months, while in the fourth year, 227–341 g per post of the above formulas is applied three to four times each year. In Malaysia, application of 100 g of NPK 15:15:15 fertilizer mixture per post at 2-month intervals is suggested, followed by 300 g per post of NPK 13:13:21 applied monthly during the flowering and fruit development stages (Department of Agriculture, 2018). In India, according to Mahorkar and Bahiram (2019), each plant should be fertilized with 10–15 kg of organic fertilizer and 70 g N, 90 g P and 40 g K per plant in the vegetative phase, and with a higher ratio of P and K and a lower ratio of N during the flowering to harvesting stages, comprising 50 g N, 50 g P and 100 g K. Application of K in the form of K_2SO_4 or KNO_3 increased fruit weight, yield, firmness and TSS compared with application of KCl or control (Hoang and Chau, 2009). However, application of KCl was reported to increase production and produce a better quality of dragon fruit (Sahu *et al.*, 2022). Applying 106.0–133.3 g K_2O in the first year and 200 g K_2O in the second and third years after planting was recommended by Fernandes *et al.* (2018).

Mizrahi (2015) reported that for fertigated pitayas grown in soil-less culture in Israel, the total requirements of NPK for 1 ha cultivation was 110 kg N, 25 kg P and 280 kg K. The needs of the plant during the various periods of the plant phenology were estimated as follows: during flowering and fruit production, N should be doubled and P should be multiplied by 5, while K should be multiplied by 2.5, in comparison with their vegetative periods.

Chemical fertilizers are usually applied to the soil around the base of the post, extending to the drip line, or the fertilizer is placed in continuous bands parallel to the row on both sides where the plant roots can easily take up the nutrients. Fertilizers should be kept at least 30–40 cm away from the stems and possibly out to a point 20–30 cm past the dripline or edge of the canopy. Watering should be done immediately after fertilizer application, or fertilizers are applied during the rainy season. Water-soluble fertilizers can be diluted with water before applying to soil, or can be applied through irrigation systems (fertigation).

8.11.3 Foliar fertilizers

Foliar fertilizers are applied directly to the branches and canopy of dragon fruit plants to improve their health and increase fertility. Application of foliar and soluble fertilizers is important to induce heavy flowering of red pitaya, increase fruit production and total yield, and improve the fruit quality and fruit size (Then *et al.*, 2014a). Spraying with a commercial foliar fertilizer mixture of 3.5% N, 5.0% P_2O_5 and 16.0% K_2O was able to increase the average fruit weight of red pitaya by 22.2% and produced 22.1% grade A fruit compared with 10.1% in the untreated plants. Spraying of KNO_3 increased the average number of fruit per post, as well as fruit weight and yield of red pitaya in Malaysia (Then, 2013, 2014a).

The effect of gibberellic acid (GA_3) at 30 ppm combined with commercial foliar fertilizers (30-10-10, 10-30-10 and 10-10-30; N-P_2O_5-K_2O) on yield and fruit quality was tested by Tuan *et al.* (2016). The results showed that spraying GA_3 plus a mixture of these three foliar fertilizers after the fruit set stage and repeating twice with a 7-day interval was the most effective treatment, enhancing TSS, total sugar and vitamin C content, and increasing fruit size, fruit weight and yield of red dragon fruit.

Applying 225 g per post of foliar fertilizers containing silicon and Mg in combination with 550 g N, 650 g P_2O_5 and 450 g K_2O increased weight, yield and fruit quality for 6-year-old white-fleshed dragon fruit planted on acrisol soil (Mai *et al.*, 2010). Calcium (Ca) also has an important role in physiological and biochemical processes in plants, and is an essential macronutrient for plants to complete their growth cycle, including fruit development (Hepler, 2005). Madani *et al.* (2016) suggested spraying $CaCl_2$ before harvest to improve fruit texture and flavour. The treatment also reduced weight loss and postharvest disease rates on papaya fruit. The firmness of red-fleshed dragon fruit (*H. polyrhizus*) increased after preharvest treatment with Ca (Ghani *et al.*, 2011). Increased Ca content in treated fruit may contribute directly to the reduction of anthracnose (*Colletotrichum gloeosporioides*) and brown rot (*Monilinia fructicola*) severity in $CaCl_2$-treated fruit (Ghani *et al.*, 2011).

8.11.4 Plant growth regulators

PGRs are applied widely in commercial dragon fruit orchards to improve branch growth, fruit yield and appearance. Dosages, combinations and phases of use vary depending on a number of factors including soil conditions, variety, climatic and weather conditions, plant health and farming system. PGRs help improve the growth, yield and appearance of all varieties grown commercially in Vietnam (Hoang, 2006; Hoang and Chau, 2008; Hang and Hong, 2016).

Application of GA_3 increased the percentage of fruit set and consequently increased the number of fruit and yield of dragon fruit (Takata *et al.*, 2016). Combined application of GA_3 and NAA at suitable concentrations can improve the yield and quality of dragon fruit. Spraying of NAA (20 ppm) and GA_3 (50 ppm) at 10 days after flower buds appear, after flower blooming, and at 5, 10, 15, 20 and 25 days after flower blooming helped to increase fruit weight and yield (Hoang, 2006). To *et al.* (2000, 2002) reported that a mixed spraying of GA_3, α-NAA and β-NAA (8, 150 and 40 ppm, respectively) on day 11 after flowering increased fruit weight by 10% and improved the fruit TSS, peel thickness and hardness of the fruit scale. Spraying of NAA (30–40 ppm) and GA_3 (40 ppm) at 10 days after flower bud appearance, and at 5, 10, 15 and 20 days after flowering increased fruit firmness, fruit weight, yield and the brightness of fruit peel on pink-fleshed dragon fruit variety 'LD5'. Combined spraying of GA_3 (40 ppm) and NAA (20 ppm) at 7, 14 and 21 days after blooming increased flesh firmness, fruit length, fruit weight and yield, and also improved the luminosity of the fruit peel of white-fleshed dragon fruit (Hang and Hanh, 2022). Boonklam *et al.* (2019) tested the effect of forchlorfenuron (*N*-(2-chloro-4-pyridinyl)-*N*-phenylurea (CPPU)) and NAA on off-season dragon fruit. The results showed that CPPU (20 mg l^{-1}), NAA (200 mg l^{-1}), or CPPU (20 mg l^{-1}) + NAA (200 mg l^{-1}) significantly enlarged the size of the dragon fruit with a slightly negative effect on fruit quality. However, fruit treated with NAA, CPPU or NAA + CPPU had lower TSS, higher titratable acidity and greater firmness.

8.12 Irrigation

Dragon fruit is a highly drought-tolerant plant (Fan *et al.*, 2014). The high drought tolerance has been attributed to its high osmotic adjustment capacity and its ability to orchestrate its enzymatic and non-enzymatic antioxidant systems, thus avoiding significant oxidative damage (Wang *et al.*, 2021). Dragon fruit thus requires less water compared with some other fruit plants and can survive well in very low rainfall regions. However, if good fruit quality and high productivity are required, an adequate and stable water supply is essential to meet the plant's needs. Sufficient soil moisture through proper irrigation is vital for commercial production of dragon fruit. It also is helpful for off-season production of dragon fruit in tropical regions.

Irrigation best management practice is essential for dragon fruit orchards to meet crop water needs and prevent drought or overwatering stress and nutrient leaching. Many types of irrigation systems are used in dragon fruit orchards, each with advantages and disadvantages related to use, infrastructure and management requirements, cost and labour for installation, operation and maintenance (Crane *et al.*, 2020). Common types of irrigation system for dragon fruit in Vietnam include hand watering, sprinkler/jet irrigation and drip irrigation.

Hand watering refers to the application of water for irrigation purposes using a hand-held hose (Fig. 8.20A). A worker holding a hose will water each plant individually. This is the conventional method of irrigation in many developing countries. It requires more water and labour and longer working hours than a localized irrigation system. It was the main irrigation method for dragon fruit in Vietnam from 1990 to 2010. It has gradually been replaced by sprinkler and drip irrigation (Mai *et al.*, 2020).

Sprinkler irrigation is a method of irrigation similar to natural rainfall (Fig. 8.21). Water is distributed to the plants through a system of pipelines, usually by pumping. The sprinklers, water supply, pipeline system, pumps and operating conditions must be designed to allow uniform watering of the plants. Jet irrigation (Fig. 8.20B, C) has similarities to sprinkler irrigation. It is commonly used for under-canopy irrigation in dragon fruit orchards in Vietnam. This system is usually more reliable, and less clogging occurs than a sprinkler system. The jets usually cost less than sprinklers (Mai *et al.*, 2020).

Drip irrigation is also called trickle irrigation. Water is conveyed through a system of pipes to the orchard, dripping slowly on to the soil through emitters or drippers. It can only water the root zone area and does not cover as large an area as the sprinkler system, thus reducing water loss through evaporation and runoff. Therefore, it is an efficient method of irrigation that eliminates uneven and excessive watering. Another advantage is that it can easily be combined with the application of fertilizers through the irrigation system (fertigation). The drip

Fig. 8.20. Irrigation systems for dragon fruit. (A) Hand watering with a hose. (B) Irrigation under the canopy with polyvinyl chloride pipes and jets. (C) Irrigation using high-density polyethylene pipes and jets.

Fig. 8.21. Overhead irrigation systems. (A) Rows of overhead sprinklers between the rows. (B) A system of both overhead sprinklers and under-canopy jets installed in the orchard.

irrigation system can be supported by a combined solar–wind–electric power generation system for sustainable use of water and energy (Muslim *et al.*, 2018). Currently, this type of irrigation system is increasingly being used for dragon fruit orchards, especially in areas where the water resource for irrigation is not abundant (Mai *et al.*, 2020). Disadvantages are that the initial cost of drip irrigation is higher than sprinkler irrigation, clogging of the nozzles can occur if the water is not filtered correctly and it requires more maintenance.

In Vietnam, many orchards are equipped with three irrigation systems: an under-canopy irrigation system, an overhead irrigation system and a system for hand watering. The hand-watering system used is mainly a water gun for washing the plants. For under-canopy irrigation, microsprinklers or microjets can be chosen. The choice is made based on the availability of water and its quality, the water resource, the trellis system in use, planting density and other factors. Microjets are commonly used in Vietnam. For a single post, two microjets or microsprinklers are used. For the fence trellis system, a microjet is usually set up along the planting row with the distance between two microjets depending on the distance the water is propelled to provide a uniform amount of water to the plant. An overhead irrigation system using sprinklers is also commonly used for irrigation of dragon fruit. Water is distributed under higher pressure through a pipeline network and can be applied to many plants. The distance between two sprinklers depends on the distance the water is propelled by the sprinkler. In Vietnam, this system is used mainly for canopy washing, for cooling the air on hot, dry days (to reduce sunburn and sooty mould disease; see Chapter 11, this volume), and for watering the plants.

The root system of dragon fruit plants is very shallow, with a depth of no more than 40 cm (Mizrahi *et al.*, 2007). Therefore, on sandy soils, irrigation should be carried out every day through a drip irrigation system with a small amount of water to keep the root system wet. In Hawaii, during dry periods, dragon fruit plants are irrigated twice a week to ensure good soil moisture and prevent oversaturation (Taitano *et al.*, 2021).

The timing of irrigation, the amount of water applied and the intervals between applications depend on the weather conditions, plant age, type of soil, development stage of the plant, season, mulching conditions and other factors. Ideally, irrigation should be carried out daily or at 2–3-day intervals for drip irrigation systems and microjet/sprinkler systems (Mai *et al.*, 2020).

The irrigation rate varies depending mainly on climatic and weather conditions, soil properties, variety, planting density, trellis system, mulching conditions, plant age and health, and expected yield. In India, irrigation is applied whenever there is a long dry spell, and the water requirement is suggested to be about 1000–1500 l per pole year^{-1} during the first and second year (Nangare *et al.*, 2020). According to Mizrahi *et al.* (2007), an irrigation rate of around 1600 m^3 of water ha^{-1} every year was not far from the optimum rate. In Vietnam, Quang (2010) stated that for drip irrigation in the arid areas of the south central coast region, the annual requirement for water is 725.4–825.6 m^3 ha^{-1} for 1-year-old plants, 1096.6–1379.7 m^3 ha^{-1} for 2-year-old plants, 1535.9–1803.0 m^3 ha^{-1} for 3-year-old plants, and 1937.5–2662 m^3 ha^{-1} for 4-year-old and older plants. Thoai *et al.* (2021) recommended an irrigation rate of 30–55 m^3 ha^{-1} during the dry season for young plants ($\leq$3 years old) with an interval of 4–5 days during the hotter and drier months of March and April, and 6–8 days for the period from November to February. During the rainy season (May–October), if the rainfall is 5 mm or less, the interval should be 4–5 days, but if the rainfall is more than 5 mm, irrigation is not required. For established plants (4 or more years old), the interval should be 2–3 days during May and June, and 4–6 days in April, July and September. During the rainy season, if rainfall is over 9 mm, irrigation is not required, but if it is less than 8 mm, watering should be carried out at a rate of 25–35 m^3 ha^{-1} the following day. In regions with scarce irrigation water, Quang (2010) and Hung *et al.* (2013) suggested the following irrigation schedule: in the first year, the irrigation rate is 25–30 m^3 ha^{-1} for a duration of 60–90 min, with two to three applications every month (700–850 m^3 ha^{-1} month^{-1}); in the second year, the irrigation rate is 35–45 m^3 ha^{-1} for 100–120 min, with three to four applications per month (1500–1800 m^3 ha^{-1} month^{-1}); in the third and subsequent years, the irrigation rate is 70–80 m^3 ha^{-1} for 120–150 min, with two applications every month (1800–2200 m^3 ha^{-1} month^{-1}). The irrigation schedule should be adjusted when the rainfall is less than 10 mm, with irrigation applied as normal. If rainfall is between 10 and 20 mm, the water amount and duration of application should be reduced to 50% compared with normal. When the rainfall is over 20 mm, irrigation is not required.

Uneven soil moisture may result in splitting of fruit (Zee *et al.* 2004). Fruit splitting is usually a result of an imbalance of soil moisture due to heavy rainfall or excessive irrigation after a prolonged drought, which is common in tropical sandy soils. Overwatering and excess rain may lead to flower drop and fruit splitting or rot (Zee *et al.*, 2004). Therefore, regular and proper irrigation during dry periods and good drainage during the rainy season will help to reduce fruit splitting when excessive rain occurs in sandy soils. Good drainage is also essential to help prevent root rot and flower and fruit drop when excessive rain occurs in heavy and poorly drained soils.

8.13 Scale Caress

Supermarkets require dragon fruit to have a bright red/pink peel with dark green scales (Fig. 8.22). In Vietnam, to improve the appearance of the fruit, a technique is used called 'scale caress', locally known as 'Vuot Tai Qua' (Mai *et al.*, 2020). The technique involves preharvest application of a mixture of PGRs and foliar fertilizers on the surface of the fruit scales. The mixture usually contains PGRs such as the registered products NAA, GA_3, kinetin and CPPU, as well as nutrients and amino acids.

The mixture for scale caress is prepared by mixing PGRs and foliar fertilizers. A volume of 1 l of mixture includes 1 g of 90% GA_3, 1.5–3.0 g of 90% NAA, and 1 g of cytokinin (diethyl aminoethyl hexanoate, 6-benzylaminopurine, CPPU or kinetin) and foliar fertilizers containing macronutrients, micronutrients and amino acids (Mai *et al.*, 2020). The mixture should be used on the day of preparation; storing the mixture is not recommended. Growers usually buy locally available products and follow the product's label and/or manufacturer's instructions for use. Some mixes are also prepared with different ingredients and proportions of foliar fertilizers and growth regulators by different growers, depending on their experience. The prepared mixture is kept in a plastic bottle (200–300 ml) with a hole in the cap. Rubber gloves are used during application to avoid skin contact, with one or more pairs of polyester or cloth gloves on top of the rubber gloves to absorb the mixture (Fig. 8.23). A few drops of the mixture are applied on the inside of the thumb and index finger and then it is gently rubbed once on both sides of the fruit scales. This should be scheduled for early morning or at cooler times of the day (Mai *et al.*, 2021). The mixture is usually applied to one-third or two-thirds of the scale from the top (Fig. 8.23).

The number of applications of the scale caress mixture varies depending on variety; a single application is suggested for white-fleshed 'Vietnam White' variety and two to three applications for the red-fleshed 'LD1' variety (Mai *et al.*, 2020). Field surveys carried out in some of the main growing regions of southern Vietnam showed that 100% of surveyed growers applied this practice, of which more than 90% of surveyed growers did one application for the

Fig. 8.22. A fruit of red-fleshed variety 'LD1' without scale caress (A) and a fruit with scale caress (after application of PGRs to the scales) (B).

Fig 8.23. Two type of gloves, rubber gloves (A) and polyester/cloth gloves (B), are widely used in Vietnam for applying plant growth regulators to fruit scale (scale caress) (C).

white-fleshed 'Vietnam White' variety and two applications for the red-fleshed 'LD1' variety. Less than 10% of the growers said they applied the mixture three times on 'LD1'. For a single application, the mixture is applied 4–7 days before harvest. For two applications, it is applied 15–16 days before harvest and 9–11 days after the first application. For three applications, it is applied 15–18 days after pollination and 4–5 days and 2–3 days before harvest. If applied to fruit at 6 or more days before harvest, the mixture is rubbed over half of the scale length from the tip (Mai *et al.*, 2020). About 2–3 days after application, the fruit scales will turn greener and become straighter, thicker and longer (Fig. 8.22B).

8.14 Pollination

Dragon fruit flowers are large, androgynous and striking (Gunasena *et al.*, 2007), usually opening during the dark hours and closing in the morning (Marques *et al.*, 2011). Species that visit the flowers include moths, bats, ants and bees. Moths and bats are better pollinators than bees for varieties with flowers that bloom at night (Merten, 2003; Crane and Balerdi, 2019). Some varieties have flowers that remain open in the morning and are also then visited by bees.

In some cases, the reason that natural pollination is unreliable is not due to lack of pollinators. Thus, manual pollination has been suggested to increase fruit weight and yield (Gunasena *et al.*, 2007). It is important to understand the flower fertility of the varieties being grown. They can be self-fertile, self-pollinating or self-sterile (cross-pollinating). Self-fertile varieties produce fruit using their own pollen. Self-pollinating varieties have self-fertile flowers with a shorter stigma, surrounded by the anthers. It is easy for the anthers to contact the stigma and release pollen on to the stigma for successful pollination. However, some self-pollinating varieties need cross-pollination by hand to ensure larger fruit. Some self-fertile varieties produce a stigma protruding far above the anthers, so that the pollen released from the anthers cannot easily contact the stigma and assistance is needed for effective pollination. Self-sterile or cross-pollinating varieties usually produce flowers with a very long stigma, protruding far from the anthers. Cross-pollination is necessary in this case to ensure the production of fruit. Pollen from a compatible variety should be used to pollinate the stigma. However, there is currently little information on pollen varieties that are compatible with commercially grown varieties.

If a new variety is unnamed or there is no available information, it is necessary to check the flower fertility. Before testing, hand cross-pollination needs to be done to ensure fruit production. After pollination, the start of fruit development is observed, characterized by desiccation of the flower remains (Marques *et al.*, 2011). The genera *Selenicereus* and *Hylocereus* are self-incompatible and not many insects or other animals are attracted to the flowers to assist in natural pollination (Mizrahi *et al.*, 2004). *Hylocereus* spp. genotypes are only 10% self-compatible and manual cross-pollination is needed to ensure fruit production (Subandi *et al.*, 2018). Crosses between red-fleshed (*H. polyrhizus*) and white-fleshed (*H. undatus*) dragon fruit increases the size, diameter, length and weight of the fruit, as well as its sugar content and yield (Subandi *et al.*, 2018). Menezes *et al.* (2015) reported that pollination of red pitaya with its own pollen allowed fructification. Flowers pollinated in April produced fruit with physical characteristics superior to flowers pollinated in January. Hand pollination is carried out by physically removing the anthers from one flower and touching them on to the stigma of another, or collecting the pollen and using a brush to pollinate multiple flowers (Merten, 2003). Pollen is most viable at the time of flower opening, but hand pollination was found to be successful well into the evening of the following day (Weiss *et al.*, 1994). In Vietnam, pollination is not usually practised in common commercial varieties. This may be because these varieties are self-compatible, and some of them are autogamous. Pollination, flowering, fruit set and fruit are discussed in detail in Chapter 9 (this volume).

8.15 Intercropping

Intercropping in fruit orchards is an effective approach for promoting sustainable agriculture and increasing land productivity. However, this practice has also been applied widely in small-scale orchards. Growers can secure a stable revenue by diversifying their income sources from dragon fruit and appropriate intercrops (Fig. 8.24).

Intercropping systems include dragon fruit/maize, coconut/tomato/dragon fruit, coconut/dragon fruit, dragon fruit/tomato and dragon fruit/papaya (Eusebio and Alaban, 2018). Yard-long bean or cowpea (*Vigna sinensis*) has also been tested as an intercrop in dragon fruit orchards (Yusoff *et al.*, 2008). The results showed that intercropping with *V. sinensis*, if done during the early planting period, was not harmful to the growth of the dragon fruit plants and provided an additional income.

In newly established orchards, growing economic crops in the interspace or under the canopy of the plants during the first few years is common practice. The intercrops also act as a ground cover in orchards. Intercropping should not be continued when dragon fruit plants develop a large canopy with less space between rows for the intercrop. Thereafter, ground-cover plants should be grown instead of an intercrop. Vegetable plants are the most popular choice for intercropping in dragon fruit orchards in the first and second years after planting, including: leafy vegetables (e.g. lettuce, Thai basil (*Ocimum basilicum* var. *thyrsiflora*), mint, coriander, culantro (*Eryngium foetidum*), Asiatic pennywort (*Centella asiatica*), leaf mustard, bok choy), solanaceous vegetables (e.g. tomato, aubergine, Vietnamese aubergine (*Solanum*

Fig. 8.24. Intercropping in dragon fruit orchard with vegetable plants in the first 2 years. (A) Bottle gourd. (B) Water spinach. (C) Pumpkin.

macrocarpon), chillies), cucurbitaceous vegetables (e.g. bitter gourd, cucumber, pumpkin), and onion and garlic (Fig 8.24) (Mai *et al.*, 2020). Shade-tolerant vegetables such as culantro, Asiatic pennywort and mint are suitable for newly established orchards as well as in established orchards of 3–4 years old or older. Papaya is usually intercropped with dragon fruit in Vietnam.

Dragon fruit needs plenty of sunlight to grow, so plants that grow tall and have a tendency towards excessive growth should not be used as intercrops. However, moringa trees (*Moringa oleifera*) are often chosen to be intercropped in dragon fruit orchards as a shade tree for preventing sunburn and branch yellowing by extreme sunlight in hot and dry regions. Their leaves, flowers and fruit can be harvested and used as a vegetable or as an alternative feed for livestock. A moringa tree is planted between every two to four dragon fruit posts. Some fruit trees are also chosen as intercrops, such as hog plum (*Spondias mombin* or *S. pinnata*).

Dragon fruit can also be grown as a secondary crop on areca palm and coconut orchards (Zee *et al.*, 2004; Thamban *et al.*, 2021). In Vietnam, only well-growing and established orchards of betel palm or coconut (>10 years old) are chosen. The trunks of coconut trees (>8 m high) are used as living posts for the vines.

Intercropping in dragon fruit orchards remains poorly researched. Little is known about orchard establishment and management, irrigation and fertilization requirements, the competition for water, nutrients and sunlight, productivity and income in such intercropping systems.

8.16 Off-season Production and Crop Regulation

8.16.1 Off-season production

Dragon fruit are harvested seasonally. The fruit are available in abundance during the harvest season but scarce during the off-season. Extending the fruiting period can solve these issues. Off-season production aims to extend the harvest time of dragon fruit earlier or later than the normal growing season (Poerwanto *et al.*, 2008; Kothiyal *et al.*, 2020; Mai *et al.*, 2022). The main objectives of off-season production are to sell fruit at harvest for a better price and to meet fruit demand in the off-season. Economically, off-season production is more efficient in terms of labour, equipment, vehicles, warehouses, etc., which are mainly used seasonally in the normal growing season, making the production system more sustainable. Off-season production is practised for many tropical fruits (Poerwanto *et al.*, 2008; Mai *et al.*, 2022), including dragon fruit. Dragon fruit is generally considered a long-day and photoperiod-dependent plant (Barbeau, 1990; Nerd and Mizrahi, 1997; Tieng, 1997; Ortiz-Hernández, 1999). The plant optimally blooms and bears fruits where the geographical condition has a critical day length of approximately 12 h (Nobel, 2002; Jiang *et al.*, 2012; Jiang, 2020). In the northern hemisphere, dragon fruit naturally flowers between May and October (Luders and McMahon, 2006; Saradhuldhat *et al.*, 2009; Jiang *et al.*, 2012). In contrast, in some growing regions in the southern hemisphere such as northern Australia, the natural fruit season is from November to April (Luders and McMahon, 2006), and is from November to May in Brazil (Rabelo *et al.*, 2018). Because the flowering period of dragon fruit is between May and October in the northern hemisphere, it is classified as a long-day plant. Yen and Chang (1997) had only limited success in promoting flowering of dragon fruit after experimentally prolonging the light period up to 2200 h between November and March in plastic house conditions. Khaimov and Mizrahi (2006) were unsuccessful in promoting dragon fruit (*Selenicereus megalanthus* and *H. undatus*) flowering in their experiments extending day length between March and July.

Several methods have been tested to induce off-season flowering of dragon fruit. Flower thinning and manipulation of radiation (using shade) showed an effect on flowering induction in Israel (Khaimov and Mizrahi, 2006). Flower thinning and application of PGRs such as CPPU and GA_3 were reported to delay both flowering and ripening of dragon fruit in Israel (Khaimov and Mizrahi, 2006). In Thailand, CPPU treatment was applied to induce off-season flowering as an alternative to using artificial light for night-breaking (Saradhuldhat *et al.*, 2016). A limitation of this method is finding proper buds under the spines on branches for the treatment. Some of the buds can be vegetative, and after treatment some vegetative shoots will grow instead of flower buds (Saradhuldhat, 2018). Many chemicals related to floral initiation have been tested, including ethephon (Ethrel), ethephon with urea, a commercial mixture of gibberellins (GA_4 and GA_7) with benzyl adenine known as Perlan, paclobutrazol, NAA, hydrogen cyanamide, 2-ethylene chlorohydrin, KNO_3 and CaC_2, but did not induce flowering in the non-productive season (Yen and Chang, 1997; Chang, 2003; Khaimov and Mizrahi, 2006). According to Ke (1997), application of KNO_3 can accelerate earlier flowering, resulting in an earlier harvest. Light intensity may affect the nutritional status of branches and thus influence the number of flowers produced, but it does not induce off-season flowering (Raveh *et al.*, 1998; Khaimov and Mizrahi, 2006). The dragon fruit is a long-day plant, and requires shorter nights to produce flowers. Thus, to create shorter dark nights, breaking the dark period with supplemental lighting can be used to induce off-season flowering. Off-season production using supplementary lighting at night to induce flowering in the short-day season has been reported and is becoming common practice in many countries (Ke, 1997; Bao, 1999; Ke *et al.*, 2000; Khaimov and Mizrahi, 2006; Saradhuldhat *et al.*, 2009; Hoang and Long, 2011; Jiang *et al.*, 2012; Xiong *et al.*, 2020).

Factors affecting flowering of dragon fruit include branch/shoot age, temperature, light and application of PGRs/foliar fertilizers (Jiang *et al.*, 2012). Other factors affecting the process include species/variety, plant age and plant health, rain pattern and application of fertilizers. In Taiwan, off-season fruit production in winter with varieties of yellow pitaya (*S. megalanthus*) are easier compared with varieties of *H. undatus* or *H. polyrhizus* (Jiang *et al.*, 2011; Lobo *et al.*, 2013). In Vietnam, the 'Vietnam White' variety needs a longer night-lighting period to induce off-season flowering than the red-fleshed 'LD1' variety (Mai *et al.*, 2020). According to Jiang *et al.* (2011), shoots of yellow pitaya (*S. megalanthus*) that sprout in spring normally flower in winter, and shoots that sprout in winter will flower the following autumn and produce fruits in winter. In southern Vietnam, new shoots and growth emerge every month on the plant so flowering induction can be carried out 10–20 days after harvest. Temperature also affects off-season flowering of dragon fruit. Regardless of day length, too high a temperature in the summer may restrict the flowering (Nerd *et al.*, 2002). Stunted growth and sunburn injuries of branches caused by extreme sunlight and temperatures during the hot season from late April to early May can reduce the flowering of light-induced dragon fruit plants in southern Vietnam. In contrast, low temperatures also strongly affect the off-season flowering response of dragon fruit, with longer night-lighting periods being needed to promote flowering during the colder season in tropical regions (Long and Hoang, 2011; Jiang *et al.*, 2016), and there may even be no flower bud formation in spring at higher-latitude subtropical regions of Israel due to low temperatures (Khaimov and Mizrahi, 2006). In subtropical regions, low air temperature in winter is one of the major limiting factors for off-season production of dragon fruit. The effect of flower induction was found to decrease if the temperature was below 18°C for a long period, especially at night (Oliveira and Tel-Zur, 2022). In the short-day winter season of Hainan Island, China, sufficient supplementary light can induce flowering in dragon fruit plants if the minimum air temperature is at least 15°C (Xiong *et al.*, 2020). Chu and Chang (2020) reported that under an inductive long-day photoperiod (14 h/10 h) in Taiwan, day/night temperatures of 29°C/19°C may be the minimum required for reproductive bud emergence, while 32°C/22°C is the optimal temperature for reproductive bud development and emergence in white-fleshed 'Vietnam White' dragon fruit induced with supplemental lighting for off-season production.

The discovery in Vietnam in 1994 that dragon fruit plants could flower off-season with the use of supplemental lighting at night was made by Dr Thien, a Vietnamese French man who returned to his hometown of My Tho and worked voluntary for the Southern Horticultural Research Institute (SOFRI) (Mai *et al.*, 2022). He accidentally discovered that dragon fruit vines planted in his front yard under an incandescent light bulb (left on at night) flowered off-season. Mr Hai Long, a dragon fruit grower from Binh Thuan province, visited the house to see for himself, and afterwards used incandescent bulbs to break the dark period and induce flowering of his dragon fruits. This lighting technique was also used by some local growers in Binh Thuan. Over time, this night-lighting practice became increasingly popular (Ke, 1997; Bao, 1999), resulting in dragon fruit production year-round (Mai *et al.*, 2020). Among the available practices for inducing off-season flowering, the use of artificial light for night-breaking is the most common and effective method used in the tropics (Table 8.3) (Mai *et al.*, 2020).

In Taiwan, Jiang *et al.* (2012) induced off-season flowering using 28 W warm white fluorescent bulbs for 6 h from 21:00 to 03:00 to break the night period. In Bangladesh, Rashid *et al.* (2021) induced off-season flowering by supplemental lighting using 100 W incandescent bulbs or 24 or 30 light-emitting diode (LED) bulbs for 6 h from 18:00 to 24:00. In Vietnam, incandescent light bulbs (75 or 100 W) were widely used initially, and gradually replaced by compact fluorescent light (CFL) bulbs to save energy. Long and Hoang (2011) reported that using 20 W CFL (yellow light) and 26 W CFL bulbs had the same effect in inducing off-season flowering as 75 W incandescent bulbs with supplemental lighting for 9 h from 21:00 to 06:00 and a duration of 13–20 nights for white-fleshed and 15 nights for red-fleshed dragon fruit (*H. undatus*). CFL bulbs are much more energy-efficient than incandescent bulbs (estimated to be ≥25% more efficient). In the Mekong Delta region, Be *et al.* (2014) observed that the same effect could be achieved using 20 W yellow-light CFL bulbs (1800 bulbs ha^{-1}) or 60 W incandescent bulbs (1000 bulbs ha^{-1}) with supplemental lighting for 10 h for 15 nights (August–September) from 20:00 to 06:00 hours, or for 9 h for 20 nights (December–January), to induce flowering between September and February. LED bulbs have also been tested for their effect on inducing flowering of dragon fruit to replace incandescent bulbs and CFL bulbs to reduce the use of electricity (Be *et al.*, 2014; Mai *et al.*, 2020). Generally, 9 or 10 W LED bulbs are now used (Mai *et al.*, 2022). Recommendations for using supplemental lighting to induce flowering of dragon fruit for the main growing regions in southern Vietnam (Mai *et al.*, 2021) are: 20–24 W CFL bulbs or 5–10 W LED bulbs for supplemental lighting for 8–10 h per night from 21:00–22:00 to 05:00–06:00 for 12–24 nights depending on season and plant health. The number of nights required is higher in second crops after the first light-induced crop.

In southern Vietnam, for the single-post trellis system with 3 × 3 m spacing, the bulbs are installed on a wire between the two rows, covering all four sides of the canopy (Fig. 8.25). A total of about 1600–1800 bulbs ha^{-1} are used for 20–24 nights in December and January, 1400–1500 bulbs ha^{-1} for 16–20 nights in October and November, 1111–1200 bulbs for 12–15 nights in August, September and February are recommended. For the fence trellis or T-bar trellis system, a 2 m distance between bulbs on the wire is recommended. The bulbs are also placed overhead on heavy-duty wires above the rows, similar to street lights. More electricity/power is required for lighting when the bulbs are placed further away from the canopy.

8.16.2 Crop regulation

Crop regulation of dragon fruit refers to encouraging crops in particular seasons in order to produce fruit with improved quality and a higher price (Mai *et al.*, 2022). Examples of fruit crops that achieve this regulation are longan, mango, citrus, guava and pomegranate (Lal *et al.*, 2017; Mai *et al.*, 2022). Crop regulation aims to force the dragon fruit plants to rest and then to produce abundant blooms and fruits during the desired season. This practice is used to adjust flowering and fruiting to the desired time of year to achieve the best price for each crop.

In growing regions of the northern hemisphere, dragon fruit starts flowering naturally from March to September (Ke, 1997; Suu *et al.*, 2017; Saradhuldhat, 2018) and is harvested from April to October in six to ten sessions per year with a peak between June and August, when the price of the fruit is usually low because of an abundant supply. In the natural flowering season, cultivation practices applied to reduce the number of fruits can be achieved by pruning, flower bud and flower thinning, and also by adjusting fertilization and irrigation regimes (Mai *et al.*, 2020).

Off-season fruit are usually sold at a higher price than those in the peak season. In southern Vietnam, three main harvest seasons are recommended for off-season production: (i) December–January; (ii) February–March; and (iii) April–May. The month/season in which the fruit is harvested can affect its price. Fruit harvested in an off-peak month/season will have a higher price due to lower supply. Regulating the appropriate harvest month/season to ensure high market value is

Table 8.3. A typical light-inducing crop production cycle and main management practices for dragon fruit production in southern Vietnam.

No. days after harvest	Enhance plant health	Night lighting	Flower buds emerge	Flower bud development	Blooming and fruit set	Fruit development	
	10–20 days	15–25 days	3–7 days	18–21 days	3–4 days	26–32 days	Harvest
	Pruning and training Hygiene Weed, water and fertilize Pest control Remove flower buds and flowers	Night lighting: 8–10 h per night from 21:00 or 22:00 for 14–24 nights, with 1000–1800 bulbs ha^{-1} Stop lighting when areoles are swollen Pest control		Flower bud thinning, leaving one to three buds per branch Water and fertilize Pest control	Remove petals	Fruit thinning, leaving one to two fruit per branch, 20–50 fruit per post Water and fertilize Pest control Scale caress	

Fig. 8.25. Use of light bulbs for supplemental lighting for flower induction in dragon fruit. (A) Compact florescent (CFL) bulbs. (B) Light-emitting diode (LED) bulbs. (C) A combination of the two types of bulbs.

important for each grower and growing region. Table 8.3 shows an estimate of the duration of night lighting for flowering induction to harvest.

8.17 Harvesting

Dragon fruit plants can produce their first fruit within 1 year of establishment from healthy cuttings. In commercial orchards, the first harvest can be 12–18 months after planting, depending on the size and quality of the cuttings at planting and the management practices followed (Mai *et al.*, 2021). The plant can continue to fruit annually for 20–30 years before it starts to decline (Crane and Balerdi, 2019). In the northern hemisphere, dragon fruit usually flowers from March to September and is harvested from April to October (Ke, 1997; Saradhuldhat *et al.*, 2009) with six to 12 cycles of harvest.

The fruit peel changes colour late in the maturation period, from green to red/pink/yellow, which is about 25–29 days after anthesis (Nerd *et al.*, 1999; Franco *et al.*, 2020). During the initial fruit development period, the outer skin of immature fruit is bright green and starts to change colour on days 25–28 after anthesis. The red colour (for red-peel varieties) on the rind develops quickly and gradually turns completely red (30 days after flowering). In Vietnam, fruits of the 'Vietnam White' variety start to change colour on days 25 and 26 after anthesis (Mai *et al.*, 2020). The acidity of fruits of *H. undatus* was found to be higher in colour-changing fruits than in fruits with advanced colouring (Nerd *et al.*, 1999; Ortiz and Takahashi, 2015; Ortiz and Takahashi, 2020).

Dragon fruit is a non-climacteric fruit and therefore should be picked when the fruit is fully mature to improve its eating quality. The maturity of fruit at harvest influences the eating quality and storage life of the fruit. Maturity, harvesting and storage are discussed in detail in Chapter 13 (this volume).

The fruit should be harvested in cool air in the early morning and/or late afternoon. Dragon fruit are usually hand-picked. Picking of dragon fruit is not easy because the peduncle is short and the fruit has thorns on some varieties. Pruning scissors or garden shears are usually used for fruit picking. Due to the short pedicel, the cut can remove part of the flesh on the branch and the entire areole. Twisting and moving the fruit in a clockwise direction can cause less damage to the fruit to avoid reducing storage life (Mai *et al.*, 2020). After picking, the fruit are usually placed in temporary field-harvesting containers (plastic crates or bamboo/plastic baskets) (Fig. 8.26). Harvesting containers should be smooth without any sharp edges that could damage the fruit. They should also not have a capacity of more than 25 kg for easy movement to the packing sheds. The harvested fruits should be stored in a dry, clean and shaded place and handled carefully at all stages after picking. Pre-sorting and packing are usually done in packing sheds established at a convenient site not far from the field. Sorting of the fruit at this stage helps save energy and

Fig. 8.26. The harvested fruit are sorted (A) and placed in bamboo or plastic baskets (B) before transportation to the packing house for cool storage.

money because culled fruit does not need to be handled, cooled, packed or transported. The harvesting equipment, tools, containers and lining papers must be in a good, clean condition and are washed and stored separately, so that they are isolated from any sources of contamination.

8.18 Conclusion

Dragon fruit is a promising fruit plant and growing dragon fruit is profitable. The areas under cultivation have increased in recent years in most of the growing countries to meet increasing demand worldwide. There has also been an increase in dragon fruit production in non-traditional areas in recent years; however, appropriate standard practices for cultivation of dragon fruit in these areas have not yet been established.

Production of dragon fruit year-round has been standardized in a few of the growing countries to satisfy market demand and also to improve income for growers. Off-season production is gradually becoming common practice in commercial production in tropical regions, and three to five crops can be obtained in a year from the same orchard. Key challenges that commercial growers may face for off-season production include: (i) a sufficient power supply for lighting for night-breaking to induce flowering; (ii) a shortage of irrigation water in some areas due to increased demand and the effects of climate change; (iii) increased risk of pest-related damage to dragon fruit during off-season production and due to climate change; and (iv) increased risk of temperature stresses caused by extreme weather, which may cause branch yellowing and sunburn at hotter temperatures during the summer and which may affect off-season flowering at colder temperatures. More research is needed to develop packages of practices for cultivation of dragon fruit for different varieties and in different ecological regions.

8.19 References

Ahmad, H., Mirana, A.S., Mahbuba, S., Tareq, S.M. and Uddin, J.A.F.M. (2016) Performance of IBA concentrations for rooting of dragon fruit (*Hylocereus undatus*) stem cuttings. *International Journal of Business, Social, and Scientific Research* 4, 231–234.

Alves, D.A., Cruz, M.C.M., Lima, J.E., Santos, N.C., Rabelo, J.M. and Barroso, F.L. (2021) Productive potential and quality of pitaya with nitrogen fertilization. *Pesquisa Agropecuária Brasileira* 56, e01882.

Arredondo, E., Chiamolera, F.M., Casas, M. and Cuevas, J. (2022) Comparing different methods for pruning pitaya (*Hylocereus undatus*). *Horticulturae* 8: 661.

Bao, D.V. (1999) [Examining the effect of artificial lighting duration on off-season flowering of dragon fruit.] BSc thesis, Nong Lam University, Ho Chi Minh City, Vietnam (in Vietnamese).

Barbeau, G. (1990) La pitahaya rouge, un nouveau fruit exotique. *Fruits* 45, 141–174.

Be, L.V., Trua, T.V., Thanh, T.Q., Thang, N.D. and Thien, N.T. (2014) [Effect of compact fluorescent lamp on offseason flowering of pitaya (*Hylocereus undatus*) in Chau Thanh district, Long An province]. *Journal of Sciences* 4, 6–13 (in Vietnamese).

Ben-Asher, J., Nobel, P.S., Yossov, E. and Mizrahi, Y. (2006) Net CO_2 uptake rates for *Hylocereus undarus* and *Selenicereus megalanthus* under field conditions: drought influence and a novel method for analyzing temperature dependence. *Phytosynthetica* 44, 181–186.

Bhardwaj, R.L. (2011) *Benchmark Survey on the Effect of Mulching Material on Crop Production*. Krishi Vigyan Kendra, Sirohi, Maharana Pratap University of Agriculture & Technology, Udaipur, India.

Boonklam, T., Abdullakasim, S., Havananda, T. and Saradhuldhat, P. (2019) The use of forchlorfenuron and naphthalenacetic acid on fruit size in off-season dragon fruit. *Journal of Science and Technology* 8, 1–11 (in Thai).

Bruand, A., Hartmann, C. and Lesturgez, G. (2005) Physical properties of tropical sandy soils: a large range of behaviors. In: *Proceedings of the International Workshop on The Management of Tropical Sandy Soils For Sustainable Agriculture, 27 November–2 December, Khon Kaen, Thailand*, pp. 148–158.

Chakma, S.P., Rashid, A.S.M.H., Roy, S. and Islam, M. (2014) Effect of NPK doses on the yield of dragon fruit (*Hylocereus costaricensis* [F.A.C. Weber] Britton & Rose) in Chittagong Hill Tracts. *American-Eurasian Journal of Agricultural and Environmental Sciences* 14, 521–526.

Chang, T. (2003) Yield and quality of pitaya (*Hylocereus undatus* Britt. & Rose) as affected by fruiting regulation. MSc thesis, National Chung Hsing University, Taichung, Taiwan.

Chhetri, S., Hasan, M.A. and Tamang, A. (2021) Influence of varying length of stem cutting and IBA concentrations on root and shoot growth in dragon fruit cv Giant White (*Hylocereus undatus*). *Environment and Ecology* 39, 1114–1118.

Chopra, M. and Koul, B. (2020) Comparative assessment of different types of mulching in various crops: a review. *Plant Archives* 20, 1620–1626.

Chu, Y.C. and Chang, J.C. (2020) Regulation of floral bud development and emergence by ambient temperature under a long-day photoperiod in white-fleshed pitaya (*Hylocereus undatus*). *Scientia Horticulturae* 271: 109479.

Crane, J.H. and Balerdi, C.F. (2019) Pitaya (Dragon Fruit): Growing in the Florida Home Landscape. UF/IFAS Extension, Gainesville, Florida. Available at: http://edis.ifas.ufl.edu/HS1068 (accessed 10 November 2023).

Crane, J., Bayabil, H., Evans, E.A. and Ballen, F. (2020) Irrigation System Descriptions for Tropical and Subtropical Fruit Crops in Florida. UF/IFAS Extension, Gainesville, Florida. Available at: https://edis.ifas.ufl.edu/publication/HS1375 (accessed 10 November 2023).

Department of Agriculture (2018) *Pitaya Plant Technology Manual.* Department of Agriculture, Kelantan, Malaysia.

Eusebio, J.E. and Alaban, M.C.S. (2018) Current Status of Dragon Fruit and its Prospects in the Philippines. FFTC Agricultural Policy Platform (FFTC-AP, Taipei, Taiwan. Available at: https://ap.fftc.org.tw/article/1295 (accessed 10 November 2023).

Fan, Q.J., Yan, F.X., Qiao, G., Zhang, B.X. and Wen, X.P. (2014) Identification of differentially-expressed genes potentially implicated in drought response in pitaya (*Hylocereus undatus*) by suppression subtractive hybridization and cDNA microarray analysis. *Gene* 533, 322–331.

Fernandes, D.R., Moreira, R.A., Cruz, M.C.M., Rabelo, J.M. and Oliveira, J. (2018) Improvement of production and fruit quality of pitayas with potassium fertilization. *Acta Scientiarum Agronomy* 40: 35290.

Franco, R.K.G., Esguerra, E.B., Tababa, J.L. and Castro, A.C. (2020) Harvest maturity affects the quality and storage behavior of white-fleshed dragon fruit [*Hylocereus undatus* (Haworth) Britton and Rose]. *Food Research* 6, 423–433.

Ghani, M.A.A., Awang, Y. and Sijam, K. (2011) Disease occurrence and fruit quality of pre-harvest calcium treated red flesh dragon fruit (*Hylocereus polyrhizus*). *African Journal of Biotechnology* 10, 1550–1558.

Gunasena, H.P.M., Pushpakumara, D.K.N.G. and Kariyawasam, M. (2007) Dragon fruit, *Hylocereus undatus* (Haw.) Britton and Rose. In: Pushpakumara, D.K.N.G., Gunasena, H.P.M. and Singh, V.P. (eds) *Underutilized Fruit Trees in Sri Lanka. Vol. 1: Asia*. World Agroforestry Centre, New Delhi, pp. 110–141.

Hang, N.T.N. and Chau, N.M. (2001) [Effect of NPK rates combined with microbial fertilizer rates on yield and quality of dragon fruit plant]. In: *Annual Research Results on Fruits*. Agriculture Publishing House, Southern Horticultural Research Institute (SOFRI), Ho Chi Minh City, Vietnam, pp. 146–152 (in Vietnamese).

Hang, N.T.N. and Hanh, T.T.M. (2022) Increase the yield and fruit quality of white flesh dragon fruit (*Hylocereus undatus*) using the combination of gibberellic acid and naphthalene acetic acid. *South Asian Journal of Agricultural Sciences* 2, 94–97.

Hang, N.T.N. and Hong, D.T.C. (2016) [Effect of plant growth regulators on fruit yield and quality of pink flesh dragon fruit variety LD5]. *Vietnam Journal of Science and Technology* 9, 30–36 (in Vietnamese).

Hariyanto, B., Mayura, E., Muas, I., Jumjunidang, J. and Octriana, L. (2023) Effects of nitrogen and phosphorus fertilizer on growth and yield of dragon fruit (*Hylocereus polyrhizus*). *Journal of Applied Agricultural Science and Technology* 7, 172–185.

Hartwig, N.L. and Ammon, H.U. (2002) Cover crops and living mulches. *Weed Science* 50, 688–699.

Hepler, P.K. (2005) Calcium: a central regulator of plant growth and development. *Plant Cell* 17, 2142–2155.

Herawati, N., Aisah, A.R., Sudarto and Hidayah, B.N. (2021) Organic fertilizers increase yield of dragon fruit in Western Lombok, Indonesia. *Advances in Biological Sciences Research* 13, 314–317.

Hieu, N.T., Thu, N.N.A., Linh, D.T., Thanh, N.T.K., Hoa, N.V. *et al.* (2018) Effect of various degree of canopy pruning on plant growth, yield, and control of canker disease of dragon fruit crop. In: *Proceedings of the International Conference on Tropical Fruit Pests and Diseases, 25–27 September, Kota Kinabalu, Sabah, Malaysia*, pp. 98–104.

Hoa, N.V., Hang, N.T.N. and Dung, D.K. (2015) [*Technical Guidance on Dragon Fruit Cultivation Using Compact Fluorescent Light Bulbs for Flowering Induction*]. Bach Khoa Publishing House and University of Agriculture Publishing House, Hanoi (in Vietnamese).

Hoang, N.H. (2006) [Effect of doses and application time of some plant regulators on yield and quality of dragon fruit]. In: *Annual Research Results on Vegetables, Flowers, and Fruit Plants*. Agriculture Publishing House, Southern Fruit Research Institute (SOFRI), Ho Chi Minh City, Vietnam, pp. 256–268 (in Vietnamese).

Hoang, N.H. and Chau, N.M. (2008) [Effect of dose and application time of growth regulators on yield and quality of white flesh dragon fruit (*Hylocereus undatus*) Cho Gao]. In: *Annual Scientific Results on Vegetables, Flower, and Fruit*. Agriculture Publishing House, Southern Horticultural Research Institute (SOFRI), Ho Chi Minh City, Vietnam (in Vietnamese).

Hoang, N.H. and Chau, N.M. (2009) [Improvement quality and sweetness of Cho Gao white flesh dragon fruit (*Hylocereus undatus*) with potassium fertilizers through basin application]. In: *Annual Scientific Results on Vegetable, Flower, and Fruit*. Agriculture Publishing House, Southern Horticultural Research Institute (SOFRI), Ho Chi Minh City, Vietnam (in Vietnamese).

Huang, X., Jiao, J., Li, Z. and Jihua, D. (2019) Effects of biogas residue organic fertilizer on the yield and quality of dragon fruit. *IOP Conference Series: Earth and Environmental Science* 371: 032069.

Hung, T., Linh, D.V., An, N.V. and Quang, D.V. (2013) [*Guide Notebook for Water-saving Irrigation Technology for Upland Crops. Vol. 1: General Department of Irrigation*]. Science and Technics Publishing House, Hanoi, Vietnam (in Vietnamese).

Ingels, C.A., Scow, K.M., Whisson, D.A. and Drenovsky, R.E. (2005) Effects of cover crops on grapevines, yield, juice composition, soil microbial ecology, and gopher activity. *American Journal of Enology and Viticulture* 56, 19–29.

Jacobs, D. (1999) Pitaya (*Hylocereus undatus*), a potential new crop for Australia. *Australian New Crops Newsletter* 29, 16.3.

Jiang, Y., Lin, T., Lee, C., Yen, C. and Yang, W. (2011) Phenology, canopy composition, and fruit quality of yellow pitaya in tropical Taiwan. *HortScience* 46, 1497–1502.

Jiang, Y.-L. (2020) Seasonal response of night-breaking on floral bud formation in red pitaya (*Hylocereus* sp.) in a noninductive period. *Scientia Horticulturae* 270: 109420.

Jiang, Y.-L., Liao, Y.-Y., Lin, T.-S., Lee, C.-L., Yang, W.-J. *et al.* (2012) The photoperiod-regulated bud formation of red pitaya (*Hylocereus* sp.). *HortScience* 47, 1063–1067.

Jiang, Y.-L., Liao Y.-Y., Lin, M.-T. and Yang, W.-J. (2016) Bud development in response to night-breaking treatment in the noninductive period in red pitaya (*Hylocereus* sp.). *HortScience* 51, 690–696.

Kader, M.A., Senge, M., Mojid, M.A. and Ito, K. (2017) Recent advances in mulching materials and methods for modifying soil environment. *Soil Till Research* 168, 155–166.

Kakade, V., Dinesh, D., Singh, D., Bhatnagar, P.R. and Kadam, D. (2019) Influence of length of cutting on root and shoot growth in dragon fruit (*Hylocereus undatus*). *Indian Journal of Agricultural Sciences* 89, 1895–1989.

Karunakaran, G. and Arivalagan, M. (2019) Dragon fruit – a new introduction crop with promising market. *Indian Horticulture* 63, 8–11.

Karunakaran, G., Tripathi, P.C., Sankar, V., Sakthivel, T. and Senthilkumar, R. (2014) Dragon fruit – a new introduction crop to India: a potential market with promising future. In: *Proceedings of a National Seminar on Strategies for Conservation, Improvement and Utilization of Underutilized Fruits, 1–3 December, Karnataka, India*, pp. 138–139.

Karunakaran, G., Arivalagan, M. and Sriram, S. (2019) Dragon fruit country report from India. In: *Dragon Fruit Network: Marketing and Whole Value Chain and Steering Committee Meeting*. Indian Institute of Horticultural Research, Bengaluru, India, pp. 105–112.

Ke, N.V. (1997) [*Dragon Fruit Plant*]. Agriculture Publishing House, Ho Chi Minh City, Vietnam (in Vietnamese).

Ke, N.V., Bao, D.N. and Thu, P.V. (2000) [Flowering induction for dragon fruit plant]. *University of Agriculture and Forestry Journal of Technology and Science of Agriculture and Forestry* 2, 15–18 (in Vietnamese).

Kek Hoe, T. (2017) Planting density of red pitaya (*Hylocereus polyrhizus*) to achieve optimum yield under Malaysia weather condition. *International Journal of Agriculture Innovations and Research* 6, 353–358.

Khaimov, A. and Mizrahi, Y. (2006) Effects of day-length, radiation, flower thinning and growth regulators on flowering of the vine cacti *Hylocereus undatus* and *Selenicereus megalanthus*. *Journal of Horticultural Science and Biotechnology* 81, 465–470.

Kleinhempel, H., Nachtigall, M., Ficke, W. and Ehrig, F. (1987) Disinfection of pruning shears for the prevention of the fire blight. *Acta Hortculturae* 217, 211–218.

Kothiyal, K., Jain, S., Rawat, H., Kiran, B. and Subhasmita, S. (2020) Off season induction of flowering and fruit production in tropical and sub-tropical fruit trees. *Indian Journal of Agriculture and Allied Sciences* 6, 15–18.

Lal, N., Sahu, N., Marboh, E.S., Gupta, A.K. and Patel, R.K. (2017) A review on crop regulation in fruit crops. *International Journal of Current Microbiology and Applied Sciences* 6, 4032–4043.

Lasmini, S.A., Wahyudi, I., Rosmini, R., Nasir, B. and Edy, N. (2019) Combined application of mulches and organic fertilizers enhance shallot production in dryland. *Agronomy Research* 17, 165–175.

Le Bellec, F., Vaillant, F. and Imbert, E. (2006) Pitahaya (*Hylocereus* spp.): a new fruit crop, a market with a future. *Fruits* 61, 237–250.

Li, Y.S., Mao, C.H., Kuo, T.Y. and Chang, Y.C. (2015) Viral diseases of pitaya and other Cataceae plants. In: Jiang, Y.L., Liu, P.C. and Huang, P.H. (eds). *Proceedings of an International Workshop on Improving Pitaya Production and Marketing*. Food and Fertilizer Technology Center, Taipei, Taiwan, pp. 119–127.

Lim, T.K. (2012) *Hylocereus undatus*. In: Lim T.K. (ed.) *Edible Medicinal and Non-medicinal Plants. Vol. 1: Fruits*. Springer, Dordrecht, Netherlands, pp. 650–655.

Lobo, R., Bender, G., Tanizaki, G., Fernandez de Soto, J. and Aguiar, L. (2013) *Pitahaya or Dragon Fruit Production in California: A Research Update*. University of California, Agriculture and Natural Resources Division, Los Angeles, California. Available at: https://ucanr.edu/sites/sdsmallfarms/files/172469.pdf (accessed 10 November 2023).

Long, N.N. and Hoang, N.H. (2011) [Study on the effect of compact fluorescent light bulbs in inducing off-season flowering of dragon fruit for saving energy]. In: Phung, M.T., Tinh, P.V. and Son, V.T. (eds) *Extension @ Agriculture Forum on Sustainable Production and Consumption of Dragon Fruit, 7 October, Binh Thuan, Vietnam*. National Agricultural Extension Center, Hanoi, Vietnam, pp. 49–63 (in Vietnamese).

Lu, Y.C., Watkins, K.B., Teasdale, J.R. and Abdul-Baki, A.A. (2000) Cover crops in sustainable food production. *Food Reviews International* 16, 121–157.

Luders, L. and McMahon, G. (2006) *The Pitaya or Dragon Fruit (*Hylocereus undatus*)*. Agnote, Northern Territory Government, Darwin, Australia.

Luo, J., Xu, M., Qi, Z., Xiong, R., Cheng, Y. *et al.* (2019) Differential responses of the soil microbial community in two pitaya orchards with different mulch types. *Scientific Reports* 9: 10413.

Luo, J., Xu, M., Liu, C., Wei, S. and Tang, H. (2021) Effects comparation of different mulching methods on soil in pitaya orchards. *International Agrophysics* 35, 269–278.

Luu, T.T.H., Le, T.L., Huynh, N. and Pablo, Q.A. (2021) Dragon fruit: a review of health benefits and nutrients and its sustainable development under climate changes in Vietnam. *Czech Journal of Food Sciences* 39, 71–94.

Madani, B., Mirshekari, A. and Yahia, E. (2016) Effect of calcium chloride treatments on calcium content, anthracnose severity, and antioxidant activity in papaya fruit during ambient storage. *Journal of the Science of Food and Agriculture* 96, 2963–2968.

Mahorkar, K.D. and Bahiram, V.K., (2019) Dragon fruit: nutritive and ruminative fruit. *Agriculture and Food: eNewsletter* 1, 52–54.

Mai, V.T., Ha, V.T. and Khoi, B.X. (2010) [Responses of fruit productivity and quality of white flesh dragon fruit variety to silica-mag fertilizers on acrisols soil in the southeast region]. In: *Annual Scientific Reports on Fruits and Vegetables*. Agriculture Publishing House, Southern Horticultural Research Institute (SOFRI), Ho Chi Minh City, Vietnam (in Vietnamese).

Mai, V.T., Ha, V.T., Van, N.T., Chu, T.A., Tran, T.V. *et al.* (2020) *Investigation of the Status of Production and Cultivation of Off-season Dragon Fruit in the Southern Vietnam*. Project Report. Southern Horticultural Research Institute (SOFRI), Ho Chi Minh City, Vietnam.

Mai, V.T., Ha, V.T., Hai, N.M., Tuyen, H.T.B. and Dien, L.Q. (2021) [*Protocol for inducing off-season flowering of dragon fruit for provinces in the Southern Vietnam*]. Southern Horticultural Research Institute (SOFRI), Ministry of Agriculture and Rural Development, Hanoi, Vietnam (in Vietnamese).

Mai, V.T., Ha, V.T., Van, N.T., An, C.T., Tran T.V. *et al.* (2022) *Research on Improvement of Practices of Off-season Production of Some Main Fruits in the South Vietnam (Dragon Fruit, Longan, Rambutan, Durian, and Mango)*. Project Master Report. Southern Horticultural Research Institute (SOFRI), Ministry of Agriculture and Rural Development, Hanoi, Vietnam (in Vietnamese).

Mansyah, E., Hendri, Muas, I. and Yuliati, S. (2019) Dragon Fruit Production and Marketing in Indonesia: Standard Quality in the Global and Regional Levels. FFTC Agricultural Policy Platform. https://ap.fftc.org.tw/article/1601 (10 November 2023).

Marques, V.B., Amato, R.A., Ramos, J.D., Araujo, N.A. and Silva, F.O.R. (2011) Fenologia reprodutiva de pitaia vermelha no município de Lavras, MG. *Ciência Rural* 41, 984–987.

McLaurin, W.J. and Gary, L.W. (1991) *Composting and Mulching: A Guide to Managing Organic Landscape Refuse.* Revised July 2022. University of Georgia Cooperative Extension Service Circular 816. University of Georgia, Athens, Georgia.

McMahon, G. (2003) *Pitaya (Dragon Fruit)*. Northern Territory Government, Darwin, Australia.

Menezes, T.P., Ramos, J.D., Bruzi, A.T. and Costa, A.C. (2015) Artificial pollination and fruit quality in red pitaya. *Bioscience Journal* 31, 801–807.

Merten, S. (2003) A review of *Hylocereus* production in the United States. *Journal of the Professional Association for Cactus Development* 5, 98–105.

Mizrahi, E., Raveh, E., Yossov, A., Nerd, A. and Ben-Asher, J. (2007) New fruit crops with high water use efficiency. In: Janick, J. and Whipkey, A. (eds) *Creating Markets for Economic Development of New Crops and New Uses*. ASHS Press, Alexandria, Virginia, pp. 216–222.

Mizrahi, Y. (2015) Thirty-one years of research and development in the vine cacti pitaya cultivation in Israel. In: Jiang, Y.L., Liu, P.C. and Huang, P.H. (eds) *Proceedings of an International Workshop on Improving Pitaya Production and Marketing*. Food and Fertilizer Technology Center, Taipei, Taiwan, pp. 1–18.

Mizrahi, Y., Mouyal, J., Nerd, A. and Sitrit, Y. (2004) Metaxenia in the vine cacti *Hylocereus polyrhizus* and *Selenicereus* spp. *Annals of Botany* 93, 469–472.

Mohamed-Yasseen, Y. (2002) Micropropagation of pitaya (*Hylocereus undatus* Britton et Rose). *In Vitro Cellular & Developmental Biology – Plant* 34, 427–442.

Muchjajib, S. and Muchjajib, U. (2012) Application of fertilizer for pitaya (*Hylocereus undatus*) under clay soil condition. *Acta Horticulturae* 928, 151–154.

Muslim, R., Widiastuti, J. and Wijayanto, D.S. (2018) Development of an innovative irrigation system for dragon fruit farming. *Journal of Mechanical Engineering and Vocational Education* 1, 65–69.

Nangare, D.D., Taware, P.B., Singh, Y., Kumar, P.S., Bal, S.K. *et al.*, (2020) *Dragon Fruit: A Potential Crop for Abiotic Stressed Areas*. Technical Bulletin No. 47. ICAR National Institute of Abiotic Stress Management, Baramati, India, p. 24.

Nerd, A. and Mizrahi, Y. (1997) Reproductive biology of cactus fruit crops. *Horticultural Reviews* 18, 321–346.

Nerd, A., Gutman, F. and Mizrahi, Y. (1999) Ripening and post-harvest behaviour of fruits of two *Hylocereus species* (Cactaceae). *Postharvest Biology Technology* 17, 39–44.

Nerd, A., Sitrit, Y., Kaushik, R.A. and Mizrahi, Y. (2002) High summer temperatures inhibit flowering in vine pitaya crops (*Hylocereus* spp.). *Scientia Horticulturae* 96, 343–350.

Nobel, P.S. (2002) *Cacti: Biology and Uses*. University of California Press, Berkeley/Los Angeles, California.

Oliveira, M.M.T. and Tel-Zur, N. (2022) Cactus pear and pitaya: fruit production and orchard management. *Acta Horticulturae* 1343, 343–354.

Ortiz, T.A. and Takahashi, L.S.A. (2015) Physical and chemical characteristics of pitaya fruits at physiological maturity. *Genetics and Molecular Research* 14, 14422–14439.

Ortiz, T.A. and Takahashi, L.S.A. (2020) Pitaya fruit quality (*Hylocereus undatus* [Haworth] Britton & Rose) according to physiological maturity. A review. *Revista Colombiana De Ciencias Hortícolas* 14, 63–75.

Ortiz-Hernández, Y.D. (1999) Pitahaya: un nuevo cultivo para-México. Editorial Limusa, Mexico City, Mexico.

Parween, T. and Hasan, A. (2019) Growth, yield, and quality of dragon fruit as influenced by NPK fertilization. *Indian Journal of Horticulture* 76, 180–183.

Pascua, L.T., Pascua, M.E. and Gabriel, M.L.S. (2015) Dragon fruit production and marketing in the Philippines: its status, constraints and prospects. In: Jiang, Y.L., Liu, P.C. and Huang, P.H. (eds) *Proceedings of an International Workshop on Improving Pitaya Production and Marketing*. Food and Fertilizer Technology Center, Taipei, Taiwan, pp. 47–63.

Poerwanto, R., Efendi, D., Widodo, W.D., Susanto, S. and Purwoko, B.S. (2008) Off-season production of tropical fruits. *Acta Horticulturae* 772, 127–133.

Quang, L.X. (2010) [Research on reasonable irrigation regime for fruit plants (dragon fruit) in the arid South Central region]. PhD thesis, Vietnam Academy for Water Resources, Hanoi (in Vietnamese).

Rabelo, J.M., Cruz, M.C.M., Alves, D.A.A., Lima, J.E., Reis, L.A.C. and Santos, N.C. (2018) Reproductive phenology of yellow pitaya in a high-altitude tropical region in Brazil. *Acta Scientiarum Agronomy* 42: e43335.

Rahad, M.A.B.K., Islam, M.A., Rahim, M.A. and Monira, S. (2016) Effects of rooting media and varieties on rooting performance of dragon fruit cuttings (*Hylocereus undatus* Haw.). *Research in Agriculture, Livestock, and Fishery* 3, 67–77.

Rashid, A.S.M.H., Azam, M.G. and Chowdhury, S.M.K.H. (2021) Influence of day-length enhancement through night-breaking by artificial lighting on off-season dragon fruit production. *Asian Journal of Advances in Agricultural Research* 17, 1–10.

Raut, U.A., Rathod, S.R., Bharad, S.G. and More, P.G. (2022) Study of different IBA levels on nursery behaviour of dragon fruit cutting. *Pharma Innovation Journal* SP-11, 346–351.

Raveh, E., Nerd, A. and Mizrahi, Y. (1998) Responses of two hemiepiphytic fruit crop cacti to different degrees of shade. *Scientia Horticulturae* 73, 151–164.

Ravikant, Verma, R.S., Singh, V., Lata, R. and Meena, D. (2022) Effect of plant growth regulators on root development of dragon fruit cuttings [*Hylocereus undatus* L. (Haworth)]. *Asian Journal of Microbiology, Biotechnology and Environmental Sciences* 24, 807–810.

Sahu, A., Kishore, K., Nayak, R.K., Dash, S.N., Sahoo, S.C. *et al.* (2022) Influence of potassium on mineral content, yield and quality attributes of dragon fruit (*Selenicereus monacanthus*) in acidic soil of Eastern tropical region of India. *Journal of Plant Nutrition* 46, 2621–2636.

Saradhuldhat, P., Boonklam, T. and Havananda, T. (2016) [Off-season floral induction in white and red flesh dragon fruit by forchlorfenuron]. *Songklanakarin Journal of Plant Science* 3, 49–53 (in Thai),

Saradhuldhat, P., Kaewsongsang, K. and Suvittawat, K. (2009) Induced offseason flowering by supplemented fluorescent light in dragon fruit (*Hylocereus undatus*). *Journal of International Society for Southeast Asian Agricultural Sciences* 15, 236–237.

Singh, S.B., Pramod, K., Prasad, K.G. and Kumar, P. (1991) Response of *Eucalyptus* to organic manure mulch and fertilizer sources of nitrogen and phosphorus. *Van Vigyan* 29, 200–207.

Son, N.H., Son, N.V., Truc, L.T.H., Quynh, L.A.N., Thu, N.N.A. *et al.* (2020) [*Protocol for Integrated Crop Management (ICM) for Dragon Fruit Production for Main Growing Regions*]. Southern Horticultural Research Institute (SOFRI), Ministry of Agriculture and Rural Development, Hanoi, Vietnam (in Vietnamese).

Son, N.V., Truc, L.T.H., Hieu, N.T., Yen, T.T.O., Hoa, N.V. *et al.* (2019) [*Protocol for Dragon Fruit Cultivation on T-bars Trellis System*] Southern Horticultural Research Institute/New Zealand Food and Crop Research Institute, Ministry of Agriculture and Rural Development, Hanoi, Vietnam (in Vietnamese).

Son, N.V., Diep, V.V. and Hang, N.T.N. (2020) Effect of NPK combined with organic fertilizers on yield and quality of pink variety LD5. *Journal of Vietnam Agricultural Science and Technology* 1, 28–34 (in Vietnamese).

Sosa, V., Guevara, R., Gutiérrez Rodríguez, B.E. and Ruiz-Domínguez, C. (2020) Optimal areas and climate change effects on dragon fruit cultivation in Mesoamerica. *Journal of Agricultural Science* 158, 461–470.

Subandi, M., Mustari, E. and Setian, A. (2018) The crossing effect of dragon fruit plant cultivars [*Hylocereus* sp.] on yield. *International Journal of Engineering and Technology* 7, 762–765.

Suu, T.D., Hoa, N.V., Thoai, V.H., Yen, T.T.O., Xuan, P.T. *et al.* (2017) [*Practices for Growing and Care of Dragon Fruit*]. Vietnam Academy of Agricultural Sciences/National Agricultural Extension Center, Hanoi, Vietnam (in Vietnamese).

Taitano, C., Ilai, A.J., Johnny, S., Acosta, M. and Tuquero, J. (2021) *Dragon fruit*. FS 5-21. College of Natural & Applied Sciences (CNAS), University of Guam, Mangilao, Guam. Available at: https://projects.sare.org/wp-content/uploads/Dragon_Fruit_Published.pdf (accessed 10 November 2023).

Takata, W.H.S., Miyake, R.T.M., Narita, N. and Ono, E.O. (2016) Effects of season and GA_3 concentrations on *Hylocereus undatus* flowering and production. *Journal of Agronomy* 15, 179–183.

Tang, W., Yang, H., Wang, W., Wang, C., Pang, Y. *et al.* (2022) Effects of living grass mulch on soil properties and assessment of soil quality in Chinese apple orchards: a meta-analysis. *Agronomy* 12: 1974.

Teviotdale, B.L., Wiley, M.F. and Harper, D.H. (1991) How disinfectants compare in preventing transmission of fire blight. *California Agriculture* 45, 21–23.

Thamban, C., Rajesh, M.K., Beegum, S., Prathibha, P.S. and Daliyamol (2021) Intercropping of fruit crops in coconut garden – vignettes from 'Chowtara thota' Meeyapadavu. *Indian Coconut Journal* 63, 8–13.

Then, T.H. (2013) The effects of foliar fertilizers on the red pitaya (*Hylocerus polyrhizus*) fruit weight. *Acta Horticulturae* 984, 227–230.

Then, K.H. (2014a) Flower induction of red pitaya by foliar fertilizer spraying under Malaysian weather conditions. *Acta Horticulturae* 1024, 193–195.

Then, K.H. (2014b) The effect of compost application to improve the red pitaya yield under various mixture fertilizer rates. *Acta Horticulture* 1024, 189–192.

Thoai, V.H., Dung, N.V., Nghi, D.Q., Kien, B.C. and Yen, D.T.P. (2021) [*Instruction Manual for the Cultivation of Dragon Fruit Adating with Climate Change*]. Agriculture Publishing House, Hanoi, Vietnam (in Vietnamese).

Thomson, P. (2002) *Pitahaya (*Hylocereus *Species): A Promising New Fruit Crop for Southern California*. Bonsall Publications, Bonsall, California.

Tieng, M.T.N. (1997) [*Dragon fruit – A Long Day Plant*]. Ho Chi Minh City Flower and Bonsai Society, Ho Chi Minh City, Vietnam (in Vietnamese).

To, L.V., Ngu, N., Duc, N.D., Trinh, D.T.K., Thanh, N.C., *et al.* (2000) Quality assurance system for dragon fruit. In: Johnson, G.I., To, L.V., Duc, N.D. and Webb, M.C. (eds) *ACIAR Proceedings 100. Quality Assurance in Agricultural Produce*. Australian Centre for International Agricultural Research, Canberra, pp. 101–114.

To, L.V., Ngu, N., Duc, N.D. and Huong, H.T.T. (2002) Dragon fruit quality and storage life: effect of harvesting time, use of plant growth regulators and modified atmosphere packaging. *Acta Horticulturae* 575, 611–621.

Torell, M. and Salamanca, A.M. (2003) Wetlands management in Vietnam's Mekong Delta: an overview of the pressures and responses. In: Torell, M., Salamanca, A.M. and Ratner, B.D. (eds) *Wetlands Management in Vietnam: Issues and Perspectives*. WorldFish Center, Penang, Malaysia, pp. 1–19.

Tri, T.M., Hong, B.T.M. and Chau, N.M. (2000) Effect of N, P, and K on yield and quality of white flesh dragon fruit. In: *Annual Research Results on Fruits, Southern Horticultural Research Institute*. Agriculture Publishing House, Ho Chi Minh City, Vietnam (in Vietnamese).

Trivellini, A., Lucchesini, M., Ferrante, A., Massa, D., Orlando, M. *et al.* (2020) Pitaya, an attractive alternative crop for Mediterranean region. *Agronomy* 10: 1065.

Tuan, H.D., Doanh, L.Q., Thanh, L.Q., *et al.* (2007) Study on utilization of wild groundnut for agriculture in upland. *Journal of Agriculture and Rural Development* 2, 21–24 (in Vietnamese).

Tuan, N.M., Huan, N.T. and Hung, N.Q. (2016) Effect of GA_3 plus Yogen foliar fertilization on yield and fruit quality of red dragon fruit at Lap Thanh district in Vietnam. *International Journal of Plant & Soil Science* 12, 1–9.

Verma, R.S., Lata, R., Ram, R.B., Verma, S.S. and Prakash, S. (2019) Effect of organic, inorganic, and biofertilizers on vegetative characters of dragon fruit (*Hylocereus undatus* L.) plant. *Pharma Innovation Journal* 8, 726–728.

Wang, A., Ma, C., Ma, H., Qiu, Z. and Wen, X. (2021) Physiological and proteomic responses of pitaya to PEG-induced drought stress. *Agriculture* 11, 7632.

Weiss, J., Nerd, A. and Mizrahi, Y. (1994) Flowering behavior and pollination requirements in climbing cacti with fruit crop potential. *HortScience* 29, 1487–1492.

Xiong, R., Liu, C., Xu, M., Wei, S.S., Huang, J.Q. *et al.* (2020) Transcriptomic analysis of flower induction for long-day pitaya by supplementary lighting in short-day winter season. *BMC Genomics* 21: 329.

Yen, C.R. and Chang, F.R. (1997) Forcing pitaya (*Hylocereus undatus* Britt. & Rose) by chemicals, controlled day length and temperature. In: Chen, Y.W. and Chang, L.R. (eds) *Proceedings of a Symposium on Enhancing Competitiveness of Fruit Industry, Vol.* 3. Taichung District Agricultural Improvement Station, Dacun, Taiwan, pp. 163–170.

Yusoff, M.M., Halim, R.A., Mohamed, M.T.M., Rastan, S.O.S. and Meon, Z. (2008) Growth, yield and fruit quality of red dragon (*Hylocereus polyrhizus*) fruit as affected by plant support system and intercropping with long bean (*Vigna sinensis*). *Journal of Food, Agriculture and Environment* 6, 305–311.

Zee, F., Yen, C.R. and Nishina, M. (2004). *Pitaya (Dragon Fruit, Strawberry Pear)*. College of Tropical Agriculture and Human Resources, University of Hawaii, Honolulu, Hawaii.

9 Flowering, Fruit Set and Fruit Development

Joanna Cho Lee Ying[1] and Phebe Ding[2*]

[1]Post-harvest Handling Programme, Horticulture Research Centre, Malaysian Agricultural Research and Development Institute (MARDI), Serdang, Selangor, Malaysia; [2]Department of Crop Science, Faculty of Agriculture, Universiti Putra Malaysia, Serdang, Selangor, Malaysia

9.1 Introduction

Dragon fruit is a common name used in Asia to indicate fruit from genera *Hylocereus* and *Selenicereus*, while the name pitaya or pitahaya is commonly used in Central America, the origin of this plant from the family Cactaceae. This plant was introduced from Central America by French priests to Indochina in the mid-19th century (Mizrahi, 2014). In 1995, Vietnam introduced pitayas to the world under the name dragon pearl fruit due to the bracts that mimic the scales of a dragon, while its symmetrical shape looks like a pearl. From this, the name dragon fruit was derived and is widely used for this plant that thrives in countries between the Tropics of Cancer and Capricorn.

9.2 Flowering Stimulation

Flowering of dragon fruit is affected by shoot age, temperature, light quantity and quality, and by application of plant growth regulators (PGRs) (Jiang *et al.*, 2012). To stimulate flowering, application of PGRs has been widely used in fruit crops. The application of PGRs such as gibberellic acid (GA_3) and *N*-(2-chloro-4-pyridinyl)-*N*-phenylurea (CPPU, also known as forchlorfenuron), a synthesized cytokinin, is able to accelerate flowering or extend the flowering period in *Hylocereus undatus* (Khaimov and Mizrahi, 2006). However, application of PGRs is not able to stimulate flowering during the off-season. This has restricted the production cycle in a year. A study carried out in Brazil revealed that application of GA_3 did not promote flowering but increased fruit set, fruit number, fruit weight and productivity in *H. undatus* (Takata *et al.*, 2016). The failure of GA_3 to promote flowering in *H. undatus* might be due to low temperatures that inhibit flower bud formation as an optimal temperature of 20–30°C is needed for flower bud formation (Khaimov-Armoza *et al.*, 2012).

Dragon fruit is a long-day plant (Jiang *et al.*, 2012), indicating it only flowers when the photoperiod exceeds a critical day length. The critical day length was found to be about 12 h for *Hylocereus* spp. (Jiang *et al.*, 2012). The flowering of a long-day plant is controlled by light signals that entrain the plant circadian clock and also by the light spectrum, especially blue, red and far-red light (Sharathkumar *et al.*, 2021). Short day lengths are experienced by countries in the northern and southern hemisphere during winter and summer, respectively. As a result, dragon fruit in the northern hemisphere blooms from May to October,

* Email: phebe@upm.edu.my

DOI: 10.1079/9781800623156.0009

while those in the southern hemisphere bloom from November to May (Arredondo *et al.*, 2022). As such, dragon fruit fails to form flower buds in certain months of the year. Therefore, artificial lighting has been used by producing countries as night-breaking treatment to prolong day length to stimulate flowering during the off-season.

In Vietnam, for off-season production, *H. undatus* plants are irradiated for 17 consecutive nights from 18:30 to 06:00, while *H. polyrhizus* plants are irradiated for 14 consecutive nights from 21:00 to 05:00 using artificial lighting (Nguyen *et al.*, 2021). Floral buds develop 7–10 days after irradiation has stopped, followed by flower blooming or anthesis 25–28 days after artificial lighting has ended. The fruit is ready to be harvested 55–60 days after artificial lighting has ended. As Taiwan is at a higher latitude than Vietnam, it requires a longer duration of night-breaking to stimulate flowering, and the duration required varies with climatic seasons. During late autumn, 4 weeks of night-breaking treatment is sufficient to promote flowering, but during winter and early spring, 3 months are required to generate a similar result (Jiang *et al.*, 2012). Growers in Taiwan found that the most effective night-breaking treatment is by irradiating plants with 75–100 W tungsten filament (incandescent) bulbs from 22:00 to 02:00 (Jiang *et al.*, 2012). Jiang (2020) further studied the seasonal response of night-breaking on floral bud formation. It was suggested that night-breaking duration in late autumn and early spring should be for 1 and 4 h, respectively.

In Bangladesh, artificial lighting of 6 h (from 18:00 to 00:00) with 100 W incandescent bulbs performs better than other treatments for off-season dragon fruit production (Rashid *et al.*, 2021). In Vietnam, improved low-power compact fluorescent lights (power capacity of 20 W) that have emission spectra focused on red and far-red regions is able to result in better growth and yield than a commercial incandescent lamp (60 W) in both *H. polyrizhus* and *H. undatus* (Nguyen *et al.*, 2021). It is suggested that the intensity of the red and far-red regions might be responsible for inducing phytochrome transformation, which is critical in the flowering process of dragon fruit.

Although artificial lighting is widely used for off-season production, the response of dragon fruit is variety or clone dependent. It was shown that 23 out of 30 genotypes (77%) of *H. undatus* and *H. polyrhizus* hybrids had a positive flowering response towards lighting treatment (Tran *et al.*, 2015a). This study also found that the duration of night-breaking treatment varied from 33 to 48 days with one or two flowering flushes. The flowering duration was 21–26 days, while the fruiting stages took 46–59 days for the 30 genotypes studied. In addition, the fruit produced during the off-season had a higher soluble solids concentration than that produced during the in-season. Recently, Malaysian growers have started using artificial lighting to promote flowering, especially during the low-season, with a positive flowering response (Fig. 9.1). However, no scientific study has yet been carried out to elucidate the efficacy of artificial lighting in

Fig. 9.1. Blooming of dragon fruit plants at night. Artificial lighting is used to stimulate dragon fruit plants to flower.

production regions with an average day length of 12 h throughout the year.

During the off-season, a pruning study was conducted to determine the yield performance of 5-year-old *H. undatus* growing in Tien Giang province, Vietnam (Hieu *et al.*, 2018). The findings showed that following pruning of unproductive and infected cladodes by 60% and 40%, the total number of emerged floral buds was 67.2 and 71.9 buds per post, respectively, 4 days after lighting had stopped. Ten days after lighting had stopped, the emergence of floral buds increased to 91.9 and 96.2 buds per post, respectively. The 60% and 40% pruned canopies also increase the number of shoots transformed from vegetative to reproductive by 15.1% and 25.4%, respectively, 4 days after lighting had stopped. This study suggested that the total number of fruit and yield per post are not affected by pruning; however, pruning tends to increase fruit weight, with canopies pruned 60% giving the heaviest fruit weight (481 g per fruit) compared to other treatments. The flower buds of dragon fruit that grow in equatorial regions appear at 2-week intervals throughout the year. However, the peak flowering season of *H. polyrhizus* growing in Malaysia is from March to September, while the low flowering season is from October to December with a mean of 82.6 flowers per trellis year^{-1}, although the latitude of Malaysia falls in the range of 1.5–6.0°N (Then *et al.*, 2020). The low flowering at the year-end in Malaysia may be due to the north-east monsoon season, which brings heavy rainfall to the country. Heavy rainfall areas are not suitable for the dragon fruit, as excessive rain causes flower and fruit drop (Karunakaran *et al.*, 2021). In India, *H. undatus* has four to five flushes of flowering during June–September, which coincides with the prevalence of the longer day length (≥13 h), rainfall, high humidity (>80%) and a moderate range of temperature (~28°C) (Kishore, 2016). In Mexico, Martinez-Ruiz *et al.* (2017) calculated the thermal constant (TC) in degree-day development (DDD), using the residual method. The reproductive phenology of three selections ('Ana', 'Juana' and 'Carlos') of *H. undatus* was studied. They observed the beginning of the reproductive period after the maximum temperature values were <34°C. For the stage of budbreak–anthesis, TC was 360 DDD, while for anthesis–fruit maturity, TC was 537 DDD; therefore, to complete from bud break to fruit maturity took 897 DDD.

9.3 Reproductive Biology

The genus *Hylocereus* is one of the members of the cactus family that produces the largest flowers (Nerd and Mizrahi, 1997; Anderson, 2001). The flower buds of *H. polyrhizus* take about 4 weeks to mature in the Malaysian climate (Cho and Ding, 2021) and flower bloom or anthesis occurs at night. The flowers begin to bloom in an upright position from 21:30 and fully bloom at 22:00 (Cho and Ding, 2022). The flowers are white and monoecious with both female and male organs (Fig. 9.2A) and remain in bloom for one night. The large flower of most *Hylocereus* spp. is characterized as having stem segments that are not flattened, with a long floral tube (27 ± 2.54 cm) that blooms after sunset (Fig. 9.2B). The flower perianth is formed by two types of tepal: the sepaloid tepals, which cover the floral tube, and the petaloid tepals, which attach around the flower cup rim (Fig. 9.2C). The sepaloids are formed by 33 long ensiform yellow tepals and 25 greenish tepals of various shapes that range from ensiform to narrowly triangular, while the petaloids are formed by 24 white lance-ovate inner tepals surrounded by five white-yellowish lanceolate outer tepals (Table 9.1).

The flowers of *H. polyrhizus* are zygomorphic, as the style and stamens are located in the ventral part of the flower (Fig. 9.3A). The basifixed yellow anthers attached to long filaments are 0.48 ± 0.04 cm long, borne on projecting yellow filaments with a mean length of 8.04 ± 0.48 cm (Fig. 9.3B; Table 9.2). The stamens are apostemonous (distinct and free) and attached to the inner surface of the floral tube, with a mean number of 1016 ± 69 stamens per flower (Fig. 9.3C). The numerous stamens are arranged in a spiral pattern attached to the surface of the floral tube (Fig. 9.3D).

The flowers contain one long style measuring 27 ± 2.54 cm long and 0.63 ± 0.07 cm wide (Fig. 9.4A, Table 9.3). The style is light yellow with a length as long as its floral tube. The style connects to the stigma at the top and the ovary at the bottom (Fig. 9.4A). The mature stigma of *H. polyrhizus* consists of 24 ± 2 stigma lobes radiating from the terminal rim of the tubular

Fig. 9.2. Flower of *Hylocereus polyrhizus* at anthesis (22:00). (A) Both male and female organs coexist in a flower. (B) Lateral view of a flower, emphasizing the spatial distance of the stigma and anthers (red circle), suggesting xenogamous behaviour. (C) Various type of sepaloid and petaloid tepal. Scale bar, 2 cm.

Table 9.1. Number, length, width and shape of sepaloid and petaloid tepals of the *Hylocereus polyrhizus* flower. Adapted from Cho and Ding, 2021.

Type of tepal	Number (mean ± SD)	Length (mean ± SD) (cm)	Width (mean ± SD) (cm)	Shape
Sepaloid				
Yellow	33 ± 6	12.28 ± 0.74	1.06 ± 0.19	Ensiform
Green	25 ± 7	12.18 ± 0.24	1.00 ± 0.19	Ensiform
		6.11 ± 0.7	1.21 ± 0.22	Narrowly triangular
Petaloid				
White	24 ± 2	10.16 ± 0.46	3.70 ± 0.44	Lance-ovate
White-yellowish	5 ± 1	10.95 ± 0.86	2.03 ± 0.38	Lanceolate

SD, standard deviation, $n = 10$.

style (Fig. 9.4B). Each stigma lobe is 1.65 ± 0.11 cm long. The opening of the style canal can be observed in the centre of the radiating stigma lobes. Dutta (1995) noted that the single long style bearing one or more stigma lobes corresponds to the number of placentae located in the ovary. The lining of the canal of the style is covered with a layer of papillae (Fig. 9.4C). The spatial difference between stigma and anthers is 1.66 ± 0.53 cm, with the stigma positioned above

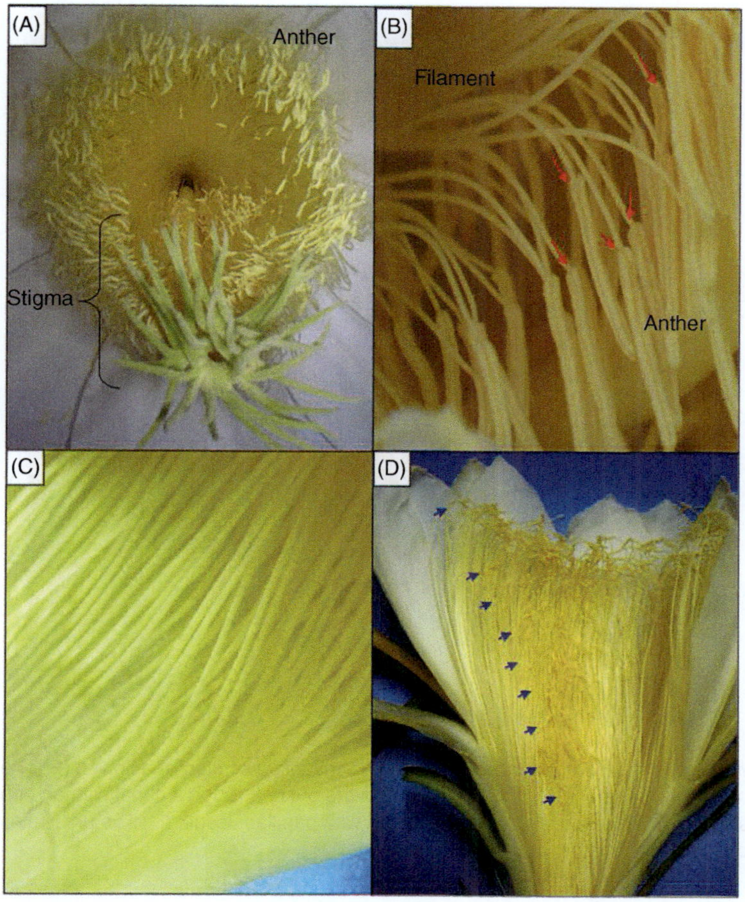

Fig. 9.3. Androecium of *Hylocereus polyrhizus*. (A) Surface view of the flower suggesting zygomorphism, as the stamens and style are located in the ventral part of the flower. (B) The basifixed anthers (arrows) are attached to long filaments. (C) The stamens are apostemonous and attached to the inner surface of the floral tube. (D) The numerous stamens are arranged in a spiral pattern (arrows).

Table 9.2. Anther and filament length, and stamen number of the *Hylocereus polyrhizus* flower. Adapted from Cho and Ding, 2021.

Anther length (mean ± SD) (cm)	Filament length (mean ± SD) (cm)	Stamen number (mean ± SD)
0.48 ± 0.04	8.04 ± 0.48	1016 ± 69

SD, standard deviation, $n = 10$.

the anthers (Fig. 9.4A). Costa *et al.* (2014) observed *H. undatus* cladodes that had borne fruit in previous years, and found that a cladode size of between 40 and 60 cm in length produced the highest percentage of flowering and fruit, and a higher number of fruit per cladode.

The pericarpel of the flower is covered with bracteoles (modified leaves) and arranged in phyllotactic spirals that subsequently becomes the peel of the developing fruit. The ovary of *H. polyrhizus* is epigynous with the perianth position above its receptacle. The unilocular ovary is filled with thousands of small, white and unfertilized ovules (Fig. 9.4D). Each ovule is borne on a funicular that attaches to the inner ovary wall (aparietal placentation) (Fig. 9.4D). Following

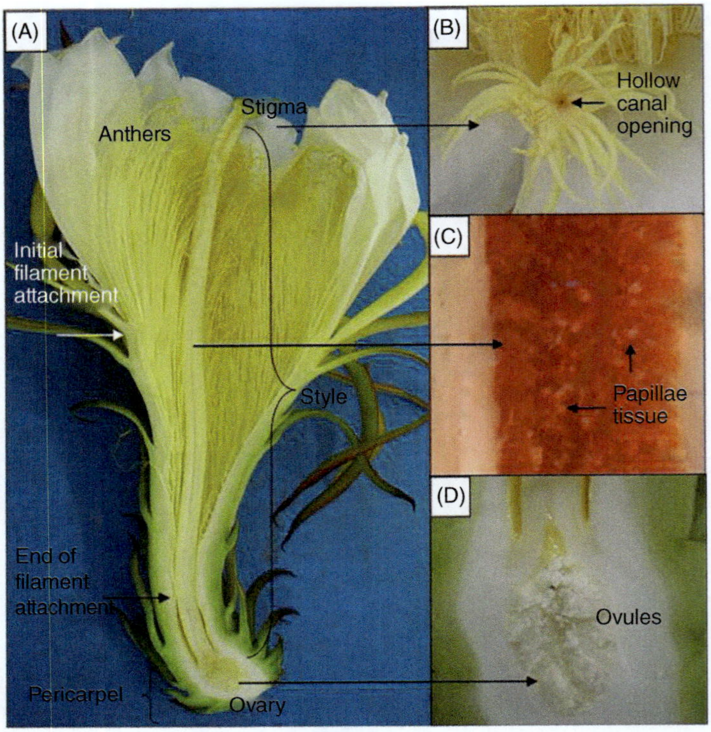

Fig. 9.4. Flower parts of *Hylocereus polyrhizus*. (A) A longitudinal section of a flower. (B) The stigma consisting of numerous stigma lobes and connected to the ovary through a canal with an opening in the centre of the stigma. (C) The canal is densely covered with papillae tissue. (D) The unilocular ovary consists of thousands of ovules in aparietal placentation.

Table 9.3. Style length, style diameter, stigma lobe number and length, distance between stigma and anther of *Hylocereus polyrhizus* flower. Adapted from Cho and Ding, 2021.

Style length (cm ± SD)	Style diameter (mean ± SD) (cm)	Stigma lobe number	Stigma lobe length (mean ± SD) (cm)	Distance between stigma and anther (mean ± SD) (cm)
27 ± 2.54	0.63 ± 0.07	24 ± 2	1.65 ± 0.11	1.66 ± 0.53

SD, standard deviation, $n = 10$.

fertilization, the ovules or eggs will become seeds that are campylotropous (each curve in upon itself so that the micropyle lies next to the funiculus) and crassinucellate (the embryo sac, containing the egg within the ovule, lies deep within nucellar tissue) (Anderson, 2001). The number of carpels in members of the Cactaceae is difficult to define due to fusion of the pericarpel with the ovary.

The nocturnal flower of *H. polyrhizus* is characterized by the typical white sepaloid tepals that facilitate visibility of the flower to night pollinators by providing maximum light reflection. The flowers also emit a strong musky scent during anthesis at night that entices night pollinators such as bats and hawkmoths (nocturnal pollinators) and bees (day-active pollinators) (Weiss *et al.*, 1994). The extension of floral anthesis to the following day in *H. polyrhizus* indicates mixed pollination syndromes that increase the probability of visits by new guilds of pollinators (Walter, 2010). A similar flowering trait has also been reported in the columnar cactus species of

Echinopsis terscheckii to ensure sexual reproduction when nocturnal pollinators vary spatially and temporally (Ortega-Baes *et al.*, 2011). In *H. monacanthus*, the sequence of floral organogenesis starts from the sepal primordial stage, develops into the petal primordial and then the stamen primordial stages, and finally into the carpel primordial stage. However, there are overlaps in time of appearance of some of the organs. The mature stigma has 24–30 shallow clefts with numerous sutures. The style has a hollow canal with an inner surface densely covered with passage cells (Guang *et al.*, 2020).

9.4 Pollination and Fertilization

After 4 days of *H. polyrhizus* flower observation at different time intervals of anthesis, the following sequence can be revealed. At pre-anthesis (19:00) (Fig. 9.5A), the flowers remain closed with clasped multiple stigma lobes that resemble a small, unopened flower (Fig. 9.5B). During anthesis, the flowers tend to bloom upright starting from 21:30 (Fig. 9.5C) and remain fully open at 23:00 (Fig. 9.5D). At postanthesis (09:00 the following morning), the flower remains open (Fig. 9.5E) but is fully closed by 12:00; however, its stigma lobe remains open, and deposited pollen grains can be found on the stigma (Fig. 9.5F). The flower will eventually start to wilt after 2 days and by day 4 after anthesis, the flower has completely wilted, and fungi are beginning to colonize the outer part of the flower (Fig. 9.5G) and the style closest to the stigma starts to turn brown (Fig. 9.5H). To ensure successful pollination and fruit set, the effective pollination period for *H. polyrhizus* should not exceed 12 h after anthesis. The lack of exudates observed on the stigmatic surface cover in papillae tissue identifies the stigma of *H. polyrhizus* as a dry type. Interaction between the dry stigma and pollen grains is highly regulated, especially in self-incompatible species such as *Arabidopsis thaliana* (Edlund *et al.*, 2004) and *Brassica oleracea* (Zheng *et al.*, 2018).

The production of many-seeded fruit in flowering cacti such as *H. polyrhizus* and the tall Chilean columnar cactus *Echinopsis chiloensis* is the result of deposition of large quantities of pollen grains on the stigma (Pimienta-Barrios and Castillo, 2002; Walter, 2010). Pollen grains of *H. polyrhizus* are released at anthesis as monads in large quantities. Their shape is large, oblate and spheroidal with a polar axis measuring 78.19 μm (Fig. 9.6A, Table 9.4) and an equatorial diameter of 80.65 μm. The polar:equatorial ratio of the grain is 0.97. The exine surface is covered with many sharply pointed, radially directed rods called echinates (Fig. 9.6B). Pollen grains viewed under a light microscope appear trizonocolpate with three apertures of colpi arrange equidistantly around the axis of the pollen grain (Fig. 9.6C) which serve as germination furrows for the emerging pollen tube during fertilization. In *H. polyrhizus*, the probability of substantial pollen deposition is enhanced by the large stigma area created by the numerous stigma lobes. As pollen grains land on the stigma, subsequent adhesion is facilitated by the highly ornamented, echinate sculpture on the pollen exine.

Pollen germination in *H. polyrhizus* occurs approximately 2 h after pollination. Upon germination on the stigmatic surface, the pollen tubes elongate underneath the papillae tissues, marking the beginning of a successful journey (Cho and Ding, 2021). Upon penetrating the papillae tissues, the pollen tubes head downwards through the transmitting tissues and subsequently grow along the style. This growth is facilitated by an extracellular matrix composed of starch granules, as observed in the adjacent cells of the transmitting tissues, and by mucilage tracts on the parenchymatous cells of the style and phenolic compounds on the papillae (Williams, 2012; Pereira *et al.*, 2014). Li *et al.* (2022) noted the highest pollen activity between 2 and 4 h after flowering, and the optimal pollination time is within 6 h of flowering. The fruit size was more affected by pollen activity than by stigma receptivity. Pollen viability is mostly affected by high temperature stress. The reduction in seed number caused by incomplete fertilization due to inferior pollen viability was found as the main cause of poor fruiting under a high temperature rather than pistil or photoinhibition (Chu and Chang, 2022).

Self-incompatibility of *H. polyrhizus* is associated with herkogamy (Barrett, 2003), as demonstrated by the spatial difference and positioning of the stigma above the anthers. The numerous stigma lobes radiating from the terminal rim of the tubular style provide a suitable landing spot, while the vector's body transports pollen

Fig. 9.5. The blooming sequence of the *Hylocereus polyrhizus* flower. (A) At 19:00, the flower remains closed with the stigma slightly protruding at the tip of the flower. (B) At 19:00, the multiple stigma lobes remain clasped, mimicking a small, unopened flower. (C) At 21:30, the flower blooms in an upright position. (D) At 23:00, the multiple stigma lobes unclasp and are ready to receive pollen grains. (E) At 09:00 (the following morning), the flower remains open with bees collecting pollen grains. (F) At 12:00 (the following midday), the stigma lobe of the flower remains open, although the flowers are fully closed. (G) By day 4 after anthesis, the flower has wilted, and fungi start to colonize the outer part of the flower. (H) By day 4 after anthesis, the style closest to the stigma starts to turn brown.

grains from previously visited flowers and ensures pollen deposition on to the stigma upon contact. This pre-pollination mechanism is a plant trait commonly found in the Cactaceae to prevent selfing, and promotes out-crossing by favouring xenogamous pollen (Almeida *et al.*, 2013; Valverde *et al.*, 2015). In fact, Weiss *et al.* (1994) previously demonstrated the self-unfruitfulness of *H. polyrhizus* grown in a controlled environment and confirmed that cross-pollination with other clones resulted in absolute (100%) fruit set with significantly larger fruit sizes. The release of pollen can be observed while the flowers are still closed at pre-anthesis, but due to the stigmas protruding above the anthers, the pollen cannot be deposited on to the stigma, which reduces the probability of self-pollination. Under such circumstances, hand cross-pollination is needed.

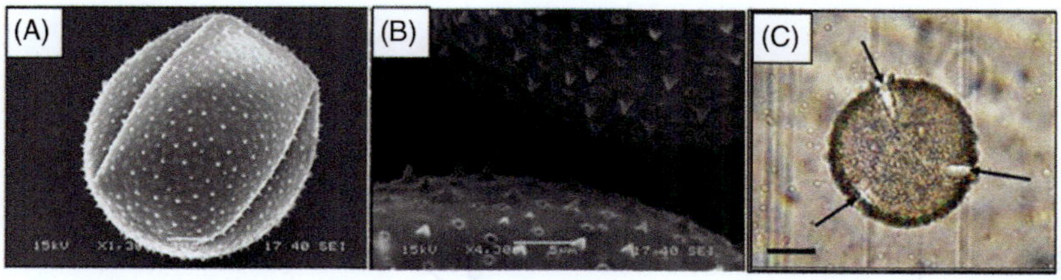

Fig. 9.6. Features of the *Hylocereus polyrhizus* pollen. (A) Scanning electron micrograph of the equatorial view of a pollen grain. Scale bar, 10 µm; 1300× magnification. (B) Side view of the echinates (arrows) on the exine surface. Scale bar, 10 µm; 1300× magnification. (C) Light microscopy of a pollen grain showing trizonocolpate features. Scale bar, 5 µm; 4300× magnification.

Table 9.4. Pollen size and polar:equatorial ratio of *Hylocereus polyrhizus* flower. Adapted from Cho and Ding, 2021.

Pollen size (µm ± SD)		
Polar axis	Equatorial diameter	Polar:equatorial ratio
78.19 ± 4.78	80.65 ± 3.84	0.97

SD, standard deviation, *n* = 25.

For self-incompatible dragon fruit varieties, hand pollination or artificial pollination is needed by collecting pollen from other flowers. Dragon fruit exhibits metaxenia where the pollen source can affect the quality, shape and ripening time of the produced fruit (Mizrahi *et al.*, 2004). In Taiwan, the pollen source (the same flower, a different flower, or a flower from a different variety) and pollination method (hand cross-pollination, hand self-pollination, automatic self-pollination and open pollination) were shown to affect fruit production in some of dragon fruit varieties (Tran *et al.*, 2015b). The distance between the anthers and stigma may be used as an indicator of whether hand pollination is necessary for a particular variety. In Indonesia, the fruit weight of *H. costaricensis* increased by 11% if the flowers were cross-pollinated using pollen from *H. undatus* and *H. polyrhizus* compared with natural pollination (Andayani *et al.*, 2016). Profitable yields of dragon fruit depend entirely on fruit weight, as the fruit are graded and priced according to their weight. It was found that the weight of dragon fruit is highly correlated with the number of seeds in a fruit (Cho, 2011; Fig. 9.7). As well as fruit weight, the number of pollen grains applied during cross-pollination can affect the respiration rate and soluble solids contents of *H. polyrhizus* fruit (Cho *et al.*, 2013). Fruit produced from a pollen load of 0.001 g was the smallest fruit (~133 g), although its carbon dioxide production and soluble solids concentration were significantly higher than fruit produced from pollen loads of 0.05, 0.10, 0.15 and 0.20 g.

The long style of *H. polyrhizus* belongs to the solid type, as most of the pollen tubes grow towards the ovary within the transmitting tissue (Cho and Ding, 2021). When observed under fluorescence microscopy, 2 h after pollination, the pollen tubes appear to elongate underneath the papillae tissues (Fig. 9.8A). There are no traces of pollen tubes detected in the middle part of the style (10 cm below the stigma) and the ovary (Fig. 9.8B). In the ovary, the ovules yield a faint fluorescent light especially at the funiculus thread region (Fig. 9.8C). Two days after pollination (DAP), most of the germinating pollen tubes are seen growing underneath the papillae tissue and growing towards the style (Fig. 9.8D). The callose plugs of pollen tubes can be detected in the transmitting tract that reaches the middle part of the style (Fig. 9.8E). A few pollen tubes reached the style end, but none is found to enter the ovary cavity at this stage (Fig. 9.8F). At 4 DAP, pollen tubes are found to be germinating at the stigma surface and still growing underneath the papillae towards the style (Fig. 9.8G). Compatible pollens develop tubes in a dense cluster and grow rapidly down the middle of the style (Fig. 9.8H) and enter the ovary cavity (Fig. 9.8I).

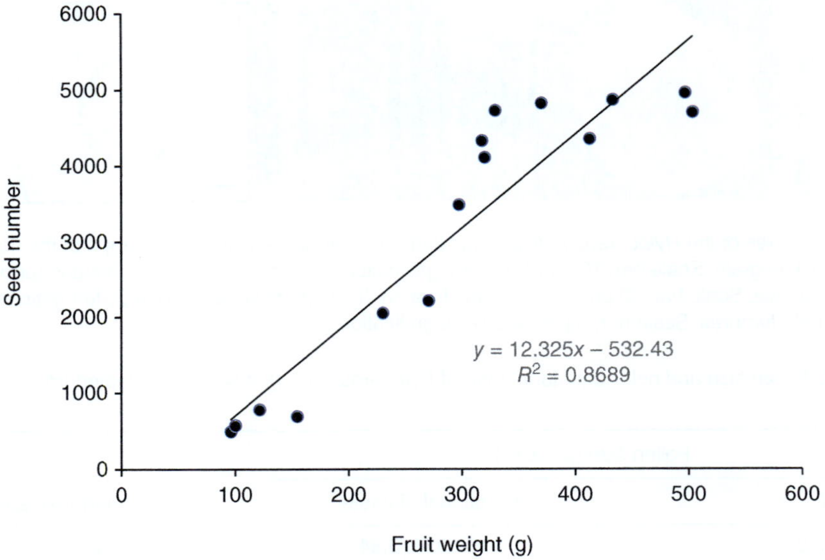

Fig. 9.7. Correlation between the numbers of seeds and fruit weight of red-fleshed dragon fruit (*Hylocereus polyrhizus*).

Compared with 2 DAP, the fluorescence properties of callose plugs are less bright at 4 DAP. At this stage, numerous ovules are fertilized by pollen tubes. With the help of fluorescing callose plugs of the growing pollen tubes, transmitting tissue is found in the centre of the style, sandwiched between the canal and the parenchyma tissue, forming a ring when viewed as a transverse section (Fig. 9.8J). The development of compatible pollen tubes in a dense cluster suggests that the style of *H. polyrhizus* is capable of supporting the growth of a large number of pollen tubes. Pollen tube growth is biphasic. In the first autotrophic phase, pollen-tube growth on the stigma depends on the reserves contained in the pollen grain because the dry-type stigma contains no resources (Cruden, 2009). As the energy reserves in the pollen grain are expended, pollen-tube growth becomes heterotrophic, using resources from the style reserves. The presence of callose plugs in the transmitting tract in the middle of the style at 2 DAP (Fig. 9.8D) indicates the transition from autotrophic to heterotrophic pollen-tube growth (Stephenson *et al.*, 2003). Although the flower starts to wilt by 2 DAP due to fungal invasion of the stigma and upper style region, most of the growing pollen tubes have successfully reached the middle style region (10 cm below the stigma) by this time.

The floral traits of *H. polyrhizus* are associated with chiropterophily pollination with bats as the main pollinators seeking pollen grains and nectar as food sources. The flowers of *H. polyrhizus* do not produce any nectar, suggesting the presence of pollinators in search of pollen only. However, natural pollination by bats/hawkmoths at night and by bees in the morning is not feasible because this could not guarantee high fruit production. A pre-pollination mechanism of herkogamy is exhibited by the spatial positioning of the stigma above the anthers in *H. polyrhizus*. This also means that growers must carry out manual pollination at night to ensure efficient and successful fertilization; however, this manual operation is tedious and the costs may be a major concern to orchard owners. In order to assure availability of pollen at all times, the pollen of dragon fruit is stored for future use. The percentage fruit set and fruit weight of *H. undatus* and *H. polyrhizus* are affected by the storage temperature of the pollen but not the storage duration (Metz *et al.*, 2000). For prolonged pollen storage, it is best stored at temperatures of less than 0°C after the moisture content of the pollen has been reduced to 5–10%. Muniz *et al.* (2019) demonstrated that *H. undatus* is capable of self-pollinating and has a high fruit set rate

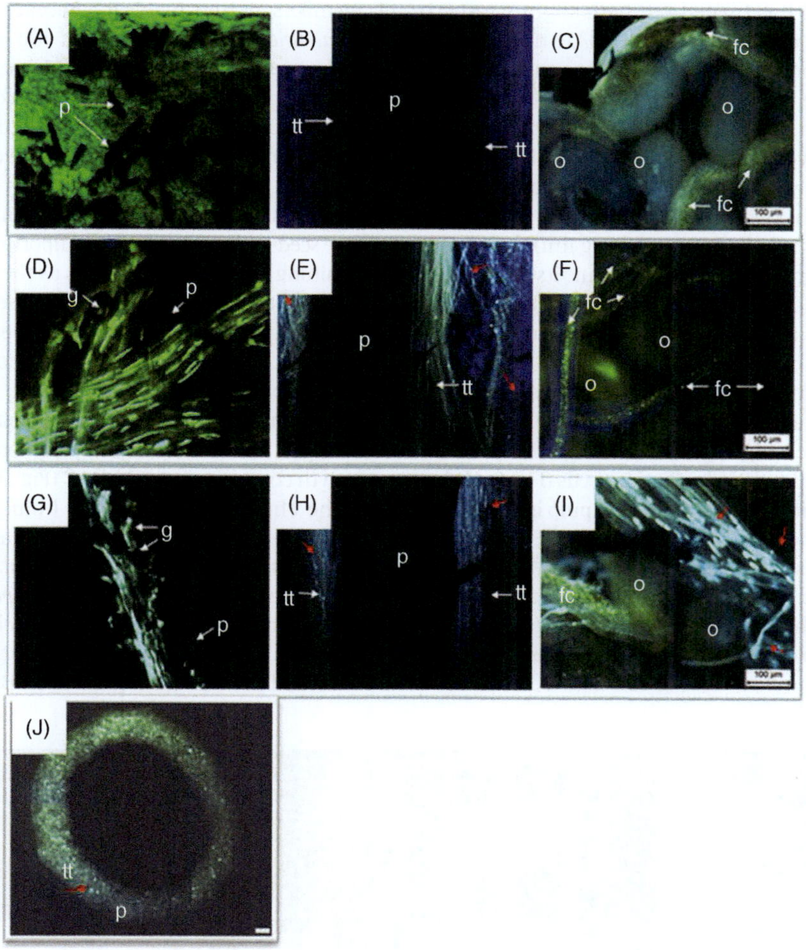

Fig. 9.8. Pollen-tube growth of *Hylocereus polyrhizus* at various days after pollination (DAP), examined under a fluorescence microscope. (A–C) 0 DAP. (A) Pollen grains with germinating tubes 2 h after pollination. (B) Pollen-tube growth is absent in the transmitting tissue (tt) in middle part of the style (10 cm below the stigma). (C) Pollen tubes are not detected in the ovary. (D–F) 2 DAP. (D) Pollen tubes growing underneath papillae towards the style. (E) Abundant pollen tubes growing down the style. (F) No pollen tubes are detected in the ovary. (G–I) 4 DAP. (G) Pollen tubes grow underneath the papillae towards the style. (H) Pollen tubes are growing abundantly down the style. Compared with 2 DAP, the fluorescence properties of callose plugs are less bright. (I) Pollen tubes are detected in the ovary and some are growing towards the ovules. (J) Transverse section of the middle part of the style showing a ring formed from fluorescing callose plugs of the growing pollen tubes. p, papillae; g, pollen grain; tt, transmitting tissue; fc, funiculus; o, ovule. Red arrows indicated the pollen tube with callose plugs.

that is not dependent on floral visitors; however, it needs the moth *Agrius cingulata* to improve yield with larger and heavier fruit. In *H. polyrhizus*, self-pollination is limited and the species requires biotic pollinators to maximize its fruit production. Pollination by bees (*Apis mellifera*) increases the size and weight of the fruit.

9.5 Fruit Development

After successful fertilization, the flower of *H. polyrhizus* wilts gradually while the fruit grows and develops as time progresses (Fig. 9.9A–D). At 25 DAP, the peel of red-fleshed dragon fruit remains green (Fig. 9.9E). At 30 DAP, the previous green peel has turned red with slightly greenish scales (Fig. 9.9F). At 35 DAP, the peel including the scales of the fruit has turned completely red (Fig. 9.9G). *H. polyrhizus* fruit exhibits a single sigmoidal growth pattern during development with fresh weight increasing steadily after successful fertilization and a decrease in weight observed from 30 to 35 DAP (Jamaludin *et al.*, 2011). During initial growth, fruit length grows faster than its diameter, but by 20 DAP, the diameter of the fruit exceeds its length and thus the shape of the fruit becomes oblong.

The fruit shape, peel, pulp and seeds of *H. polyrhizus* also change with DAP (Cho, 2011). At 5 DAP, the pulp region of *H. polyrhizus* fruit is creamy white with developing seeds (Fig. 9.10A). As the fruit develops, the fleshy pulp is gradually formed and turns white (Figs. 10B, C). At 20 DAP, the seeds have matured with prominent black seed coats and are dispersed among the white fleshy pulp (Fig. 9.10D). At 25 DAP, the fleshy pulp close to matured seeds turns red-violet, while the peel remains green (Fig. 9.10E). Fruit of members of the family Cactaceae, such as *Opuntia*, *Selenicereus* and *Pereskia* are generally fleshy berries, as the pulp of the fruit originates from the funiculi of the developing seeds in the ovary (Almeida *et al.*, 2018). During fruit maturation, the funiculi become longer and thicker, and become one of the major edible parts of the fruit (Nerd and Mizrahi, 1997). Hence, the number of mature seeds in the fruit plays an important role in determining the final fruit weight as the pulp develops from the funiculi of developing seeds. The red-violet pigment of *H. polyrhizus* is a betacyanin (Phebe *et al.*, 2009) and its development occurs much earlier in the pulp than in the peel. By 30 DAP, both the pulp and peel of the fruit are red-violet (Fig. 9.10F) and the colour gets more intense as growth continues up to 35 DAP (Fig. 9.10G).

The change of peel colour from green to red-violet is due to chlorophyll degradation and biosynthesis of betacyanin (Phebe *et al.*, 2009). The peel betacyanin content of *H. polyrhizus* fruit is lesser than that of its pulp. The change of dragon fruit peel colour from green to red-violet is often used as a harvesting index. In Brazil, the

Fig. 9.9. Fruit development of *Hylocereus polyrhizus* at intervals of 5 days after pollination (DAP). (A) At 5 DAP, the young fruit begin to develop after successful fertilization. (B) At 10 DAP, the flower has completely dried and wilted. The size of the fruit continues to expand in terms of length and diameter. (C) At 15 DAP, the length and diameter of the fruit continue to increase. (D) At 20 DAP, the length of the fruit reaches its final size. (E) At 25 DAP, the peel remains green. (F) At 30 DAP, the peel turns red. The diameter of the fruit reaches its final size. (G) At 35 DAP, the peel and scales of the fruit turn completely red.

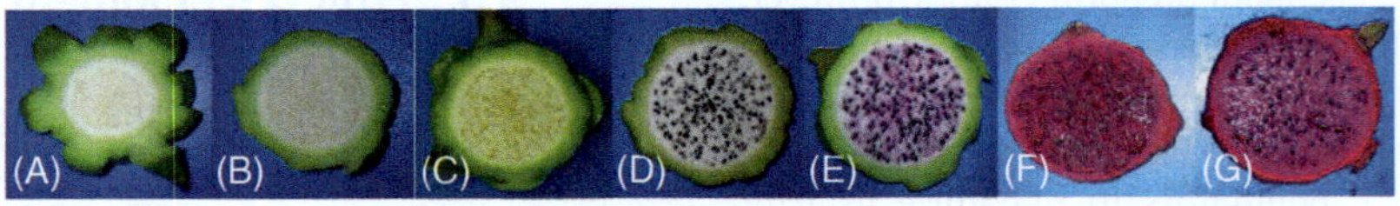

Fig. 9.10. Development changes in the pulp and seed within *Hylocereus polyrhizus* at increasing days after pollination (DAP). (A) 5 DAP. (B) 10 DAP. (C) 15 DAP. (D) 20 DAP. (E) 25 DAP. (F) 30 DAP. (G) 35 DAP.

harvesting stage of white-fleshed dragon fruit is recommended to be carried out when the fruit peel has 25–75% of its red colour to obtain a higher yield and quality than at other stages (Rangel *et al.*, 2021). In Malaysia, the fruit of *H. polyrhizus* is harvested at about 28 DAP to avoid fruit splitting due to delayed harvesting. In Brazil, harvesting of *H. polyrhizus* (Magalhaes *et al.*, 2019a) and *H. undatus* (Magalhaes *et al.*, 2019b) is recommended at 34–36 and 34–38 days after anthesis (DAA), respectively. After 40 days, the fruit of white-fleshed pitaya (*H. undatus*) becomes less visually attractive because of scale wilting and loss of the intense pink colour of the skin, although they possess adequate internal quality to be consumed up to 42 days and may also be used for processing (Magalhaes *et al.*, 2019b). Under the semi-arid conditions of north-western India, the optimum harvesting stage of red-fleshed dragon fruit is about 35–40 DAA, depending on market distance (Singh *et al.*, 2022). The harvesting stage of dragon fruit is affected by production region and species.

During *H. polyrhizus* fruit development, no respiration peak or ethylene is detected at any growth stage, and hence dragon fruit is defined as a non-climacteric fruit (Jamaludin *et al.*, 2011). The firmness of *H. polyrhizus* fruit decreases from 4.8 to 0.9 N, while its soluble solids concentration increases from 1.9% to 12.6% from 5 to 35 DAP (Cho, 2011). The reducing sugars content of *H. polyrhizus* fruit increases significantly as it develops and achieves a maximum value of 121.8 g kg^{-1} at 32 DAA (Trong *et al.*, 2022). In contrast, the starch content decreases from its highest value of 172.8 g kg^{-1} at 18 DAA to 61.2 g kg^{-1} at 32 DAA. This finding corroborates scanning electron microscope observations where the density of starch granules in *H. polyrhizus* pulp cells decreases as ripening progresses, as reported by Cho (2011) (Fig. 9.11). The vitamin C content of Vietnamese *H. polyrhizus* fruit increases as it ripens, achieving 26.18 g kg^{-1} of vitamin C at 32 DAA (Trong *et al.*, 2022).

The activity of acid invertase and neutral invertase in white-fleshed pitaya increases during fruit development before 25–27 DAA (Huang and Lin, 2010). Acid invertase activity was found to be greater than neutral invertase activity. Therefore, acid invertase might be the key enzyme associated with rapid accumulation of glucose in the fruit. α-Amylase activity reduced gradually during fruit maturation up to 29 DAA. Maleic dehydrogenase activity also declined during fruit development. Maleic acid, citric acid and ascorbic acid in the pulp reached the highest peak on 25 DAA, while the concentration of oxalic acid reduced gradually during fruit maturation. Acid invertase was considered to be the key enzyme associated with a rapid accumulation of soluble sugars. Red-fleshed pitaya (*H. polyrhizus*) treated with 100 ppm CPPU on the day of flowering (on 1 cm flower buds) can increase the thickness of bracts and delay the occurrence of yellowing without affecting fruit hardness, soluble solids, titratable acid and weight loss percentage during storage (Sheng *et al.*, 2019).

The physical, chemical and physiological changes in the development of baby pitaya (*Selenicereus setaceus*) at different fruit growth stages was studied by Rodrigues *et al.* (2022). The days required from anthesis until abscission of fruit was 70 days when the inflorescence occurred in late August and the first ripe fruit was harvested in October. The pattern of cumulative growth of the fruit followed a double-sigmoidal model, characterized by two phases of pronounced growth, and the final size was obtained at 63 DAA. The baby pitaya collected at 63 DAA has a completely red peel (L^* = 46.52, a^* = 28.32, chlorophyll = 0.79 mg 100 g^{-1}, betacyanin = 1.28 mg 100 g^{-1}). The pulp showed marked

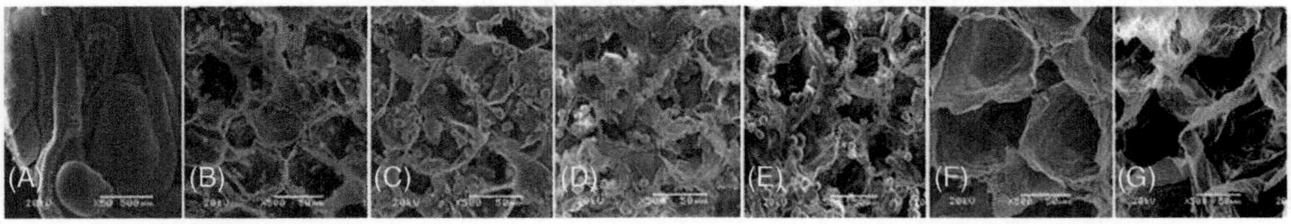

Fig. 9.11. Development changes in starch content within *Hylocereus polyrhizus* at increasing days after pollination (DAP) using scanning electron microscopy. (A) 5 DAP. (B) 10 DAP. (C) 15 DAP. (D) 20 DAP. (E) 25 DAP. (F) 30 DAP. (G) 35 DAP.

changes in some chemical constituents, resulting in reduced contents of starch and titratable acidity and high concentrations of soluble solids and total sugars. Respiration and ethylene production displaying behaviours were similar to those in non-climacteric fruit.

9.6 Conclusion

Although dragon fruit originated from Central America, it is now more popular in tropical and subtropical countries of Asia. This fruit, especially the red-fleshed species, contains betacyanin, a type of betalain pigment that appears in beetroots. It is an important phytochemical in human diets. Although the fruit cycle of dragon fruit is short, the flowering of this plant is affected by photoperiod. Furthermore, some dragon fruit varieties demonstrate herkogamy with spatial separation of the anthers and stigma. In addition, the metaxenia characteristic causes dragon fruit quality to be affected by pollen source. Hence, there is still room for researchers to explore this crop in order to produce consistent yield and a high quality of fruit.

9.7 References

Almeida, O.J.G., Cota-Sánchez, J.H. and Paoli, A.A.S. (2013) The systematic significance of floral morphology, nectarines, and nectar concentration in epiphytic cacti of tribes *Hyloceraceae* and *Rhipsalideae* (Cactaceae). *Perspectives in Plant Ecology, Evolution and Systematics* 15, 255–268.

Almeida, O.J.G., Souza, L.A., Paoli, A.A.S., Davis, A.R. and Cota-Sánchez, J.H. (2018) Pericarp development in fruits of epiphytic cacti: implications for fruit classification and macro-morphology in the Cactaceae. *Botany* 96, 621–635.

Andayani, N., Latifah, L.N. and Astuti, L.M. (2016) The effort to increase production of Super Red Dragon Fruit (*Hylocereus costaricensis*) by artificial pollination. In: *Proceeding of the 1st IBSC: Towards the Extended Use of Basic Science Health, Environment, Energy and Biotechnology for Enhancing, 26–27 September, University of Jember, Indonesia*, pp. 45–46.

Anderson, E.F. (2001) *The Cactus Family*. Timber Press, Portland, Oregon.

Arredondo, E., Chiamolera, F.M., Casas, M. and Cuevas, J. (2022) Comparing different methods for pruning pitaya (*Hylocereus undatus*). *Horticulturae* 8: 661.

Barrett, S.C. (2003) Mating strategies in flowering plants: the outcrossing-selfing paradigm and beyond. *Philosophical Transactions of the Royal Society of London B: Biological Sciences* 358, 991–1004.

Cho, J.L.Y. (2011) Floral structure and pollen load in relation to fruit development and physico-chemical characteristics of dragon fruit [*Hylocereus polyrhizus* (Weber) Britton & Rose] at harvest. MSc thesis, Universiti Putra Malaysia, Serdang, Malaysia.

Cho, J.L.Y. and Ding, P. (2021) Floral morphology and pollination process of red-fleshed dragon fruit (*Hylocereus polyrhizus*) grown under open field. *Horticultural Science & Technology* 39, 277–293.

Cho, J.L.Y. and Ding, P. (2022) Floral characteristics of red-fleshed dragon fruit (*Hylocereus polyrhizus*) grown under Malaysia climate. *Acta Horticulturae* 1342, 139–142.

Cho, J.L.Y., Ding, P., Abd. Razak, A.R. and Wahab, Z. (2013) Pollen load affects quality of red-fleshed dragon fruit (*Hylocereus polyrhizus*). *Acta Horticulturae* 1012, 253–258.

Chu, Y.C. and Chang, J.C. (2022) Heat stress leads to poor fruiting mainly due to inferior pollen viability and reduced photosystem II efficiency in "Da Hong" pitaya. *Agronomy* 12: 225.

Costa, A.C., Ramos, J.D., Silva, F.O.R. and Durate, M.H. (2014) Flowering and fruitification in different types of cladodes of red pitaya in Lavras. *Revista Brasileira de Fruticultura* 36, 279–284.

Cruden, R.W. (2009) Pollen grain size, stigma depth, and style length: the relationships revisited. *Plant Systematics and Evolution* 278, 223–238.

Dutta, A.C. (1995) *Botany*. Oxford University Press, Kolkata, India.

Edlund, A.F., Swanson, R. and Preuss D. (2004) Pollen and stigma structure and function: the role of diversity in pollination. *Plant Cell* 16, 84–98.

Guang, S.P., Hao, C.Z., Jun, S.C., Xia, G.S., Gang, G. *et al.* (2020) Floral organogenesis of red pitaya (*Hylocereus monacanthus*). *Fujian Journal of Agricultural Science* 35, 943–949.

Hieu, N.T., Thu, N.N.A., Linh, D.T., Thanh, N.T.K., Hoa, N.V. and Rangaswamy, M. (2018) Effect of various degree of canopy pruning on plant growth, yield, and control of canker disease of dragon fruit crop. In:

Proceedings of the 2018 International Conference on Tropical Fruit Pests and Diseases, 25–27 September, Kota Kinabalu, Malaysia, pp. 98–104.

Huang, S. and Lin, H. (2010) Changes in the activities of sugar synthesis enzymes and malate dehydrogenase in *Hylocereus undatus* fruit during later period of development. *Journal of Taiwan Society for Horticultural Science* 56, 81–91.

Jamaludin, N.A., Ding, P. and Hamid, A.A. (2011) Physico-chemical and structural changes of red-fleshed dragon fruit (*Hylocereus polyrhizus*) during fruit development. *Journal of the Science of Food and Agriculture* 91, 278–285.

Jiang, Y.L. (2020) Seasonal response of night-breaking on floral bud formation in red pitaya (*Hylocereus* sp.) in a noninductive period. *Scientia Horticulturae* 270: 109420.

Jiang, Y.L., Liao, Y.Y., Lin, T.S. and Lee, C.L. (2012) The photoperiod-regulated bud formation of red pitaya (*Hylocereus* sp.). *HortScience* 47, 1063–1067.

Karunakaran, G., Arivalagan, M. and Sriram, S. (2021) Cultivation and characterization of dragon fruits in India. In: Fullerton, R.A. and Lee, W.L. (eds) *From Production to Market.* Food and Fertilizer Technology Center for the Asian and Pacific Region, Taipei, Taiwan, pp. 131–150.

Khaimov, A. and Mizrahi, Y. (2006) Effects of daylength, radiation, flower thinning and growth regulators on flowering of the vine cacti *Hylocereus undatus* and *Selenicereus megalanthus*. *Journal of Horticultural Science and Biotechnology* 81, 465–470.

Khaimov-Armoza, A., Novak, O., Strnad, M. and Mizrahi, Y. (2012) The role of endogenous cytokinins and environmental factors in flowering in the vine cactus *Hylocereus undatus*. *Israel Journal of Plant Sciences* 60, 371–383.

Kishore, K. (2016) Phenological growth stages of dragon fruit (*Hylocereus undatus*) according to the extended BBCH-scale. *Scientia Horticulturae* 213, 294–302.

Li, J., Shi, H., Dai, H., Wang, Y., Zhao, J. *et al.* (2022) Pollen germination and hand pollination in pitaya (*Selenicereus* spp.). *Emirates Journal of Food and Agriculture* 34, 369–387.

Magalhaes, D.S., da Silva, D.M., Ramos, J.D., Pio, L.A.S., Pasqual, M. *et al.* (2019a) Changes in the physical and physico-chemical characteristics of red-pulp of dragon fruit during its development. *Scientia Horticulturae* 253, 180–186.

Magalhaes, D.S., Ramos, J.D., Pio, L.A.S., Boas, E.V.B.V., Pasqual, M. *et al.* (2019b) Physical and physicochemical modifications of white-fleshed pitaya throughout its development. *Scientia Horticulturae* 243, 537–543.

Martinez-Ruiz, E.R., Tijerina-Chavez, L., Becerril-Roman, A.E., Rebolledo-Martinez, A., Velasco-Cruz, C. and del Angel-Perez, A.L. (2017) Phenology and thermal constant of dragon fruit (*Hylocerus undatus* Haw. Britt. & Rose). *Agroproductividad* 10, 3–8.

Metz, C., Nerd, A. and Mizrahi, Y. (2000) Viability of pollen of two fruit crop cacti of the genus *Hylocereus* is affected by temperature and duration of storage. *HortScience* 35, 199–201.

Mizrahi, Y. (2014) Vine-cacti pitayas: the new crops of the world. *Revista Brasileira de Fruticultura* 36, 124–138.

Mizrahi, Y., Mouyal, J., Nerd, A. and Sitrit Y. (2004). Metaxenia in the vine cacti *Hylocereus polyrhizus* and *Selenicereus* spp. *Annals of Botany* 93, 469–472.

Muniz, J.P., Bomfin, I.G.A., Correa, M.C.M. and Freitas, B.M. (2019) Floral biology, pollination requirements and behavior of floral visitors in two species of pitaya. *Revista Ciencia Agronomica* 50, 640–649.

Nerd, A. and Mizrahi, Y. (1997) Reproductive biology of cactus fruit crops. *Horticultural Reviews* 18, 321–346.

Nguyen, Q.T., Ngo, M.D., Truong, T.H., Nguyen, D.C. and Nguyen, M.C. (2021) Modified compact fluorescent lamps improve light-induced off-season floral stimulation in dragon fruit farming. *Food Science & Nutrition* 9, 2390–2401.

Ortega-Baes, P., Saravia, M., Sühring, S., Godínez-Alvarez, H. and Zamar, M. (2011) Reproductive biology of *Echinopsis terscheckii* (Cactaceae): The role of nocturnal and diurnal pollinators. *Plant Biology* 13, 33–40.

Pereira, A.M., Masiero, S., Nobre, M.S., Costa, M.L., Solis, M.T. *et al.* (2014) Differential expression patterns of arabinogalactan proteins in *Arabidopsis thaliana* reproductive tissues. *Journal of Experimental Botany* 65, 5459–5471.

Phebe, D., Chew, M.K., Suraini, A.A., Lai, O.M. and Abdullah, J.O. (2009) Red-fleshed pitaya (*Hylocereus polyrhizus*) fruit colour and betacyanin content depend on maturity. *International Food Research Journal* 16, 233–242.

Pimienta-Barrios, E. and Castillo, R.F. (2002) Reproductive biology. In: Nobel, P.S. (ed.) *Cacti: Biology and Uses*. University of California Press, Berkeley, California, pp 75–90.

Rangel, I.V.J., Magalhaes, D.S., Rodrigues, F.A., Pasqual, M. and Pio, L.A.S. (2021) Fruit quality and harvest point determination in white-fleshed dragon fruit. *Research, Society, and Development* 10: e11810716287.

Rashid, A.S.M.H., Azam, M.G. and Chowdhury, S.M.K.H. (2021) Influence of day-length enhancement through night-breaking by artificial lighting on off-season dragon fruit production. *Asian Journal of Advances in Agricultural Research* 17, 1–10.

Rodrigues, L.J., Tomas, M.G., de Paula, N.R.F., Pinto, D.M., Boas, E.V. and Costa, C. (2022) Physicochemical and chemical modifications of baby pitaya during its development. *Scientia Horticulturae* 306: 111460.

Sharathkumar, M., Heuvelink, E., Marcelis, L.F.M. and Leperen, W. (2021) Floral induction in the short-day plant chrysanthemum under blue and red extended long-days. *Frontiers in Plant Science* 11: 610041.

Sheng, C.C., Lu, J.Y., Ya, C.L., Cha, L.T. and Tsang, C.P. (2019) The effect of forchlorfenuron (CPPU) application at different flowering stage on fruit quality of red-fleshed pitaya (*Hylocereus polyrihizus*). *Journal of Taiwan Society for Horticultural Science* 65, 183–194.

Singh, A., Swami, S., Panwar, N.R., Kumar, M., Shukla, A.K. *et al.* (2022) Development changes in the physicochemical composition and mineral profile of red-fleshed dragon fruit grown under semi-arid conditions. *Agronomy* 12: 355.

Stephenson, A.G., Travers, S.E., Mena-Ali, J.I. and Winsor, J.A. (2003) Pollen performance before and during the autotrophic–heterotrophic transition of pollen tube growth. *Philosophical Transactions of the Royal Society B: Biological Sciences* 358, 1009–1018.

Takata, W.H.S., Miyake, R.T.M., Narita N. and Ono, O.E. (2016) Effects of season and GA_3 concentrations on *Hylocereus undatus* flowering and production. *Journal of Agronomy* 15, 179–183.

Then, K.H., Faiz, O.M. and Norshafiqah, K. (2020) The flowering pattern and fruit production of red pitaya (*Hylocereus polyrhizus*) under Malaysian growing condition. *International Journal of Agriculture Innovations and Research* 8, 2319–1473.

Tran, D.H., Yen, C.R. and Chen, Y.H. (2015a) Flowering response of a red pitaya germplasm collection to lighting addition. *International Journal of Agricultural and Biosystems Engineering* 9, 175–179.

Tran, D.H., Yen, C.R. and Chen, Y.K.H. (2015b) Effect of pollination method and pollen source on fruit set and growth of red-peel pitaya (*Hylocereus* spp.) in Taiwan. *Journal of Horticultural Science and Biotechnology* 90, 254–258.

Trong, L.V., Thuy, L.T., Chinh, H.V. and Thinh, B.B. (2022) Physiological and biochemical changes of red-fleshed dragon fruit (*Hylocereus polyrhizus*) during development and maturation. *Journal of Food and Nutrition Research* 61, 139–145.

Valverde, P.L., Jiménez-Sierra, C., López-Ortega, G., Zavala-Hurtado, J.A., Rivas-Arancibia, S. *et al.* (2015) Floral morphometry, anthesis, and pollination success of *Mammillaria pectinifera* (Cactaceae), a rare and threatened endemic species of Central Mexico. *Journal of Arid Environments* 116, 29–32.

Walter, H.E. (2010) Floral biology of *Echinopsis chiloensis* ssp. *chiloensis* (Cactaceae): evidence for a mixed pollination syndrome. *Flora* 205, 757–763.

Weiss, J., Nerd, A. and Mizrahi, Y. (1994) Flowering behavior and pollination requirements in climbing cacti with fruit crop potential. *HortScience* 29, 1487–1492.

Williams, J.H. (2012) Pollen tube growth rates and the diversification of flowering plant reproductive cycles. *International Journal of Plant Sciences* 173, 649–661.

Zheng, Y.Y., Lin, X.J., Liang, H.M., Wang, F.F. and Chen, L.Y. (2018) The long journey of pollen tube in the pistil. *International Journal of Molecular Science* 19: 3529.

10 Pests and Nematodes

Haibo Long[1]*, Liangde Tang[2] and Huadong Li[3]
[1]*Environment and Plant Protection Institute, Chinese Academy of Tropical Agricultural Sciences, Haikou, China;* [2]*National Key Laboratory of Green Pesticide, Guizhou University, Guiyang, China;* [3]*Sinochem (Hainan) Agricultural Ecology Limited Company, Haikou, China*

10.1 Introduction

Dragon fruit or pitaya is an emerging tropical fruit grown in a number of countries. In the last 10 years, the area of dragon fruit under cultivation has increased markedly in Asia. The plants are subject to attack by various insects, but only a few insect species populations have increased to levels high enough to be considered pests. Major pests may require immediate attention and control. Minor pests should be monitored but will normally remain under biological control by beneficial insects and good cultivation practices (Carrillo *et al.*, 2021). This chapter focuses on the major pests and nematodes occurring on dragon fruit.

10.2 Thrips

10.2.1 Damage

Thrip is the general term for members of the order Thysanoptera. Thrips are tiny sucking insects, which prefer to inhabit and damage flowers and young tissues. They frequently infest dragon fruit and are regarded as one of the most important insect pests of this plant. There are usually dozens if not hundreds of thrips gathered on tender branches and fruit, and their feeding induces the formation of chlorosis and scars on infested plant parts. Heavy infestation often results in a considerable reduction in fruit quality (Figs 10.1–10.3).

10.2.2 Species

The flower thrips, *Frankliniella intonsa* (Trybom), are the most common thrips species occurring on dragon fruit in China (Fig. 10.4). Both sexes are fully winged. The body and legs are variable in colour, mainly brown with the head and pronotum often paler than the abdomen. The tibiae and tarsi are largely yellow, the major setae are dark, the antennal segments II–IV are yellow with shaded apices, and the forewings are pale with dark setae. The antennae have eight segments; segments III and IV have a forked sense cone, and segment VIII is as long as segment VII. The head is wider than it is long, and three pairs of ocellar setae are present, with pair III slightly longer than the side of the ocellar triangle, arising on the anterior margins of the triangle; postocular setae pair I are present, and pair IV are short, no longer than the diameter of the hind ocellus. The pronotum is marked with five pairs of major setae; the anteromarginal setae are shorter than the anteroangulars, and one pair of minor setae is present medially between the posteromarginal

* Email: hb_long@yeah.net

DOI: 10.1079/9781800623156.0010

Fig. 10.1. Damage caused by thrips to tender branches of dragon fruit.

Fig. 10.2. Damage caused by thrips to flower buds of dragon fruit.

submedian setae; the upper surface of the fore coxae bears microtrichia on sculpture lines. The metanotum has two pairs of setae at the anterior margin, and the campaniform sensilla is absent. The fore wing bears two complete rows of veinal setae. Abdominal tergites V–VIII bear paired ctenidia on tergite VIII anterolateral to the spiracle; a posteromarginal comb is present on tergite VIII and complete, but microtrichium is tiny, slender and arises from the triangular bases. Sternites III–VII are devoid of discal setae.

The male is usually smaller and paler than the female, and tergite VIII has no posteromarginal comb; tergite IX has posterolateral setae and is stout in larger males. Each of the sternites III–VII has a transverse pore plate.

This species is closely similar to *Frankliniella occidentalis*, but the postocular setae pair IV is considerably shorter, and only very rarely has metanotal campaniform sensilla.

10.2.3 Distribution

This species is widely distributed across Europe, from Scandinavia to southern regions, as well as from

Fig. 10.3. Unmarketable dragon fruit due to scarring from thrips.

Fig. 10.4. Adult male (left) and female (right) flower thrips (*Frankliniella intonsa* (Trybom)).

Iran to northern India, southern China and various other Asian countries. It can also be found in Oceania and North America (Ullah and Lim, 2015).

10.2.4 Biology

Thrips have a short generation time and a high reproductive potential, allowing them to increase their population size rapidly. These characteristics make thrips a typical 'r-strategist' insect. *F. intonsa* is a highly polyphagous insect (Wang *et al.*, 2010; Lim *et al.*, 2013) and can survive wide environmental variations in temperature. Its reproduction may be amphimictic or parthenogenic. All offspring that develop by parthenogenesis are males. They are paurometabolous insects, meaning that their life cycle contains egg, nymph, pseudo-pupa and adult stages. The eggs are inserted into plant tissues. Post-embryonic development involves two nymphal instars and two pupal (pre-pupa and pupa) stages that pupate either in the soil or in the leaf litter.

10.2.5 Management

Both the nymphs and adults of *Frankliniella* spp. hide in a relatively closed space, such as buds, flowers and axils, and thus it is difficult to reach their habitat when spraying with insecticides. A good thrips management programme relies on a combination of methods such as agronomic measures, physicochemical lure control, biological control and chemical control.

Agronomic measures

Cultural strategies include intercropping, trap cropping, soil improvement and weed control. *F. intonsa* is a highly polyphagous insect and has a particular preference for flowers, so a reduction in thrips density can be achieved by intercropping with flowering plants. Soil improvement by supplementation with phosphate fertilizer, which allows the plants to obtain enough nutrients, can have a compensatory effect on pest damage. As a part of agronomical measures, the use of row covers or reflective mulch should be explored in thrips management.

Physicochemical lure control

Vision (colour) and olfaction (odour) are important in the development of insect attractants/repellents for monitoring and/or controlling insect pests. Sticky card trap colour tests show that *F. intonsa* is more sensitive to blue and white traps in different legume crop fields.

Pheromone trapping is species specific, active at very low concentrations and safe for the environment (Sampson and Kirk, 2013). Aggregation pheromones can attract both sexes and have more potential for trapping. The aggregation pheromones of *F. intonsa* have been identified (Zhang *et al.*, 2011).

Biological control

The effect of biological control agents on controlling thrips is considered to be limited due to their high intrinsic population growth, high reproductive rate and fast development time. Recent evidence has demonstrated the potential of several generalist predators in regulating thrips populations. Predatory mites and *Orius* spp. are the most important groups of predators effective against sucking pests such as thrips (Wu *et al.*, 2017).

Beyond natural enemies, entomopathogenic fungi are another important biological control agent of thrips species. The entomopathogenic fungi used for thrips control are mainly *Beauveria bassiana*, *Metarhizium anisopliae*, *Isaria fumosorosea* and *Lecanicillium lecanii* (formerly *Verticillium*) (Wu *et al.*, 2018).

Chemical control

Insecticides used to control thrips widely belong to the organophosphorus, pyrethroid, neonicotinoid and spinosyn families. However, most organophosphorus insecticides against vegetable pests have been phased out in China. The adult and larval thrips are best controlled by the application of spinetoram (Tang *et al.*, 2022), while spirotetramat is the most effective insecticide against eggs.

Some strategies have been used to improve insecticide efficacy, such as the synergism of adjuvants mixed with insecticides for the management of thrips (Tang *et al.*, 2022; Wang *et al.*, 2022).

10.3 Mealybugs

10.3.1 Damage

Mealybugs (Hemiptera: Pseudococcidae) are small, phloem-sucking insects and can cause considerable economic damage to a variety of fruit crops, including citrus, guava, grape, mango and dragon fruit (Sartiami *et al.*, 2020; Subramanian *et al.*, 2021). Damage caused by mealybugs in dragon fruit include a dry skin surface, fruit shrinkage, a wrinkled skin and honeydew, which causes sooty moulds (Mani and Shivaraju, 2014; Nurhafizah *et al.*, 2020). They can harm the flowers, stems and fruits of the dragon fruit. They often gather in the cracks between the scales of the fruit (Fig. 10.5). The mealybug damage is related to their consumption of sap. They use a stylet to extract sap from the plant, which decreases plant vigour. They expel wax and sticky honeydew, which lowers the quality of the plant and fruit, particularly when black sooty mould develops on the honeydew. The fruit loses marketability due to the ability of the sooty mould to prevent photosynthesis and the difficulty in removing it (Fig. 10.6). At times, mealybugs prodice toxins and act as vectors of certain viruses detrimental to plant life (Tsai *et al.*, 2010). Mealybugs thrive in hot, humid weather, and damage is common during the dry season.

10.3.2 Species

Striped mealybugs (*Ferrisia virgata* (Cockerell)) frequently infest the flowers and fruits of dragon fruit in China. *F. virgata* is identified by the presence of two dark, dorsal stripes that run longitudinally down its body (Fig. 10.7). These stripes are visible on the mealybug cuticle through bare patches in the waxy covering. The length of the adult female body is approximately 2.0–4.5 mm. This mealybug also has slender, clear, hair-like secretions protruding through the fluff, as well as two noticeable, white, tail filaments. Males have two wings, long antennae, well-developed legs and a pair of long, white filaments at the rear (Kaydan and Gullan, 2012). Mealybug species commonly found on dragon fruit in Indonesia are *F. virgata* (Cockerell), *Phenacoccus solenopsis* Tinsley, *Planococcus minor* (Maskell), and *Pseudococcus jackbeardsteyi* Gimpel & Miller (Rezeki *et al.*, 2021). In Egypt, Youssef *et al.* (2021) identified four pseudococcid species on dragon fruit, *Phenacoccus solenopsis* Tinsley, *F. virgata* (Cockerell), *Planococcus citri* (Risso) and *Maconellicoccus hirsutus* (Green).

10.3.3 Distribution

F. virgata is a small, highly polyphagous insect. It is of neotropical origin and spread around the world in only about 10 years after first being described in Jamaica. Its polyphagous nature has

Fig. 10.5. Mealybugs in the cracks between the scales of dragon fruit.

Fig. 10.6. Mealybugs on fruit, which decreases their value.

Fig. 10.7. An adult mealybug (*Ferrisia virgata* (Cockerell)). Photograph courtesy of Han Dongyin.

facilitated its spread by human transport of infested plants. Nowadays, it can be found in tropical and temperate climates across the world (Kaydan and Gullan, 2012). Its small size and cryptic habits make it difficult to detect and identify in plant quarantine inspection laboratories. The increase in international trade of fresh plant material in recent years is likely to facilitate its continued spread (CABI Compendium, 2016).

10.3.4 Biology

The striped mealybug is sexually reproductive, with each female mating only once. The longevity of the female is about 50 days and the fecundity ranges from 300 to 700 eggs per female. The eggs are deposited in groups, rarely singly, and are usually concealed under the body. As the eggs are laid in an advanced stage of embryonic development, they hatch soon after oviposition and are therefore seldom seen. The egg period lasts about 28 min. Female and male nymphs moult three and four times, respectively, and their development times range are 26–47 and 31–57 days, respectively. The complete life cycle takes about 6–7 weeks at 30–35°C and 65% relative humidity, and 13–14 weeks at 16.6°C (CABI Compendium, 2016). They complete three to five generations a year depending on the availability of the host plant and favourable climatic conditions.

10.3.5 Management

Cultural control

Striped mealybugs can spread rapidly and therefore any infested plant part should be removed (pruned) and burnt as soon as an outbreak is detected. Attention must be paid to parts such as young shoots, flowers and fruit where the bugs often hatch. The area around the dragon fruit tree should be cleaned on a regular basis by collecting and disposing of leftover fruit to help with ventilation of the area, creating favourable conditions for plants to grow and develop, and limiting the habitat of the mealybugs. Additionally, balanced fertilization and watering are required to limit the mealybug growth. It is recommended that a water pump with a strong spray is used to water dragon fruit plants to wash the places where the mealybugs may be attached, and to create moisture on the tree and reduce the density of mealybugs.

Biological control

A commercial biocontrol agent that can be released in greenhouses is the mealybug destroyer (*Cryptolaemus montrouzieri*) (McCorquodale and Hodges, 2016; Carrillo, *et al.*, 2021). In Egypt, three species of Coccinellidae, two species of Neuroptera and three parasitoid wasp species (*Leptomastix* sp. and two species of *Tetrastichus*) were reported as attacking *F. virgata* (Ammar *et al.*, 1979). As mentioned earlier, Youssef *et al.* (2021) identified four mealybug species in Egypt affecting dragon fruit. They also identified natural enemies for the species: *Aenasius arizonensis* (Giraut), family Encyrtidae, and *Chartocerus dactylopii* (Ashmead), family Signiphoridae, on *Phenacoccus solenopsis* Tinsley; *Coccidoxenoides peregrinus* (Timberlake) and *Leptomastidea abnormis* (Giraut), family Encyrtidae, on *Planococcus citri* (Risso); *Anagyrus kamali* Moursi and *Leptomastix dactylopii* Howard, family Encyrtidae, on *Maconellicoccus hirsutus* (Green); and *Blepyrus insularis* (Cameron), family Encyrtidae, on *F. virgata* (Cockerell). In India, a coccinellid predator (*Scymnus* sp.) has been reported attacking *F. virgata*, controlling a population kept in culture (Ghose and Paul, 1972). Additionally, it is necessary to protect the natural enemies of mealybugs such as weaver ants, ladybugs, yellow-eyed beetles, red bugs and parasitic wasps. Weaver ants are the most effective natural enemy, as they prey on larvae and adult mealybugs, and can repel and prevent adult bugs from harming the plant and laying eggs. Another consideration when using biological control agents for mealybug management is the presence of ants. Honeydew is very attractive to ants as a source of food. Some species of ant will defend mealybugs from predation and parasitism to protect their access to the honeydew. If it is suspected that ants are contributing to a mealybug infestation, it is recommended that insecticide baits, traps or granules are used combined with barriers such as sticky tape to prevent ants from hindering predation or other mealybug control efforts (McCorquodale and Hodges, 2016).

Chemical control

Spraying insecticides works best at the stage when the dragon fruit tree is flowering or bearing young and developing fruits. The younger stages of the insects should be targeted when using contact insecticides. This is because mealybugs in their first and second instars are not shielded by the thick waxy covering that forms in the later instars. Spraying of 20% acetamiprid soluble powder or 10% imidacloprid suspension concentrate at 15-day intervals effectively controls mealybug infestation. Effectiveness has also been demonstrated for systemic insecticides applied to the soil and absorbed by the roots of the plant. However, due to the overlapping generations of mealybugs, repeated, frequent applications are often required, which could speed up the emergence of insecticide resistance. Regular pesticide usage might also harm beneficial insects.

10.4 Aphids

10.4.1 Damage

Aphids (Hemiptera: Aphididae) are small, soft-bodied, pear-shaped insects with specialized mouths to suck the sap from tender stems, flowers and fruit. Nymphs and adults of aphids congregate in the dragon fruit flowers and scales, resulting in drying of the flowers and production of shrivelled fruit (Fig. 10.8). Aphids also produce honeydew, which leads to the development of sooty mould on flower and fruit surfaces (Fig. 10.9). The sooty mould is difficult to clean and diminishes the market value of the fruit (Carrillo *et al.*, 2021). Honeydew can also attract other insects, such as ants, that will feed on the honeydew (Fig. 10.10). Ants can aggressively defend aphids from predators and parasites, and transport them to new sites where these aphids start new colonies. More aphids survive with the help of their bodyguard ants and further damage the dragon fruit. Some ants are also attracted to the sweet fruit and may damage it (Fig. 10.11) (Carrillo *et al.*, 2021).

10.4.2 Species

The cotton or melon aphid (*Aphis gossypii* (Glover)) has been found feeding on the flowers and fruit of dragon fruit plants in Hainan, China. Like other aphids, both winged (alate) and wingless (apterous) forms exist for *A. gossypii*. The wingless parthenogenic females are 1–2 mm in length, and blackish green or dark green mottled with lighter green. In hot conditions or when crowded, they are smaller, and these dwarf forms are a very pale whitish-yellow. The legs are pale with the tips of the tibiae and tarsi black. The cornicles also are black. Winged parthenogenic females measure 1.1–1.7 mm in length.

Fig. 10.8. Flowers and young fruit infested by aphids.

Fig. 10.9. Young fruit heavily infested with aphids and covered in black sooty mould.

Fig. 10.10. Aphids produce honeydew, which attracts ants, causing damage to the flowers.

Fig. 10.11. Ant damage to sweet fruits.

The head and thorax are black, and the abdomen yellowish-green except for the tip, which is darker. The wing veins are brown. When deposited, the eggs are yellow but quickly turn shiny black. The nymphs are smaller versions of the non-winged adults and can be tan, grey or green in colour (Capinera, 2021).

10.4.3 Distribution

A. gossypii is highly polyphagous and occurs in tropical and temperate regions throughout the world, except for the northern-most areas. In temperate climates, it is found in glasshouse cultivation.

10.4.4 Biology

In tropical regions, *A. gossypii* reproduces asexually (parthenogenically), giving birth to live young without mating and egg laying. As a consequence of this type of reproduction, populations are composed solely of females and there are no males present. The optimal temperature for reproduction is reported to be about 21–27°C. Viviparous (giving birth to live young) females produce a total of about 70–80 offspring at a rate of 4.3 day^{-1}. Thus, it is possible for *A. gossypii* to have as many as 50 generations a year. These generations also occur as frequently as every 5–7 days under optimum conditions (Ebert and Cartwright, 1997; Saha *et al.*, 2016).

10.4.5 Management

Cultural control

Timely pruning of infested branches, flowers and fruits, and burning to reduce the insect population should be carried out. Aphids prefer the colour yellow, so yellow sticky cards can be hung in orchards to attract adult insects. Reflective mulches also affect aphid populations in open-field conditions. Light-coloured mulches (e.g. silver, white) reflect short-wave light and repel flying aphids. Fewer aphids land on crops growing in fields with reflective mulch (compared with bare soil).

Biological control

There are several naturally occurring beneficial insects that feed on aphids, such as lady beetles, lacewings, hoverflies and parasitic wasps. Natural enemies of aphids can be artificially reared and released in the orchard, but these must be protected from foliar applications of broad-spectrum insecticides. Growers should try to use fewer broad-spectrum pesticides and avoid applying pesticides during peak natural enemy activity (Carrillo *et al.*, 2021).

Chemical control

A chemical insecticide may be needed if large numbers of aphids are found. Acetamiprid, imidacloprid, spiromesifen and thiamethoxam are useful in controlling aphids (Cloyd and Bethke, 2011). To protect natural enemies, indiscriminate spraying of insecticides should be avoided.

10.5 Snails

10.5.1 Damage

Snails are one of the most destructive pests commonly found in dragon fruit orchards. They consume young branches, flowers and fruit, creating irregular holes with smooth edges on stems or fruit by scraping with their rasp-like tongues (Fig. 10.12). Small, succulent plant parts are easily clipped by snail feeding. As a result, both the yield and quality of dragon fruit are reduced. Snail damage can be confused with feeding by other chewing insects. The appearance of silvery mucous trails confirms that snails have caused the damage, rather than other pests.

10.5.2 Species

The Asian tramp snail (*Bradybaena similaris* (Férussac)) is the most common snail species occurring on dragon fruit in Hainan, China. It is moderately sized, measuring about 12–16 mm in diameter at maturity and 9–11 mm in height. The shell has five to five-and-a-half whorls. The umbilicus is pronounced when viewed from below. The colour is variable, often brownish, yellowish or tan, usually with a narrow brown stripe on the perimeter of the whorl. The mouth of the shell is slightly flared, and has a white edge (Capinera and White, 2011).

10.5.3 Distribution

B. similaris occurs in all regions where dragon fruit is cultivated. It is originated in eastern Asia and is now spread throughout the tropics and subtropics.

10.5.4 Biology

All land snails are hermaphrodites, so are able to lay eggs after mating with another individual.

Fig. 10.12. Snail damage to stems and young dragon fruit.

Snails reach maturity after about 3–6 months, depending on the species. It takes about 6 months for *B. similaris* to reach sexual maturity and begin producing eggs. Mating occurs most often at dusk, at night or on rainy days. A study conducted in Brazil found that snails produced, on average, 30 eggs over the length of their life, but egg production was highly variable, and up to 115 eggs could be found in a single clutch (Carvalho *et al.* 2008). *B. similaris* may live more than a year.

Snails prefer a dark, humid environment, and are most active on rainy days or after watering; they can engage in day and night activities, including feeding. Most snails begin to move and feed after 18:00, peaking between 20:00 and 23:00. After midnight, their food intake gradually decreases, and in the morning, they stop feeding and go into the soil or seek shelter. On sunny days, they seek hiding places out of the heat and bright light. Often, the only clues to their presence are their silvery trails and plant damage. Snails cause the most damage from April to May and from September to October, and the damage is more severe in the autumn than in the spring (Capinera and White, 2011).

10.5.5 Management

Snails have shells in which they hide in most of the time, and they do most of their damage on rainy days when it is difficult to reach them by spraying with pesticides. A good snail management programme relies on a combination of methods, include cultural practices, hand-picking, traps and chemical baits (Sagers and Rood, 2011).

Cultural practices

Snails favour high humidity. Therefore, elimination of mulch, ground cover, wood and stones will deny them a moist, sheltering environment. Dense ground cover and turf are ideal hiding places. Weeds and unnecessary foliage should be removed so that the soil surface dries out rapidly. Switching from sprinkler irrigation to drip irrigation will reduce humidity and moist surfaces, making the habitat less favourable for these pests. Irrigating near sunrise will reduce the amount of time that the foliage and ground are moist.

Hand-picking

Hand-picking is a time-consuming but effective method of controlling snails when done regularly. When suffering heavy damage, growers should look for snails daily. Special attention should be paid to places where they like to hide. After the population has noticeably declined, weekly hand-picking can be sufficient.

Traps

A common snail trap is a beer pan. Simply fill a shallow pan with beer and leave it out overnight. The snails will be attracted to the beer and will drown in it. The beer will need to be replaced every few days to remain effective.

Chemical control

Several types of snail and slug bait products (molluscicides) are available. They may kill by ingestion of the bait or by contact. Snail and slug baits can be effective when used properly and in conjunction with a cultural programme that incorporates the other methods discussed above. Baits alone will not effectively control snails or slugs in the long term. Baits are also toxic to all snails and slugs, including the predatory decollate snail and native species.

Metaldehyde has been the standard bait used for many years and is most effective in dry, warm weather. It does not kill pests directly but paralyses them and causes them to froth and lose large amounts of water. During warm weather or on dry days, the pests die of desiccation. It is less effective in rainy weather or in cool areas because the pests recover a few days after eating the bait. Baits containing the active ingredient metaldehyde are very toxic to cats and dogs. Some metaldehyde products are formulated with carbaryl to increase the spectrum of pests controlled. Carbaryl is toxic to pets, wildlife and beneficial insects, including honey bees.

Iron phosphate ($FePO_4$) is a newer, registered molluscicide that kills slugs and snails. Iron phosphate is normally thought of as a fertilizer, but it shows control activity equal to or better than most other slug and snail baits. It is usually safe for children, pets, birds and fish. Iron phosphate baits are more effective than metaldehyde baits under moist conditions. After ingesting iron phosphate, the slugs and snails usually hide before they die, so the empty shells or carcasses are usually not visible.

10.6 Green Weevils

10.6.1 Damage

The green weevil (*Hypomeces squamosus* (Fabricius)), also known as the gold dust beetle, is a common polyphagous pest that damages a variety of agricultural crops and forest trees, including cotton, tea, rubber, sugarcane, mango and citrus (CABI Compendium, 2022b). The larvae live and pupate in the soil, feeding on living plant roots. The adult weevils are long-lived, doing damage to foliage and sometimes defoliating young bushes and nursery trees. When they feed on dragon fruit plants, the adult weevil makes cuts or holes on the young stems (Fig. 10.13). The feeding pattern is usually from the stem edge inwards, forming a semicircle.

10.6.2 Species

H. squamosus belongs to the true weevil family Curculionidae. The adult weevil measures about 14–18 mm in length and is covered with green or yellow metallic scales, giving the beetle a dust-like appearance. The head has a broad, long snout, clubbed antennae, large black eyes and chewing mouth parts. The larvae are legless grubs that live in the soil (Hill, 2008).

10.6.3 Distribution

H. squamosus is native to South-east Asia, its range extending from Pakistan and India to Japan, the Philippines and Indonesia. It is also widely distributed in southern tropical and subtropical regions of China (CABI Compendium, 2022b).

Fig. 10.13. Damage to young dragon fruit branches caused by green weevils (*Hypomeces squamosus*).

10.6.4 Biology

Adult green weevils have limited flying ability but excel at crawling. They are active and feed during the day, particularly from late afternoon to dusk on the leaves or stem surface, and frequently hide on the back of the leaves at other times. When adults are disturbed, they pull in their legs, fall to the ground and feign death. Adults can mate multiple times; the majority of the eggs are scattered on the leaves, and the oviposition period lasts over 80 days, with each female laying more than 80 eggs. After hatching, the larvae burrow into the soil at a depth of 10–13 cm to feed on weeds or roots (Khoo *et al.*, 1991).

10.6.5 Management

Green weevils can be hand-picked, or sprayed with an organic insecticidal spray containing neem as an antifeedant or chemicals such as carbaryl, cypermethrin or malathion, which act as a contact stomach poison. Cultural practices such as tilling the soil to destroy weevil eggs and larvae, applying fertilizer and providing adequate space for root growth will ensure healthy tree growth.

10.7 Cotton Bollworms

10.7.1 Damage

The cotton bollworm moth (*Helicoverpa armigera* (Hubner)) (Lepidoptera: Noctuidae), has been found damaging young stems and flower buds of dragon fruit in Hainan, China (Fig. 10.14). *H. armigera* is highly polyphagous and is known to attack a large number of crops, such as cotton, maize, fruit and vegetable crops. Damage occurs when the larvae bore into the host's flowers and fruits and feed within the plant; the larvae may also feed on the leaves or stems of the host plant (Lammers and MacLeod, 2007). Although this moth has not been found in large numbers on dragon fruit in Hainan, it has the potential to become a significant pest.

10.7.2 Distribution

H. armigera is cosmopolitan in its distribution (Haile *et al.*, 2021). The pest status of *H. armigera* is due in part to the broad host range of its larvae, its feeding preference for reproductive stages of plants, its high fecundity, its high

Fig. 10.14. Damage to young branches and dragon fruit caused by cotton bollworms (*Helicoverpa armigera*).

mobility, and its ability to enter facultative diapause and thus adapt to different climates (Cleary *et al.*, 2006).

10.7.3 Biology

H. armigera takes about 4–6 weeks to develop from an egg to an adult in tropical Hainan. The adult is stout bodied, brown or yellowish-brown, with a V-shaped speck on light brown forewings and a dark border on the white hind wings. It lays about 500 spherical, yellowish-white eggs singly on tender parts of the plant. The eggs hatch in 4–5 days. The larval stage, comprising six larval instars, is completed in 12–15 days. The larvae initially feed on tender stems and later bore into developing spikes by making holes and feeding on buds and flowers. Pupation takes place in the soil, and the pupal period lasts for 12–15 days. Adults emerge from the ground, mate and disperse to lay eggs. The moths consume liquids and nectar as food but are not damaging to plants (Lammers and MacLeod, 2007; Haile *et al.*, 2021; CABI Compendium, 2022a).

10.7.4 Management

Cultural control

Cultural control relies on timely removal of weeds from the orchard. Collecting and killing of fully grown larvae reduces further damage and decreases the population. At the peak of pupation, ploughing of the soil combined with irrigation can drown most of the pupae.

Physical control

Phototaxis of adults can be used to trap and kill them using black light at the peak stage. When the moth population begins to increase, a mixture of sweet and sour wine can also be used to induce killing. The sugar:vinegar:wine:water ratio for the mixture is 3:4:1:2. After mixing, a small amount of insecticide is also added.

Biological control

Various predacious invertebrates and parasitic wasps occur naturally, and successful control of *H. armigera* has been achieved by the release of egg-parasitic wasps in the genus *Trichogramma* and by *H. armigera* nuclear polyhedrosis virus in China (Luo *et al.*, 2014). Entomopathogenic fungi such as *Beauveria* spp., *Metarhizium* spp. and *Nomuraea* spp. have also been used to control *H. armigera* in Europe, Africa and Asia (Tang, 2003; Nguyen *et al.*, 2007).

Chemical control

Emamectin benzoate and pyrethroid insecticides have been used to control larvae during the early stages of development. Emamectin benzoate, indocarb, chloranobenzamide and other agents can be used to treat large larvae above the third instar.

10.8 Oriental Fruit Flies

The oriental fruit fly (*Bactrocera dorsalis* (Hendel)) (Insecta: Diptera: Tephiritidae) is a destructive pest of fruit in areas where it occurs. It is native to large parts of tropical Asia and has become established over much of sub-Saharan Africa. The adult is larger than a house fly (about 8.0 mm in length; the wing is ~7.3 mm). It is the most important and destructive pest of dragon fruit in Vietnam. It can cause significant losses in the field and is a quarantine pest for many importing countries, requiring vapour heat or irradiation treatment of fruit before export. A second species, guava fruit fly (*Bactrocera correcta*) can infest dragon fruit but is of relatively minor importance compared with *B. dorsalis* (Hien *et al.*, 2021). It is also reported from many dragon fruit-growing countries. The close proximity of dragon fruit orchards to other fruit crops provides year-long populations of fruit.

By using a combination of insecticide-impregnated fibre blocks or cotton containing the powerful male attractant methyl-eugenol and the sterile insect (sterile male) technique, Japan eradicated the oriental fruit fly in 1985 (Weems *et al.*, 2019). An area-wide integrated pest management strategy for fruit flies by reducing the number of males (using the male annihilation technique (MAT)) in the area was suggested by Hien *et al.* (2021) A combination of MAT, protein baits for killing females and orchard hygiene (removing and burning of all dropped fruit) successfully minimized the population (Hien *et al.*, 2021).

10.9 Stink Bugs

Stink bugs are another pest affecting dragon fruit. The bugs suck sap from the fruit at all stages of fruit development, causing lumps on the fruit, which become unmarketable. Six stink bug species have been reported to damage dragon fruit: *Amblypelta* sp.; *Mictis longicornis*, *Physomerus grossipes*, *Celtus* sp., *Megymenum* sp. and *Nezara viridula*. Hien *et al.* (2021) suggested bagging of fruit to limit access of the bugs. Spraying in the late afternoon with abamectin or flubendiamide reduces populations of bugs.

10.10 Minor Pests

There are two pests affecting dragon fruit that are currently considered minor pests. These are: (i) the beet armyworm (*Spodoptera exigua* (Hubner)) (Lepidoptera: Noctuidae), which damages flowers; and (ii) scale insect (*Philephedra tuberculosa* Nakahara & Gill) (Hemiptera: Coccidae), which feeds on plant sap and produces honeydew.

10.11 Root-knot Nematodes

10.11.1 Damage

Root-knot nematodes (*Meloidogyne* spp.) are global pests of more than 3000 plant species and cause an average 10% yield loss in crops. Symptoms associated with root-knot nematodes on dragon fruit are easily diagnosed by the presence of distinctive galls on the root systems (Fig. 10.15). The formation of these galls hinders the water- and nutrient-conducting abilities of the roots. Other symptoms include chlorotic yellowish stems, stunted unhealthy growth, root decay and poor fruit yield (Fig. 10.16). Root-knot nematodes can also form disease complexes with other soil-borne pathogens such as *Fusarium* or *Pythium* spp. and further hinder plant growth and increase yield loss.

10.11.2 Species

Two root-knot nematode species, *Meloidogyne enterolobii* and *M. javanica* (Tylenchida: Heteroderidae),

Fig. 10.15. Severe root gall and root decay due to infestation of root-knot nematodes.

Fig. 10.16. Above-ground symptoms of root-knot nematode infestation, showing yellow stems. Right panel from Long *et al.*, 2022, CC BY.

have been reported from dragon fruit in Hainan, China (Long *et al.*, 2022). *M. enterolobii* is considered one of the most damaging species, due to its wide host range, pathogenicity, and ability to develop and reproduce on several crops carrying resistance genes (Castagnone-Sereno, 2012). Adult females of *M. enterolobii* are pearly-white and pear-shaped to globular without a posterior protuberance (Fig. 10.17D). The perineal pattern is usually oval-shaped with distinct phasmids and tail tip, a squared or rounded dorsal arch, smooth and coarse striae, and mostly indistinct lateral lines (Fig. 10.17E). The second-stage juvenile has large, triangular lateral lips and a broad, bluntly rounded tail tip (Fig. 10.17A–C).

10.11.3 Distribution

M. enterolobii belongs to the group of tropical root-knot nematodes and was originally described from a population that caused severe damage to the pacara earpod tree on Hainan Island in China (Yang and Eisenback, 1983). Following the initial description, the distribution of *M. enterolobii*

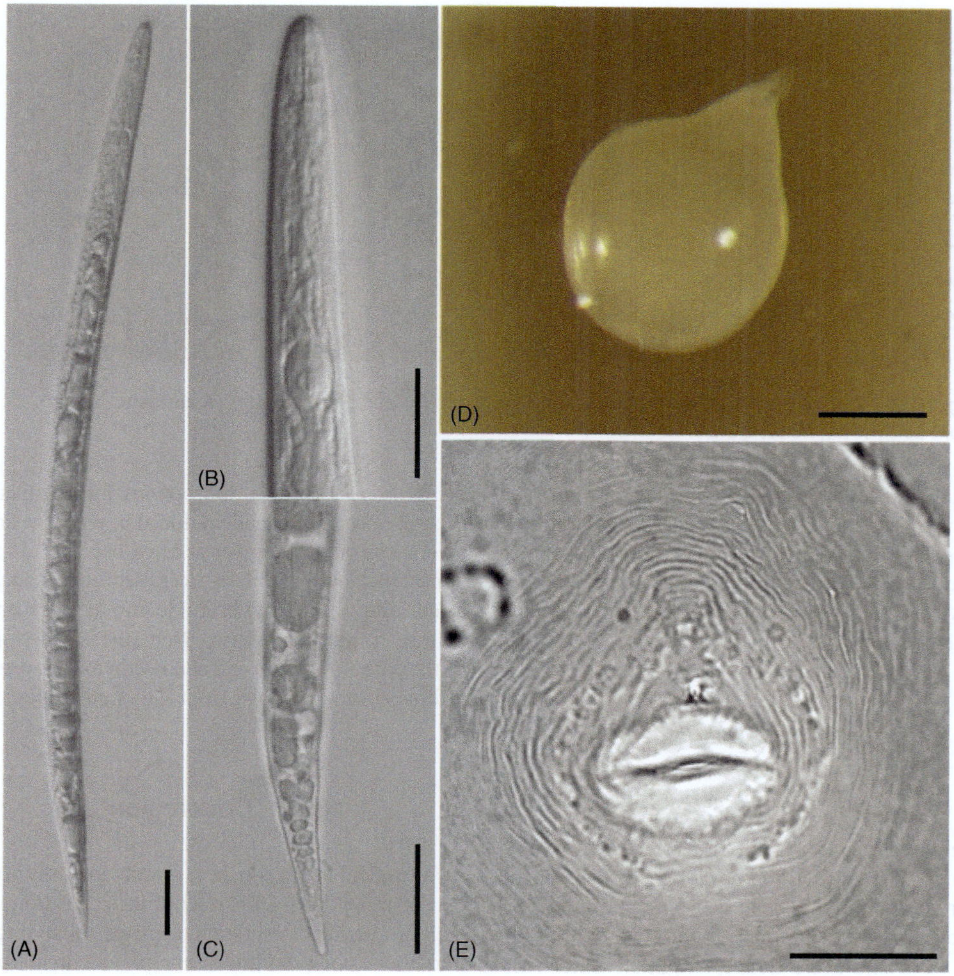

Fig. 10.17. Morphology of *Meloidogyne enterolobii.* (A–C) Second-stage juveniles. Scale bar, 20 µm. (D) Adult female. Scale bar, 100 µm. (E) Perineal pattern of the female. Scale bar, 20 µm. Panel (E) from Long *et al.*, 2022.

expanded into other regions of the world. Due to its thermal requirements, *M. enterolobii* is largely distributed in tropical climates in Africa, Asia, Europe, Central and South America, and the Caribbean (Castagnone-Sereno, 2012).

10.11.4 Biology

Root-knot nematodes are obligate sedentary endoparasites that require nutrients from host plants to complete their life cycle, which consists of eggs, four juvenile stages, and the adult male or female (Karssen and Moens, 2006). The adult females lay eggs in masses in a gelatinous matrix on the surface of the root, containing up to 1000 eggs. Development of the first-stage juvenile occurs within the egg where the first moult takes place and the second-stage juvenile forms. The second-stage juvenile comes out of the egg to find and invade the plant root just behind the growing tip (the zone of elongation) and migrate into the vascular system. Second-stage juveniles will then become sedentary at their chosen feeding site and induce the formation of four to seven giant cells to provide nutrients for growth and reproduction of the nematode (Fig. 10.18). After three moults at the feeding site, they develop into adult females or males.

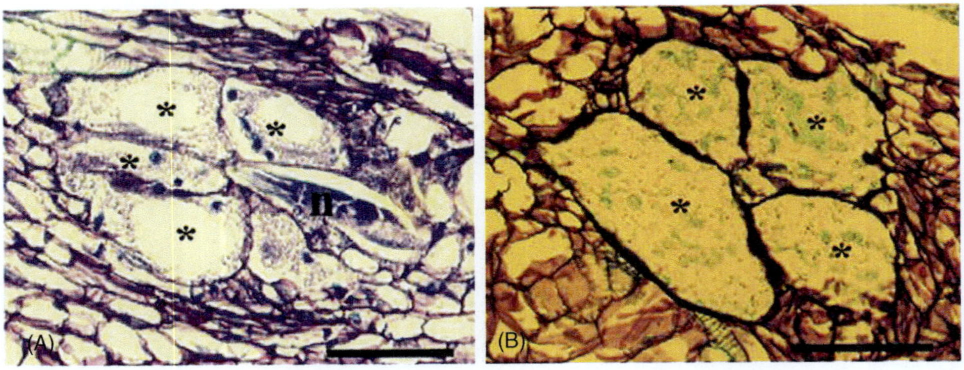

Fig. 10.18. (A, B) Root slices showing swollen nematodes (n) and giant cells (*). Scale bars: 100 μm. From Long *et al.*, 2022, CC BY.

The females lay egg masses on the galled tissue of the host roots to cause repeated infections in one planting season, or may remain inactive until the host crops are planted in a new growing season (Karssen and Moens, 2006; Siddique and Grundler, 2018).

10.11.5 Management

Sanitation

Nematodes are usually introduced into new areas with infested soil or plants. To prevent the spread of nematodes, growers should avoid moving plants and soil from infested areas of the orchard and avoid letting irrigation water run-off from plants that are infected. Tools should be well cleaned and surface sterilized before being used in the orchard, as nematodes can be found in soil adhering to equipment and tools that have been used elsewhere.

Chemical control

For the past 50 years, root-knot nematodes have largely been controlled using soil fumigants and nematicides. However, the options for chemical control have been greatly reduced in recent years as environmental concerns have become more apparent. Some highly toxic nematocides such as carbofuran, methyl bromide and aldicarb are banned in many countries. Nematicides such as fosthiazate, fluopyram and abamectin may be used before planting to control root-knot nematodes.

10.12 Conclusion

As a tropical fruit crop, pest management is one of the major problems encountered in dragon fruit production. Pests such as thrips, mealybugs and root-knot nematodes are prevalent in the orchard and cause severe economic losses. In China, control of dragon fruit pests relies heavily on the use of chemical pesticides. In the future, it will be necessary to further improve growers' scientific understanding of pests, establish a sound detection and monitoring system, and adopt integrated control strategies to reduce the use of pesticides in order to improve dragon fruit production and quality and promote sustainable development of the dragon fruit industry.

10.13 References

Ammar, E.D., Awadallah, K.T. and Rashad, A. (1979) Ecological studies on *Ferrisia virgata* CKLL. on *Acalypha* shrubs in Dokki, Giza (Homoptera, Pseudococcidae). *Deutsche Entomologische Zeitschrift* 26, 207–213.

CABI Compendium (2016) *Ferrisia virgata* (striped mealybug). CAB International, Wallingford, UK. Available at: https://doi.org/10.1079/cabicompendium.2398 (accessed 10 November 2023).

CABI Compendium (2022a) *Helicoverpa armigera* (cotton bollworm). CABI International, Wallingford, UK. Available at: www.cabidigitallibrary.org/doi/10.1079/cabicompendium.26757 (accessed 10 November 2023).

CABI Compendium (2022b) *Hypomeces squamosus* (green weevil). CAB International, Wallingford, UK. Available at: https://doi.org/10.1079/cabicompendium.27783 (accessed 10 November 2023).

Capinera, J.L. (2021) Melon Aphid or Cotton Aphid, *Aphis gossypii* glover (Insecta: Hemiptera: Aphididae). University of Florida Institute of Food and Agricultural Sciences, Gainesville, Florida. Available at: https://edis.ifas.ufl.edu/publication/IN330 (accessed 10 November 2023).

Capinera, J.L. and White, J. (2011) Terrestrial Snails Affecting Plants in Florida. University of Florida Institute of Food and Agricultural Sciences, Gainesville, Florida. https://entnemdept.ufl.edu/creatures/misc/gastro/terrestrial_snails.htm (accessed 10 November 2023).

Carrillo, D., Duncan, R. and Pena, J.E. (2021) Pitaya (Dragon Fruit) (*Hylocereus undatus*) Pests and Beneficial Insects. University of Florida Institute of Food and Agricultural Sciences, Gainesville, Florida. Available at: https://edis.ifas.ufl.edu/publication/IN1292 (accessed 10 November 2023).

Carvalho, C.M., Bessa, E.C. and d'Ávila, S. (2008) Life history strategy of *Bradybaena similaris* (Fèrussac, 1821) (Mollusca, Pulmonata, Bradybaenidae). *Molluscan Research* 28, 171–174.

Castagnone-Sereno, P. (2012) *Meloidogyne enterolobii* (= *M. mayaguensis*): profile of an emerging, highly pathogenic, root-knot nematode species. *Nematology* 14, 133–138.

Cleary, A.J., Cribb, B.W. and Murray, D.A.H. (2006) *Helicoverpa armigera* (Hübner): can wheat stubble protect cotton from attack. *Australian Journal of Entomology* 45, 10–15.

Cloyd, R.A. and Bethke, J.A. (2011) Impact of neonicotinoid insecticides on natural enemies in greenhouse and interiorscape environments. *Pest Management Science* 67, 3–9.

Ebert, T. and Cartwright, B.O. (1997) Biology and ecology of *Aphis gossypii* Glover (Homoptera: Aphidiae). *Southwestern Entomologist* 22, 116–153.

Ghose, S.K. and Paul, P.K. (1972) Observations on the biology of the mealybug, *Ferrisia virgata* (Cockerell) (Pseudococcidae: Hemiptera). *Proceedings of the Zoological Society (Calcutta)* 25, 39–48.

Haile, F., Nowatzki, T. and Storer, N. (2021) Overview of pest status, potential risk, and management considerations of *Helicoverpa armigera* (Lepidoptera: Noctuidae) for U.S. soybean production. *Journal of Integrated Pest Management* 12, 1–10.

Hien, N.T.T., Hoat, T.X., Thanh, H.M., Liem, N.V., Ngoc, T.B. *et al.* (2021) Pest and disease management in dragon fruit in Vietnam. In: Fullerton, R.A. and Lee, W.L. (eds) *Dragon Fruit – From Production to Market*. Food and Fertilizer Technology Center for the Asian and Pacific Region, Taipei, Taiwan, pp. 241–253.

Hill, D.S. (2008) *Pests of Crops in Warmer Climates and Their Control*. Springer, Dordrecht, Netherlands.

Karssen, G. and Moens, M. (2006) Root-knot nematodes. In: *Plant Nematology*. CAB International, Wallingford, UK, pp. 59–90.

Kaydan, M.B. and Gullan, P.J. (2012) A taxonomic revision of the mealybug genus *Ferrisia* Fullaway (Hemiptera: Pseudococcidae), with descriptions of eight new species and a new genus. *Zootaxa* 3543, 1–65.

Khoo, K.C., Ooi, P.A.C. and Ho, C.T. (1991) *Crop Pests and Their Management in Malaysia*. Tropical Press, Bangsar, Kuala Lumpur.

Lammers, J.W. and MacLeod, A. (2007) Report of a pest risk analysis: *Helicoverpa armigera* (Hübner, 1808). Department for Environment, Forestry and Rural Affairs, London. Available at: https://pra.eppo.int/pra/de9a63ea-789b-46ef-90a0-d8b5f935f297 (accessed 10 November 2023).

Lim, U.T., Kim, E. and Mainali, B.P. (2013) Flower model traps reduced thrips infestations on a pepper crop in field. *Journal of Asia-Pacific Entomology* 16, 143–145.

Long, H., Chen, Y., Pei, Y., Li, H., Sun, Y. and Feng, T. (2022) Occurrence and identification of root-knot nematodes on red dragon fruit (*Hylocereus polyrhizus*) in Hainan, China. *Agronomy* 12: 1064.

Luo, S., Naranjo, S.E. and Wua, K. (2014) Biological control of cotton pests in China. *Biology Control* 68, 6–14.

Mani, M. and Shivaraju, C. (eds) (2014) *Mealybugs and Their Management in Agricultural and Horticultural Crops*. Springer, New Delhi.

McCorquodale, A. and Hodges, A. (2016) Striped mealybug *Ferrisia virgata* Cockerell (Insecta: Hemiptera: Pseudococcidae). University of Florida Institute of Food and Agricultural Sciences, Gainesville, Florida. Available at: https://edis.ifas.ufl.edu/in1164 (accessed 10 November 2023).

Nguyen, T.H.N., Borgemeister, C., Poehling, H. and Zimmermann, G. (2007) Laboratory investigations on the potential of entomopathogenic fungi for biocontrol of *Helicoverpa armigera* (Lepidoptera: Noctuidae) larvae and pupae. *Biocontrol Science and Technology* 17, 853–864.

Nurhafizah, A.Y., Widians, J.A. and Budiman, E. (2020) Dragon fruit pest identification expert system. *Jurti* 4, 11–18.

Rezeki, M.S., Harahap, I.S., Sartiami, D., Ah, I. and Watson, G.W.(2021) Identification key to nymphal and adult mealybugs (Hemiptera: Pseudococcidae) associated with dragon fruit in Indonesia. *Biodiversitas* 22, 3113–3118.

Sagers, L.A.and Rood, K. (2011) Controlling Slugs and Snails in Utah. Utah State University, Logan, Utah. Available at: https://extension.usu.edu/pests/research/slugs-snails (accessed 10 November 2023).

Saha, J., Chakraborty, K. and Chatterjee, T. (2016) Biology of cotton aphid *Aphis gossypi*i Glover. *Journal of Global Biosciences* 5, 4467–4473.

Sampson, C. and Kirk, W.D.J. (2013) Can mass trapping reduce thrips damage and is it economically viable? Management of the western flower thrips in strawberry. *PLOS ONE* 8: e80787.

Sartiami, D., Saptayanti, N., Syahputra, E., Mardiasih, W.P. and Desmawati (2020) Mealybugs (Hemiptera: Pseudococcidae) associated with dragon fruit in Indonesia. In: *Proceedings of the International Conference and the 10th Congress of the Entomological Society of Indonesia, 6–9 October, Bali, Indonesia*, pp. 29–34.

Siddique, S. and Grundler, F.M. (2018) Parasitic nematodes manipulate plant development to establish feeding sites. *Current Opinion in Microbiology* 46, 102–108.

Subramanian, S., Boopathi, T., Nebapure, S.M., Yele, Y. and Shankarganesh, K. (2021) Mealybugs. In: Omkar (ed.) *Polyphagous Pests of Crops*. Springer, Singapore, pp. 231–272.

Tang, L. (2003) Potential application of the entomopathogenic fungus, *Nomuraea rileyi*, for control of the corn earworm, *Helicoverpa armigera*. *Entomologia Experimentalis et Applicata* 88, 25–30.

Tang, L.D., Guo, L.H., Ali, A., Desneux, N. and Zang, L.S. (2022) Synergism of adjuvants mixed with spinetoram for the management of bean flower thrips, *Megalurothrips usitatus* (Thysanoptera: Thripidae) in cowpeas. *Journal of Economic Entomology* 115, 2013–2019.

Tsai, C.W., Rowhani, A., Golino, D.A., Daane, K.M. and Almeida, R.P.P. (2010) Mealybug transmission of grapevine leafroll viruses: an analysis of virus-vector specificity. *Phytopathology* 100, 830–834.

Ullah, M.S. and Lim, U.T. (2015) Life history characteristics of *Frankliniella occidentalis* and *Frankliniella intonsa* (Thysanoptera: Thripidae) in constant and fluctuating temperatures. *Journal of Economic Entomology* 108, 1000–1009.

Wang, C.L., Lin, F.C., Chiu, Y.C. and Shih, H.T. (2010) Species of *Frankliniella* Trybom (Thysanoptera: Thripidae) from the Asian-Pacific Area. *Zoological Studies* 49, 824–838.

Wang, X.N., Wang, S.W., Chang, H., Gu, Y.P. and Liu, Y.P. (2022) Synergistic effect of two spray adjuvants on thiamethoxam in the control of *Megalurothrips usitatus* in cowpea. *Chinese Journal of Pesticide Science* 24, 368–375.

Weems, H.V., Heppner, J.B., Nation, J.L. and Steck, C.J. (2019) Oriental fruit fly, *Bactrocera dorsalis* (Hendel) (Insecta: Diptera: Tephritidae). University of Florida Institute of Food and Agricultural Sciences, Gainesville, Florida. Available at: https://edis.ifas.ufl.edu/publication/IN240 (accessed 10 November 2023).

Wu, S.Y., Tang, L.D., Zhang, X.R., Xing, Z.L., Lei, Z.R. and Gao, Y.L. (2017) A decade of a thrips invasion in China: lessons learned. *Ecotoxicology* 27, 1032–1038.

Wu, S.Y., Tang, L.D., Fang, F., Li, D.G., Yuan, X. *et al.* (2018) Screening, efficacy and mechanisms of microbial control agents against sucking pest insects as thrips. *Advances in Insect Physiology* 55, 199–217.

Yang, B. and Eisenback, J.D. (1983) *Meloidogyne enterolobii* n. sp. (Meloidogynidae), a root-knot nematode parasitizing pacara earpod tree in China. *Journal of Nematology* 15, 381–391.

Youssef, A.S., Salloum, W. and Asfour, M. (2021) Survey of the pseudococcid mealybugs (Hemiptera: *Pseudococcidae*) infesting dragon fruit and their natural enemies in some new reclaimed land, Egypt. *Egyptian Journal of Plant Protection Research Institute* 4, 452–457.

Zhang, P.J., Zhu, X.Y. and Lu, Y.B. (2011) Behavioural and chemical evidence of a male-produced aggregation pheromone in the flower thrips *Frankliniella intonsa*. *Physiological Entomology* 36, 317–320.

11 Diseases

Mai Van Tri[1,2]* and Dinh Thi Yen Phuong[3]
[1]*Former Deputy Director, Southern Horticultural Research Institute (SOFRI), Tien Giang, Vietnam;* [2]*Director, Southeast Horticultural Research Center, Ba Ria-Vung Tau, Vietnam;* [3]*Applied Biotechnology Institute, Tan Phu, Ho Chi Minh City, Vietnam*

11.1 Introduction

Extreme weather, acreage expansion and cropping intensity along with increased densities, excessive use of agrochemicals and low genetic diversity caused by intensive farming practices increase the risk of the spread of diseases and threaten the yield and quality of dragon fruit. Dragon fruit diseases can be caused by fungi, fungal-like organisms, oomycetes, bacteria and viruses. Balendres and Bengoa (2019) summarized 25 species of plant pathogens currently infecting dragon fruit, which are mostly fungal diseases of the stem, fruit and flowers, with only two bacterial and one viral disease recorded in this study. In recent years, there has been an increase in the number of diseases caused by bacterial, fungal and viral infections. Sunburn is also an important disorder caused by excessive heat and sunlight. In this chapter, we review the major diseases that limit commercial dragon fruit production.

11.2 Diseases Caused by Fungi

There are many fungal pathogens reported to attack the stem/branch, flower and fruit of dragon fruit plants. Balendres and Bengoa (2019) compiled a list of 21 fungal species causing diseases in this plant, with fruit and stem rot (*Bipolaris cactivora*) and anthracnose diseases (*Colletotrichum* spp.) reported most frequently. In Vietnam, stem canker (*Neoscytalidium dimidiatum*), *Bipolaris* rot (*B. cactivora*), sooty mould (caused by some fungi) and anthracnose (*Colletotrichum* spp.) are the major fungal diseases of dragon fruit.

11.2.1 Stem canker

Stem canker, caused by *Neoscytalidium dimidiatum*, has been reported in Taiwan (Chuang *et al.*, 2012), mainland China (Lan *et al.*, 2012; Yi *et al.*, 2015), Israel (Ezra *et al.*, 2013), Malaysia (Masratul Hawa *et al.*, 2013; 2015b), Vietnam (Hien and Oanh, 2014; Hieu *et al.*, 2014a; Oanh *et al.*, 2014), Thailand (Athipunyakom *et al.*, 2015), Indonesia (Muas and Jumjunidang, 2015), the USA (Sanahuja *et al.*, 2016), Puerto Rico (Serato-Diaz and Goenaga, 2021), India (Salunkhe *et al.* 2022) and Ecuador (Espinoza-Lozano *et al.*, 2023). The disease is also locally called white spot, brown spot, gecko spot and ghost spot disease in Vietnam (Mai *et al.*, 2020; Hieu *et al.*, 2022). Stem canker is considered a major disease of dragon fruit. This disease was not reported in commercial production areas for a long time but has quickly become a limiting

* Email: maivantri2017@gmail.com

DOI: 10.1079/9781800623156.0011

factor in the production of dragon fruit in many growing areas. Stem canker is a highly destructive disease that, if not controlled, has the potential to destroy orchards (Fullerton *et al.*, 2018). Canker disease incidence can be as high as 70% in commercial orchards, and can have a major and constant impact on pitahaya production (Sanahuja *et al.*, 2016; Hong *et al.*, 2020). It is considered a major pathogen with significant economic impact on the pitaya industry in Asia (Ezra *et al.*, 2013; Masratul Hawa *et al.*, 2013; Fullerton *et al.*, 2018).

Symptoms

Symptoms of stem canker have been described by several authors in various countries (Lan *et al.*, 2012; Masratul Hawa *et al.*, 2013; Xu *et al.*, 2018; Hong *et al.*, 2020). Symptoms are usually easy to detect on the branch or cladodes, beginning with whitish-yellow spots that turn orange to brown over time, and which can coalesce to form larger spots or lesions (Lan *et al.*, 2012; Masratul Hawa *et al.*, 2013; Hong *et al.*, 2020). These roundish, orange spots with a small darkened spot in the centre and formation of an incomplete chlorotic halo may or may not occur depending on weather conditions and other factors. Gradually, these spots turn reddish-brown in colour with a complete chlorotic halo. As the disease progresses, beige lesions with a reddish halo and beige-greyish lesions appear, both with a bulging and cracked centre, and an expanded and irregular chlorotic halo forms around them (Lan *et al.*, 2012; Masratul Hawa *et al.*, 2013; Hong *et al.*, 2020). In advanced stages, the coalescence of several lesions occurs and they tend to detach, leaving a hole in the stem (Lan *et al.*, 2012; Masratul Hawa *et al.*, 2013; Hong *et al.*, 2020). However, some authors do not describe these orange or reddish-brown sunken spots with the apparent necrotic water-soaked halo affecting the cladodes/branches that expand and form a lesion in advanced stages (Hong *et al.*, 2020). Fullerton *et al.* (2018) made field and laboratory observations about the symptoms of stem canker and the sequence of symptom expression (Fig. 11.1).

N. dimidiatum also causes fruit rot disease of dragon fruit (Nerd *et al.*, 1999; Ezra *et al.*, 2013; Yi *et al.*, 2015). The fungus caused internal rot, resulting in brown rot of the fruit flesh, and the peel colour of diseased fruit changes from green to red several days earlier than the normal fruit (Nerd *et al.*, 1999). The brown rot first appears below the persistent style, and gradually spreads and develops into a rotten, water-soaked area (Nerd *et al.*, 1999; Ezra *et al.*, 2013; Yi *et al.*, 2015). Conspicuous grey-white mycelia develop on the diseased area under hot and humid conditions. The pathogen also attacks flower buds, young fruit and fruit at harvest with similar symptoms to those on the stem/branch (Fig. 11.2).

Conidiogenous cells (pycnidial anamorph) of *N. dimidiatum* are characterized as hyaline and intermingled with paraphyses; the conidia are ellipsoid to ovoid, rod shaped or round, hyaline with an acutely rounded apex and truncate base, and initially aseptate, becoming brown with two septa at maturity with the central cell darker than the end cells. The hyphae are branched, septate, brown and disarticulated into aseptate or one-septa arthrospores (Masratul Hawa *et al.*, 2013).

Epidemiology

This fungus has a wide geographical distribution and a wide host range. Besides dragon fruit, it has also been reported on a number of trees including plum (*Prunus* sp.), almond (*Prunus dulcis*), mango (*Mangifera indica*), sisal (*Agave sisalana*), guava (*Psidium guajava* L.), pineapple (*Ananas comosus*), tomato (*Solanum lycopersicum*), white spider lily (*Hymenocallis littoralis*), royal poinciana (*Delonix regia*), *Citrus* sp., *Musa* spp., *Populus* sp. and *Ficus* spp. (Reckhaus, 1987; Farr *et al.*, 1989; Ray *et al.*, 2010; Nadiah *et al.*, 2017; Al-Bedak *et al.*, 2018; Hajlaoui *et al.*, 2018; Nouri *et al.*, 2018; Türkölmez *et al.*, 2019; Xie *et al.*, 2020; Ismail *et al.*, 2020; Taguiam *et al.*, 2020a; Al Raish *et al.*, 2020; Kuruppu *et al.*, 2021). The inoculum can come from these crops if they are intercropped or planted near dragon fruit plants.

According to Fullerton *et al.* (2018), *N. dimidiatum* commonly produces two types of spores in lesions: phragmospores, formed by the disassociation of hyphal cells in the dead tissue, and pycniospores, produced in shallow pycnidia just below the surface of younger cankers. Because the phragmospores are trapped within the hard necrotic tissues of the lesion and are not readily available for dispersal, their role in the epidemiology of canker is probably limited. Therefore,

Fig. 11.1. Symptoms caused by *Neoscytalidium dimidiatum* on the stems/branches of dragon fruit. (A) A small, sunken, whitish-yellow spot with a brownish-red fleck at the centre. (B) The spots enlarge, with the centre turning red, brown or grey over time. (C) The spot expands with a hard grey to brown scab in the centre, with pycnidia on the surface. (D) A chlorotic halo forms around or to one side of the lesion with expansion of the lesion. (E) Successive expansion leads to large affected lesions destroying large areas of the branch. (F) Rotten parts of old lesions drop off, leaving large holes in the branch.

Fig. 11.2. Symptoms of stem canker in flower buds (A), young fruit (B) and fruit at harvest (C).

pycniospores are the main means for dispersal and infection of stem canker. Pycniospores are released from pycnidia during wet weather and spread by rain splashes (Fullerton *et al.*, 2018). *N. dimidiatum* needs high humidity to induce sporulation and germination of pycniospores. Spread and infection will be much more severe during the wet season when infection conditions occur frequently. In addition, irrigation using overhead sprinklers can also increase the humidity during periods of no rain, which can increase the risk of disease spread and infection. The young tissues at the tips of rapidly growing branches are very susceptible to infection (Fullerton *et al.*, 2018), especially in wet conditions. Thus, branches are susceptible to diseases during the elongation and maturation stage but become more resistant before flower bud emergence, when the tissue on the branch becomes older (Ni *et al.*, 2013; Fullerton *et al.*, 2018). The spores of *N. dimidiatum* produced on diseased branches/cladodes during the early season may contribute to secondary infections on developing fruits (Crane and Balerdi, 2006). The majority of dragon fruit growing areas are located in a tropical monsoon climate. Seasonal heavy rainfall and high levels of humidity in the rainy season are favourable for development of stem canker. The severity of stem canker is high during the wet weather of the rainy season (Mai *et al.*, 2020). Additionally, higher incidence has been found in densely planted, poorly drained and poorly managed orchards.

Causal agent

Based on morphological, pathological and molecular characterization, *N. dimidiatum* has been confirmed as the cause of stem canker on dragon fruit (Yi *et al.*, 2015; Sanahuja *et al.*, 2016; Retana-Sánchez, 2019). *N. dimidiatum* was first described in 1933 by Rolland Marshall Nattrass as *Hendersonula toruloidea* responsible for causing dieback disease of plum, apricot and apple trees in Egypt (Machouart *et al.*, 2013). Decades later, it was determined to be a causative agent of human dermatomycosis-like infections and foot infections predominantly in tropical areas. *Neoscytalidium* spp. (formerly *Scytalidium*) are ascomycetous fungi that cause phaeohyphomycosis. Some can infect humans via direct contact, sometimes traumatic, with colonized plants or vegetable residues (Hay, 2002). *N. dimidiatum* causes onychomycosis and dermatomycosis. *N. dimidiatum* and its hyaline mutant, *N. dimidiatum* var. *hyalinum*, can cause clinical lesions in nail and skin similar to those produced by dermatophytes (Avner *et al.*, 2005). A newer name, *Scytalidium dimidiatum*, was used for the synanamorph *Nattrassia mangiferae* (Howard, 2003), otherwise known as *Neofusicoccum mangiferae*. Substantial confusion has arisen in the literature about this species resulting from the use of multiple different names including *Torula dimidiata*, *Fusicoccum dimidiatum* (André, 2018), *Scytalidium dimidiatum* and *Hendersonula toruloidea* (Avner *et al.*, 2005). *Scytalidium lignicola* and *Scytalidium lignicolum* are also considered earlier names of *N. dimidiatum* (Hay, 2002). The fungus *N. dimidiatum* has been isolated from various substrates including human skin and nails, plant tissues and soil, and is also known as a plant pathogen (Punithalingam and Waterston 1970; Crous *et al.*, 2006; Masratul Hawa *et al.*, 2013).

On potato dextrose agar (PDA) medium, *N. dimidiatum* produces hairy, olive-green to greyish colonies with dark grey to black pigmentation and a *Scytalidium*-like fungus; the hyphae are brown, branched, septate and constricted into spore chains before disarticulation into arthroconidia; the arthroconidia are ellipsoid to ovoid in shape and hyaline to dark brown with thick walls and septate arthrospores. Pycnidia rarely develop on PDA after 4 weeks of incubation; the pycnidial conidia are aseptate, ellipsoidal to nearly fusiform, and 8.5–15.5 μm × 3.5–5.7 μm in size (Dy *et al.*, 2022).

N. dimidiatum has been reported to cause internal black rot in pitahaya fruits in Israel and China (Ezra *et al.*, 2013; Yi *et al.*, 2015). Although the symptoms have not been documented in other regions, the possibility that fruit rot disease causes losses in local commercial production should not be excluded.

Management

Controlling stem canker disease after it has become established is not easy. Integrated management approaches that include cultural, biological and chemical controls are recommended (Mai *et al.*, 2020).

Management of stem canker disease aims to avoid the favourable conditions that promote growth of *N. dimidiatum* and uses practices that promote soil and plant health. Preventing the introduction and spread of the pathogen into dragon fruit orchards is an important measure. Using free-disease planting materials is necessary to prevent the disease entering new planting areas (Fig 11.3).

For new plantings, disease-free cuttings should always be used (Mai *et al.*, 2020). In the orchard, the inoculum must be reduced or eliminated for disease management. Diseased tissue and diseased parts on the branches/canopy should be removed (with a small, sharp knife) in the dry season to reduce the level of disease in the rainy season. Older, dried lesions will become activated when wet conditions occur repeatedly (Fig. 11.4). Removing lesions on the canopy is an important practice to reduce disease pressure and prevent the spread of the pathogen to younger plant parts (Fig 11.5). Branches or parts of branches that are less than 30 cm above the ground should be pruned. After pruning, the pruned material is removed from the orchard and destroyed.

It is crucial to keep the orchard as disease free as possible when young branches (or cladodes) are emerging and developing. Maintaining a low inoculum threshold in a diseased orchard and effectively managing stem canker during the early season to minimize disease pressure on the fruit are important strategies for disease management (Fullerton *et al.*, 2018; Hong *et al.*, 2019; Mai *et al.*, 2020). In the late dry season in monsoon tropical regions, the young branches and branch tips should be trimmed or cut off 3–5 cm from the top to accelerate the aging process of branch tissue and reduce the chance of infection.

Management practices that promote soil and tree health should be carried out, such as correcting soil pH and deficiencies, applying balanced and complete fertilization and irrigation, applying organic manure, mulching and regular mowing of weeds in orchards to the proper height (Fullerton *et al.*, 2018; Hieu *et al.*, 2018; Mai *et al.*, 2022). These practices are described in detail in Chapter 8 (this volume).

Choosing an irrigation method that does not result in foliar wetness is important. Reducing the duration of foliar wetness is beneficial for disease management. Using drippers instead of overhead sprinklers for irrigation reduces disease pressure in the orchard. As well as higher water-use efficiency, drip irrigation helps reduce the risk of diseases that thrive in wet conditions including stem canker. In the wet season, irrigation should be done early in the day so that the upper plant parts dry off rapidly, or water can be

Fig. 11.3. Using infected cuttings is one of the causes of disease spread to new planting areas. (A) A stem canker lesion in a cutting that has been prepared for a new orchard. (B) A lesion developing in a cutting 2 months after planting.

applied directly at the root zone to avoid or reduce splashing water up on to the canopy (Mai *et al.*, 2020).

Plants that are 4 or more years old need to have old branches and branches hidden in the canopy pruned to increase ventilation and reduce humidity, which helps reduce disease pressure and the level of primary inoculum. Infected lesions (Fig 11.5) and heavily infected branches are removed to prevent the spread of pathogens (Suu *et al.*, 2017; Fullerton *et al.*, 2018; Hieu *et al.*, 2018; Mai *et al.*, 2022). This is done when the plants are dry and no rain is forecast for that day. Infected tissues or parts of branches and plant are removed as soon as they are noticed (known as rogueing). This is usually done when the number of lesions on the plant or the number of diseased plants in the orchard is not high (Mai *et al.*, 2020). A sharp knife is used to cut into the healthy tissue around the lesion and remove the lesion from the branch. All wounds should be disinfected afterwards with a

Fig. 11.4. Old lesions become dry during the dry period (A) but are activated under wet conditions (B).

Fig. 11.5. Removal of infected parts before the flowering stage is an important measure for management of stem canker. (A) A lesion on a branch. (B) Removal of the branch. (C) Lesions that are not removed from the branches can easily infect young tissues or flower buds.

copper-based fungicide. Pruning tools should be sterilized after use to prevent the spread of pathogens to healthy plants. This type of pruning should be done in the dry season, and pruned branches should be removed from the orchard and destroyed; they should not be left in the orchard (Fig 11.6A). Pruning of plants on living posts, cutting/mowing of cover weeds/grasses/plants to increase ventilation and reduction of humidity in the orchard during the wet season are also recommended for disease management.

Management of stem canker disease should be practised at all stages of production. Stem canker is one of the most important diseases of dragon fruit. If left unmanaged in the orchard, it will cause significant damage and become difficult to eliminate (Fig 11.6B).

As mentioned above, all pruned plant parts should be destroyed to prevent the spread of stem canker and also other important diseases of dragon fruit. These removed plant parts can be used as materials for making compost or organic fertilizer (Suu *et al.*, 2017; Mai *et al.*, 2020). To avoid the risk of spreading the pathogens, care should be taken and microbial antagonists and cellulose-degrading microorganisms should be added during the composting process. A protocol for making compost using removed branches of dragon fruit and a number of antagonistic microbes has been developed by researchers at the Southern Horticultural Research Institute (SOFFRI) in Vietnam (Truc, 2022). The protocol can be summarized as follows: (i) collect the pruned materials, pile them up where they are to be composted and cover with plastic tarpaulin until ready to start the process; (ii) cut the branches into 3–5 cm segments and stack them in 1000 kg piles; (iii) evenly spray the compost pile with a solution containing 15 kg of product containing *Bacillus subtilis* (Truc, 2016) diluted with water (ratio of 1:1); (iv) cover the pile again with plastic tarpaulin and leave to compost for about 2 weeks (v) evenly spray 15 kg of product containing *Lactiplantibacillus plantarum* diluted in water (ratio of 1:1) on the compost pile and cover it again with plastic tarpaulin for a further 3 weeks; (vi) mix the composted product with farmyard compost at a ratio of 2:1, cover again with plastic tarpaulin and leave to compost for another week. The finished product can now be used as an organic fertilizer.

During the wet season, the growth of young shoots on the plant should be limited and they

Fig. 11.6. (A) Poor sanitation in an orchard, with infected branches left on the ground. (B) A neglected, diseased and unmanaged orchard of dragon fruit.

should be pruned from the branches. The wounds should be disinfected immediately with a copper-based fungicide solution or sprayed with a registered fungicide. Pruning or harvesting should not be done when the plants are wet, after rain or before rain is predicted. In heavy soil, growers should be especially alert for excessive moisture of the soil under the covers/mulches after rain and/or irrigation (Mai *et al.*, 2020).

In the late dry season, during the weeks of transition from the dry to the wet season, young shoots/branches should be removed and young tips of mature branches should be cut off, 3–6 cm from the top, to accelerate the ageing process of branches and reduce the risk of pathogen infection in the following wet season (Suu *et al.*, 2017; Mai *et al.*, 2020).

Cultivation of resistant varieties is the most effective control measure. Unfortunately, the three main dragon fruit species, *Hylocereus monacanthus*, *H. undatus* and *H. megalanthus*, were found to be susceptible to stem canker in laboratory trials (Taguiam *et al.*, 2020b). In Vietnam, field investigations showed that the red-fleshed variety 'LD1' is more susceptible to stem canker than the white-fleshed 'Vietnam White' variety', and tolerant varieties such as 'Vietnam White' should be used in regions with high disease pressure, replacing the more susceptible varieties (Mai *et al.* 2020).

Biological control is a sustainable and environmentally friendly management method. Some microorganisms have been found to inhibit *N. dimidiatum*. *Bacillus subtilis* is a promising biocontrol agent for protecting dragon fruit plant from infection by *N. dimidiatum* (Taguiam *et al.*, 2020a; Ratanaprom *et al.*, 2021). *B. subtilis* supernatant combined with sodium bicarbonate protected dragon fruit plants against stem canker disease during plant development in the field through pathogenic fungal inhibition and the induction of defence response mechanisms (Ratanaprom *et al.*, 2021). Dung *et al.* (2018) isolated and evaluated the antagonistic activity of some of these local isolates of *Bacillus* spp. in Vietnam and suggested that *Bacillus atrophaeus* has the potential to prevent stem canker disease on dragon fruit. Luong *et al.* (2016) evaluated ten strains of locally isolated microorganisms and identified two strains of *Streptomyces fradiae* and *Bacillus polyfermenticus* with high inhibitory activity against *N. dimidiatum*. An *et al.* (2020) isolated and tested 69 local soil bacterial isolates in Binh Thuan (Vietnam) and identified a strain of *Bacillus amyloliquefaciens* with high inhibitory ability against both spore germination and mycelial growth of *N. dimidiatum* causing stem canker disease on dragon fruit. Hai and Thuy (2021) isolated some isolates from some of the main dragon fruit growing regions in southern Vietnam and found two strains of *B. subtilis* and *B. amyloliquefaciens* with high inhibitory capacity against *N. dimidiatum*. Al Hamad *et al.* (2021) confirmed that the actinobacterial isolate *Streptomyces griseorubens* UAE2 can be used as biological control agent against *N. dimidiatum*, while Al Raish *et al.* (2021) evaluated and confirmed that isolate *S. antibioticus* UAE1 was more effective than other *Streptomyces* isolates under the experimental conditions. *Streptomyces lydicus* WYEC108, a naturally occurring bacterium that is commonly found in soil environments, showed strong *in vitro* antagonism to various fungal plant pathogens. *S. lydicus* WYEC 108 can be applied to soil against a range of root-decay fungi and can also be applied to plant foliage to control some plant diseases of the canopy. *S. lydicus* is packaged and presented as a biological fungicide in the USA and other countries including Vietnam (Yuan and Crawford, 1995; Hong *et al.*, 2019).

Trichoderma spp. are important biocontrol agents in the management of plant diseases. They are being studied extensively to enhance their effectiveness as biocontrol agents for controlling diseases of fruit crops. *Trichoderma* spp. have been screened for use in controlling several diseases on dragon fruit stems (Hung *et al.*, 2018; Astuti *et al.*, 2020). Le *et al.* (2022) isolated local strains in Binh Thuan (Vietnam), and identified and tested the isolate *Trichoderma asperellum* T10 as a promising probiotic strain against *N. dimidiatum* causing stem canker disease on dragon fruit under greenhouse conditions. Xuyen *et al.* (2017) found some promising local strains of *Trichoderma* spp. in Long An (Vietnam) for controlling stem canker caused by *N. dimidiatum*. The product Trico-DN (DHCT) has been developed and released by Can Tho University researchers to inhibit the stem canker pathogen on pruned dragon fruit branches before they are used to make compost/organic fertilizer, or for periodic spraying to prevent and control stem canker in the field. The disadvantages of these biological products are higher costs compared with chemical control, the need for repeated applications, and variations in their effectiveness depending on application conditions. To date, there are few products using biological agents that are commercially registered and widely used in dragon fruit production.

Chemical control is a popular choice for management of the disease (Hoa *et al.*, 2011; Hieu *et al.*, 2014b). However, there are not many suitable or effective fungicides for the different levels of disease and for different development stages of the plant (Hong *et al.*, 2019; Mai *et al.*, 2020). In Taiwan, there are nine synthetic fungicides that can be used in dragon fruit production. Testing on culture medium in the laboratory showed that mycelial growth of *N. dimidiatum* was effectively inhibited by cyprodinil + fludioxonil, azoxystrobin + difenoconazole and tebuconazole, while spore germination was inhibited by metiram, trifloxystrobin, pyraclostrobin, azoxystrobin, azoxystrobin + difenoconazole and iminoctadine (Ni *et al.*, 2013). These fungicides have been recommended to control stem canker disease in the field.

In Florida, registered chemical fungicides include azoxystrobin, cyprodinil + fludioxonil and copper sulfate pentahydrate (Hong *et al.*, 2019). Other fungicides include an extract of *Swinglea glutinosa*, garlic oil, hydrogen peroxide and peroxyacetic acid, and *Streptomyces lydicus* WYEC108 (as a biofungicide product). The active ingredients of azoxystrobin (Fungicide Resistance Committee Action (FRAC) group 11) and cyprodinil + fludioxonil (FRAC group 9 + FRAC group 12) are limited to a total of four applications per year, making it difficult to develop a disease-management programme that effectively protects dragon fruit throughout the season in Florida (Hong *et al.*, 2020).

In Vietnam, some chemical fungicides were tested for control of stem canker. However, some of the active ingredients that were effective in disease control are no longer allowed for use. The active ingredients recommended for controlling stem canker include copper-based fungicides, mancozeb, propineb, dimethomorph + mancozeb, metiram complex, pyraclostrobin, difenoconazole and difenoconazole + azoxystrobin (Hoa *et al.*, 2011; Hieu *et al.*, 2018). Mixtures of azoxystrobin (200 g l^{-1}) and difenoconazole (124 g l^{-1}) have been recommend by the Ministry of Agriculture in Indonesia for controlling stem canker using three applications at 10-day intervals, with a preharvest interval of 7 days (Noegrohati *et al.*, 2019).

Judicious, careful and rational use of fungicides is recommended (Hoa *et al.*, 2011). In conditions of high disease pressure (in the wet period, young-shoot period, and period of young flower buds up to 10 days old), the orchards should be monitor regularly and closely for the disease situation, undertaking timely spraying when necessary using systemic fungicides. Areas with severe disease need to be sprayed immediately after a prolonged rainstorm ends to reduce the primary inoculum before shoots appear on the canopy, and this can be repeated if necessary. The mix of difenoconazole + azoxystrobin is recommended for use in high disease pressure periods (Hieu *et al.*, 2018; Mai *et al.*, 2020). Spraying of fungicides at the preharvest stage should be careful, paying attention to the preharvest interval when choosing the appropriate fungicide. Rotation of fungicides among different groups during reproductive growth gives better protection of the flowers and fruits, and also reduces the risk of the development of fungicide-resistant pathogen populations (Hong *et al.*, 2020; Mai *et al.*, 2020). The efficacy of fungicide application is influenced by many factors, including soil and weather conditions, variety, plant health and development stages, the occurrence of disease, and most importantly the growers' ability and willingness to implement disease-management strategies. In conditions that are conducive for disease (wet weather, susceptible varieties, many young branches/shoots formed and/or high disease pressure), the effectiveness of fungicide spraying is often limited.

Management of stem canker disease is a challenge (Hong *et al.*, 2019). Disease management must be based on integrated application using appropriate and effective cultural measures and applying fungicides at the appropriate time. In Vietnam, to successfully restore a highly diseased and unproductive orchard, a combination of strict field sanitation, management of leaf wetness duration and effective fungicide programmes should be implemented (Fullerton *et al.*, 2018; Mai *et al.*, 2020).

A programme to restore a stem canker-infected orchard was suggested as follows (Mai *et al.*, 2020). The programme should start during the dry season, ideally early in the season. First, remove all heavily infected branches and remove all lesions on lightly infected branches. Trim the tips of the branches, leaving 40–80 cm from the top of the trellis/support post. The trimming level depends on the level of infection. Then proper weeding, mulching, watering, fertilization and pruning need to be practised to keep the plants healthy. A spray of copper-based fungicide should be applied after pruning. Regularly monitor the plants, remove and destroy any lesions that appear on branches (Fig 11.5), and apply two to four sprays of effective systemic fungicide rotated with a protectant fungicide to protect the branch tips where necessary, especially in wet conditions. The fungicides can be used alone or mixed with products with different modes of action. Reducing the duration of leaf wetness, watering in the morning and use of registered products of copper- and silver-based nanoparticles can all help to reduce the use of chemical fungicides.

11.2.2 *Bipolaris* rot

Stem rot, flower rot and fruit rot caused by *Bipolaris cactivora* are becoming common and serious diseases as a result of increased planting area and the number of harvests per year in hot and humid conditions. The pathogen attacks the stem, flower and fruit causing rot. *Bipolaris* stem rot is one of the most important diseases in grafted cactus farms, sometimes devastating a large area of the plantations (Hyun *et al.*, 2001; Kim *et al.*, 2004). Severe *Bipolaris* stem rot of cactus occurred in up to 77% of the fields in Koyang and Kimcheon (South Korea) from 1996 to 1997 (Chang *et al.*, 1998). Without proper care, the disease can cause up to 44% economic loss of pitaya yield (Taba *et al.*, 2007).

The disease has been reported in many countries and regions including Korea (Chang *et al.*, 1998; Hyun *et al.*, 2001), Taiwan (Wang and Lin, 2005), Japan (Taba *et al.*, 2007), Thailand (Athipunyakom and Likhitekaraj, 2010), Florida (Tarnowski *et al.*, 2010), Israel (Ben-Ze'ev *et al.*, 2011), Vietnam (Hieu *et al.*, 2016) and China (Qiu *et al.*, 2021). In addition to fruit and stem rot, the diseases caused by *B. cactivora* are also known by a number of other names, such as black spot (Taba *et al.*, 2007), fruit decay (Deng *et al.*, 2016), fruit blotch (Ben-Ze'ev *et al.*, 2011) and shattering (Chase, 1982).

Symptoms

On fruit, the symptoms are small, light brown, depressed spots, initially observed on the surface. Gradually, the spots develop into water-soaked, depressed, olive to black, dense, powdery lesions. These lesions rapidly enlarge and gain a light yellow halo, and become shrunken, water-soaked black lesions with a powdery dark brown to black conidial mass in the centre, ultimately resulting in complete rot of the fruit (Fig. 11.7B,C) (Chang *et al.*, 1998; Kim *et al.*, 2004; Tarnowski *et al.*, 2010; Hieu and Hoa, 2015; Mai *et al.*, 2020).

On the flower and flower bud, symptoms initially are small, brown to black spots on the surface, enlarging rapidly into ellipsoid sunken

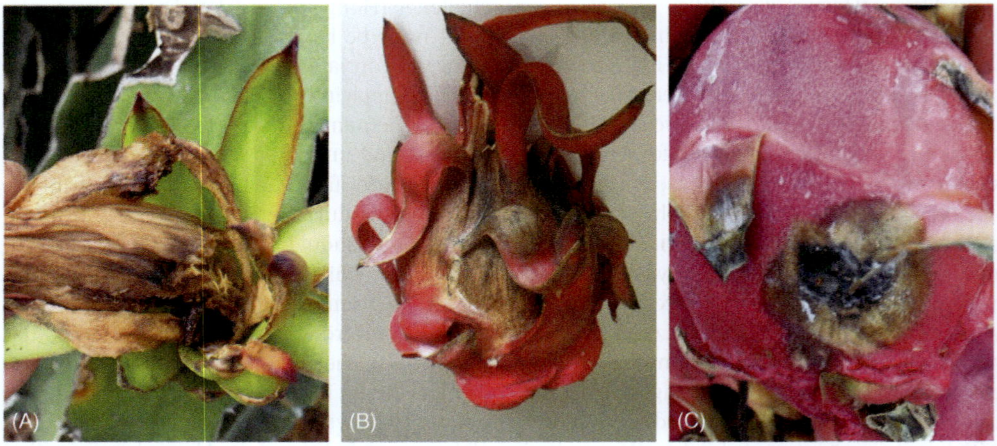

Fig. 11.7. Symptoms of *Bipolaris* rot on dragon fruit flowers (A), preharvest fruit (B) and fruit at harvest (C).

lesions with a light brown margin, and brown to black conidial masses covering the central portion of the lesion (Hieu and Hoa, 2015; Sanoamuang and Oeurn, 2018). Under favourable conditions, the diseased areas enlarge rapidly, causing the flower buds and flower to turn water-soaked black and covered with felty, black fungal growth (12.7A). Diseased flower buds and flowers will dry and remain attached to the branch (Sanoamuang and Oeurn, 2018; Mai *et al.*, 2020). The infected flower buds fail to open.

On stems and branches, the symptoms are initially light yellow, water-soaked, round lesions that subsequently turn light brown and dry out, causing death of the tissue (Chang *et al.*, 1998). According to Kim *et al.* (2004), symptoms on grafted cacti are variable. On the rootstock, the symptoms are initially pale yellow, water-soaked lesions, which turn light brown in colour and later rot. Sometimes, the whole stem is destroyed. On the scion, initial symptoms are small spots with an irregular margin that enlarges over time. The scions turn light brown and gradually dry out and die. Sometimes, they rot with the top turning black.

Causal agent

The disease is caused by the fungus *B. cactivora*, also known as *Drechslera cactivora*. As well as being found on *H. undatus*, as described by Balendres and Bengoa (2019), *B. cactivora* also causes disease in many other dragon fruit species. *B. cactivora* is considered a necrotrophic pathogen that causes a top or basal rot on various cactus species including *Cephalocereus mezcalaensis*, *Cereus peruvianus*, *Cereus tetragonus*, *Chamaecereus silvestrii*, *Gymnocalycium mihanovichii*, *H. undatus*, *H. trigonus* and *H. costaricensis* (Chase, 1982; Hyun *et al.*, 2001; Taba *et al.*, 2007; Hieu and Hoa, 2015; Qiu *et al.*, 2021).

Colonies on PDA plates are composed of pale olive to black aerial mycelium and are more exuberant and fluffier on PDA than on other media. On oat sucrose agar medium, the colonies are whitish to brownish, and appressed with little aerial mycelium, while on V8A medium, the aerial mycelium is white to light grey (He *et al.*, 2012). According to Kim *et al.* (2004), the conidiophores are erect, elongated, brown, rarely branched, irregularly nodulose and light coloured at the tip. The conidia are formed singly and are hyaline at first, later becoming thick walled and dark olive to chocolate brown, with one to five septa (usually three). For the most part, these are cylindrical and straight with rounded ends, but occasionally there are bifurcate types. Their size is 25–50 × 7–11 μm. The conidial morphology varies somewhat among isolates, but they can be differentiated into two types. Type A is less dark and somewhat conical at the terminal end, while type B is darker and/or more round at the terminal end.

The rapid spread of pathogens in host tissue, supported by a large number of potentially destructive spores (Tarnowski *et al.*, 2010), is one factor responsible for making the disease widespread and is important in commercial production of dragon fruit.

Epidemiology

According to Taba *et al.* (2007), *B. cactivora* usually overwinters in the asexual state as conidia on grass, weeds and debris, although sometimes the mycelia will also survive on debris or weeds. The pathogen normally begins to infect when the host starts to flower and the temperature increases above 25°C.

Tarnowski *et al.* (2010) reported that *B. cactivora* causes flower and fruit rot on dragon fruit but does not seriously affect mature plant stems/branches. However, the pathogen can attack young tissues of shoots causing stem rot (Hyun *et al.*, 2001; Hieu and Hoa, 2015). According to Tarnowski *et al.* (2010), after infecting the petals, the pathogen does not cause any visible symptoms until the fruit begins to grow and ripen within the flower. However, several reports have indicated that the pathogen can infect and form typical symptoms on flower buds, flowers and fruit causing flower bud rot, flower rot and fruit rot as mentioned above. This difference may be due to the different environmental conditions surveyed. The hot and humid conditions in tropical and subtropical regions are conducive to growth and spread of the pathogen, making it a more serious disease in these regions.

Prolonged hot and humid conditions are favourable for development of the disease. Warm temperature and high humidity over the whole growing season of grafted cacti in greenhouses are favourable for development and spread of the pathogen, and promote the serious level of this disease in Korea. Spread of the fungus is mediated by wind, irrigation and rain. Planting at high density and prolonged wetness of the plants following irrigation or rain are conducive to the growth and spread of *B. cactivora* (Choi *et al.*, 2010).

Oeurn *et al.* (2015) observed that when the fruit ripens, the conidia germinate, forming mycelia. The mycelia then produce hyphae that penetrate the natural openings in the plant tissues and cause symptoms. When the temperature rises to around 35°C, the mycelia start to mature into fruiting bodies and release ascospores. These are distributed by wind, rain and irrigation. Some will infect and survive in a viable host and form conidia on the surface. The conidia will then infect additional hosts that are in close proximity, or may wait for the next flowering period for infection. The conidia are dispersed primarily by wind and rain.

Management

The management practices used for stem canker and anthracnose diseases in the orchard will also contribute to management of *Bipolaris* rot. Where appropriate, use of tolerant varieties is suggested for areas with high disease pressure.

Cultivation practices are important for disease management, such as increasing ventilation of the canopy and orchard through an appropriate planting density, pruning and weed management; reducing disease pressure and preventing spread of the inoculum by removing infected plant parts (stems, branches, flower buds, flowers and fruit), low branches and non-productive branches; orchard sanitation; and removing nearby plant and weed hosts (Mai *et al.*, 2020). Reducing the amount of time that the canopy is wet by using a drip irrigation system, watering in the morning to allow the plants to dry, pruning the petals 2–3 days after flowering and early removal of infected flowers are also suggested (Mai *et al.*, 2022).

When a conidial suspension of *B. cactivora* was sterilized at 121°C for 20 min, it lost its virulence and had an inhibitory effect on the development of stem diseases caused by other pathogens. The method helps reduce disease development on grafted cacti and was nearly equivalent to the application of the fungicide iminoctadine triacetate under experimental conditions (Choi *et al.*, 2010).

There are currently no bioproducts specifically for controlling *Bipolaris* rots on dragon fruit. Two species, *Bacillus subtilis* and *Bacillus amyloliquefaciens*, were reported to have good potential to be developed as biocontrol agents for *Bipolaris* stem rot on grafted cacti (Bae *et al.*, 2013; Kim *et al.*, 2016a). In Vietnam, biofungicides based on chitosan are being evaluated for control of *Bipolaris* rot in dragon fruit (Hieu and Hoa, 2015).

In Vietnam, some fungicides are recommended and a fungicide spray schedule has been suggested as the following (Suu *et al.*, 2017): (i) the first application occurs when the flower bud reaches about 7 days old and involves spraying with copper-based fungicides, validamycin A or products based on chitosan; (ii) the second application is on flower buds that are 12–14 days old and uses validamycin A, iprodione, difenoconazole or diniconazole; (iii) the third application is to 20–21-day-old flower buds and uses iprodione, difenoconazole or diniconazole; (iv) the final spray occurs after removing the petals (2–4 days after flowering) and uses iprodione, difenoconazole, diniconazole or validamycin. The choice of fungicide and number of sprays depends on the disease pressure and weather conditions. A combination of fungicide spraying and cultural practices increases the efficacy of control and reduces the cost and impact of spraying (Mai *et al.*, 2020).

11.2.3 Anthracnose

Anthracnose caused by *Colletotrichum* spp. is a major disease of dragon fruit. The disease has been considered the most common disease of dragon fruit in Malaysia (Masratul Hawa *et al.*, 2015a). Anthracnose has been reported from many countries (Awang *et al.*, 2010; Masyahit *et al.*, 2013; Udayanga *et al.*, 2013; Guo *et al.*, 2014; Ma *et al.*, 2014; Suzianti *et al.*, 2014; Meetum *et al.*, 2015; Hieu *et al.*, 2016; Uyen *et al.*, 2018; Zhao *et al.*, 2018; Abirami *et al.*, 2019; Balendres and Bengoa, 2019; Nascimento *et al.*, 2019; Evallo *et al.*, 2021a). It is a destructive disease of dragon fruit and can cause up to 50% yield loss (Ma *et al.*, 2014). In São Paulo (Brazil), the disease has been reported to cause a loss of up to 5% on yellow pitaya (Takahashi *et al.*, 2008). However, the importance of anthracnose on dragon fruit varies among geographical and ecological regions.

Anthracnose affects the branches, flower buds, flowers and fruit at pre- and postharvest stages. It can attack fruit in the field, and during transport and storage, leading to a shorter shelf life. Postharvest infection can cause losses of dragon fruit in terms of yield and quality, and presents a serious threat to commercial production and export of dragon fruit.

Symptoms

Anthracnose caused by *Colletotrichum gloeosporioides* is characterized by water-soaked, reddish-brown lesions with yellow chlorotic halos on the stems and fruit (Fig 11.8B). The lesions have a brown centre, and eventually coalesce and rot (Masyahit *et al.*, 2009a). The acervuli can be found as dark, concentric circles on infected branches. Symptoms can increasingly develop on postharvest fruit under high-humidity conditions during storage (Masyahit *et al.*, 2009a). On postharvest fruit, the symptoms usually appear first on wounds (e.g. cutting sites) and fruit scales during storage. Water-soaked sunken spots of a light brown colour are found on the surface of infected fruit. Under conditions of high humidity, mycelial growth with a greyish to pinkish-white spore mass can be found on some lesions (Lin *et al.*, 2017). The lesions expand rapidly and coalesce, and can completely rot an entire fruit in 1 week. In addition to the common symptoms mentioned above, some lesions of infected fruit produce black microsclerotia, setae and acervuli, and some lesions are light to dark brown (Lin *et al.*, 2017). The symptoms also occur on flower buds, with a light brown lesion on the bud, which will gradually expand under wet conditions (Fig 11.8A).

Symptoms can vary slightly depending on the infecting species and plant variety. As well as the typical symptoms of brown, sunken and severely water-soaked lesions caused by *C. gloeosporioides*, *Colletotrichum truncatum* causes black spots and forms microsclerotia in the inoculation site, while *Colletotrichum boninense* causes brown to dark brown lesions with mild water-soaking (Lin *et al.*, 2017). Evallo *et al.* (2021a) observed sunken, necrotic, dark brown lesions with faint chlorotic halos on both *H. undatus* and *H. monacanthus*. Moreover, the presence of light to dark brown pycnidia is also observed on the marginal linings of lesions on infected stems of *H. undatus* and *H. monacanthus*.

The variation in anthracnose symptoms can be attributed to differences in variety, plant age, pathogen, environment conditions and management regimes (Balendres and Bengoa, 2019; Mai *et al.*, 2020).

Causal agent

C. gloeosporioides was reported on *H. megalanthus* in Brazil (Takahashi *et al.*, 2008) and on *H. undatus*

Fig. 11.8. Lesions caused by anthracnose disease on a flower bud (A) and on a stem (B).

in Japan (Taba *et al.*, 2006) and also in Florida (Palmateer *et al.*, 2007). *C. gloeosporioides* has also been reported to affect the stems and fruit of *H. polyrhizus*, *H. undatus* and *H. megalanthus* in Malaysia (Masyahit *et al.*, 2009a). Latiffah (2021) referred to this pathogen as *C. gloeosporioides sensu lato.* Later, Ma *et al.* (2014) reported *C. gloeosporioides* affecting the stems of *H. undatus* in China. Zhao *et al.* (2018) recorded *Colletotrichum siamense* as the cause of stem anthracnose in *H. polyrhizus* in China. In Thailand, Meetum *et al.* (2015) reported two species, *Colletotrichum aenigma* and *C. siamense*, causing anthracnose on the stems and fruit of *H. undatus*. In Taiwan, Lin *et al.* (2017) recorded three species, *C. gloeosporioides*, *C. truncatum* and *C. boninense*, causing anthracnose in three species of dragon fruit, *H. polyrhizus*, *H. undatus* and *H. costaricensis*. In Brazil, *Colletotrichum karstii* was reported to cause anthracnose in the stems of *H. undatus* (Nasimento *et al.*, 2019). *C. siamense* was also reported to cause anthracnose in the fruit of *H. undatus* in India (Abirami *et al.*, 2019). *C. truncatum* was found as the cause of anthracnose in the stems of *H. polyrhizus* in Malaysia (Suzianti *et al.*, 2014) and in the fruit of *H. undatus* in markets in China (Guo *et al.*, 2014). This species was also reported to cause anthracnose in the stems and fruit of *H. undatus* in Vietnam (Uyen *et al.*, 2018). Balendres and Bengoa (2019) compiled a list of *Colletotrichum* spp. causing anthracnose on dragon fruit, namely *C. gloeosporioides*, *C. aenigma*, *C. siamense*, *C. karstii* and *C. truncatum*. Thereafter, Zakaria (2021) listed the following *Colletotrichum* spp. as causal agents of anthracnose: *C. aenigma*, *C. boninense*, *C. gloeosporioides*, *C. gloeosporioides sensu lato*, *C. karstii*, *C. siamense*, *C. tropicale* and *C. truncatum*. Based on this approach, an updated list of *Colletotrichum* spp. has been produced, which includes: *C. aenigma*, *C. boninense*, *C. fructicola*, *C. gleosporioides*, *C. gleosporioides sensu lato*, *C. karstii*, *C. siamense*, *C. tropicale* and *C. truncatum* (Table 11.1).

Colletotrichum spp. have differences in colonial morphology. Colonies of *C. gloeosporioides* are greyish-white to dark grey, form aerial mycelia initially and later produce massive orange conidia. Colonies of *C. truncatum* are greyish to greenish-white and cottony to floccose. Colonies of *C. boninense* are smooth and flat, and white to salmon in colour (Lin *et al.*, 2017). Masyahit

Table 11.1. *Colletotrichum* spp. associated with dragon fruit (*Hylocereus* spp.) reported from in different countries.

Country	*Colletotrichum* spp.	Dragon fruit species	Diseased parts	Reference(s)
Brazil	*C. gloeosporioides*	*Hylocereus megalanthus*	Stem and fruit	Takahashi *et al.* (2008)
China	*C. gloeosporioides*	*H. undatus*	Stem	Ma *et al.* (2014)
	C. truncatum	*H. undatus*	Fruit	Guo *et al.* (2014)
	C. siamense	*H. polyrhizus*	Stem	Zhao *et al.* (2018)
India	*C. siamense*	*H. undatus*	Stem	Abirami *et al.* (2019)
Japan	*C. gloeosporioides*	*H. undatus*	Fruit	Taba *et al.* (2006)
Malaysia	*C. gloeosporioides*	*H. polyrhizus*, *H. undatus*, *Selenicereus megalanthus*	Stem and fruit	Masyahit *et al.* (2009a)
	C. truncatum	*H. polyrhizus*	Stem	Suzianti *et al.* (2015)
Philippines	*C. fructicola*, *C. tropicale*	*H. undatus* and *H. monacanthus*	Stem	Evallo *et al.* (2021b), Evallo *et al.* (2022)
Taiwan	*C. gloeosporioides*, *C. truncatum*, *C. boninense*	*H. polyrhizus*, *H. undatus* and *H. costaricensis*	Fruit	Lin *et al.* (2017)
Thailand	*C. gloeosporioides*, *C. aenigma*, *C. siamense*, *C. truncatum*	*H. undatus*	Stem and fruit	Meetum *et al.* (2015)
USA	*C. gloeosporioides*	*H. undatus*	Stem	Palmateer *et al.* (2007)
Vietnam	*C. gloeosporioides*, *C. truncatum*	*H. undatus*	Stem and fruit	Uyen *et al.* (2017, 2018)

et al. (2009a) stated that, on potato dextrose broth, *C. gloeosporioides* develops whitish-orange colonies, with septate hyphae and circular acervuli bearing conidia on conidiogenous cells and capsule-like conidia (6–10 × 2–2.5 μm) containing one cell.

Epidemiology

Colletotrichum spp. that cause anthracnose on dragon fruit have a wide range of hosts, and these plant hosts, when cultivated in a cropping system, lead to serious cross-infection problems in the field and at postharvest stages (Bordoh *et al.*, 2020). *C. fructicola* has a diverse host range, and has been reported on avocado in Australia, papaya in India, coffee in Thailand, cacao and *Tetragastris* spp. in Panama, apple and strawberry in the USA and Brazil, fig in Germany, sea lavender in Israel, Asian pear in Japan, bean and cowpea in Iran, and yam in Nigeria (Prihastuti *et al.*, 2009; Weir *et al.*, 2012; Atghia *et al.*, 2015; Saini *et al.*, 2016; Fuentes-Aragon *et al.*, 2018; Evallo *et al.*, 2021a). It is also common to find multiple hosts infected by a single *Colletotrichum* sp. and a single host infected by multiple species of the pathogen (Bordoh *et al.*, 2020).

Evallo *et al.* (2021a) stated that *C. fructicola* caused stem anthracnose in *H. undatus* and *H. monacanthus* in the Philippines. However, *C. fructicola* does not produce symptoms in the stem of *H. megalanthus* using an *in vitro* pathogenicity test. Evallo *et al.* (2021a) reported that *C. tropicale* caused anthracnose in *S. undatus* and *S. monacanthus*, but this species does not exhibit symptoms in *S. megalanthus* and they recommend use of *S. megalanthus* as a source of resistance to anthracnose in breeding programmes developing disease-resistant varieties.

A field survey by Masyahit *et al.* (2009a) on the occurrence of anthracnose disease in Peninsular Malaysia showed that the most severe disease occurs in orchards that have poor sanitation and in areas with prolonged high monthly rainfall that not be suitable for plant growth but creates favourable environmental conditions for the spread of diseases. The disease occurs widely in dragon fruit growing areas in Peninsular Malaysia with a monthly temperature range of 26.5–28.1°C and relative humidity of 77.5–86.6%. In Vietnam, a field survey in the main growing regions found that anthracnose was more severe in orchards with a higher planting density and poor drainage, that are unpruned or incorrectly pruned, and that have poor sanitation, imbalanced fertilization, excessive application of nitrogen and use overhead sprinklers for irrigation (Mai *et al.*, 2020).

Colletotrichum spp. is one of the major cosmopolitan phytopathogens causing postharvest anthracnose in dragon fruits. The pathogen attacks flower buds, flowers and fruit in the field, as well as fruit during long-distance transport and cold storage, leading to a shorter shelf life. Traditionally, the plants are sprayed with synthetic fungicides, which is a common approach for controlling diseases in general and anthracnose in particular for dragon fruit production. Due to the demand for safe storage methods for consumers and concerns about health and the environment, non-chemical alternative approaches to managing postharvest anthracnose should be applied (Mai *et al.*, 2020). Management of anthracnose in storage conditions is described in Chapter 13 (this volume).

Management

Management practices used for control of stem canker in the orchard will also help to prevent or reduce anthracnose disease. Where appropriate, the use of tolerant/resistant varieties is suggested for areas with high disease pressure.

Cultural measures to reduce favourable conditions for the disease include: appropriate planting density, correct pruning, orchard sanitation, good drainage, avoiding the use of overhead sprinklers during wet periods, and removal and destruction of low branches (<30 cm from the ground) and heavily diseased plant parts (Mai *et al.*, 2020). Dragon fruit plants should be maintained in a good growing condition by preventing physical injury caused by insects and animal pests and by the judicious use of fertilizers, organic manures and mulches, and appropriate weeding management. The petals should be removed 2–3 days after flowering to avoid infection with anthracnose; removing infected plant parts from the canopy also helps to reduce the primary inoculum on fruit pre- and postharvest (Mai *et al.*, 2022).

Evallo *et al.* (2022a) recommend using *S. megalanthus* as a source of resistance to anthracnose in breeding programmes. However, there has been no research on screening the resistance of dragon fruit varieties and species to develop anthracnose-resistant varieties for commercial production.

Tuong *et al.* (2018) tested and suggested some strains of *Actinomyces* spp. that have a high ability to reduce *C. truncatum* causing anthracnose on dragon fruit. *Trichoderma harzianum* and *T. asperellum* were reported to show strong antagonistic activity against the mango anthracnose fungus *C. asianum* in Vietnam (Thu *et al.*, 2021). The product Trico-DN (DHCT), described earlier, containing antagonistic *Trichoderma* spp. has been used to control anthracnose on dragon fruit branches (Xuyen *et al.*, 2017).

Chemical control is an important measure in anthracnose management. Treatment with synthetic fungicides is crucial where there is high disease pressure. Chemical fungicides traditionally used for control of anthracnose include propiconazole, difenoconazole, carbendazim, benomyl, maneb, captan, strobirulin, dithiocarbamate, benzimidazole and triazole compounds, together with other chemicals, such as prochloraz, imazalil and chorothalonil (Wharton and Diéguez-Uribeondo, 2004; Ali *et al.*, 2016; Bordoh *et al.*, 2020). However, many are currently not on the registered list for use.

In Vietnam, treatments for control of anthracnose often include active ingredients such as mancozeb, iprodione, difenoconazole, azoxystrobin, chlorothalonil and propineb, and mixtures of these (Mai *et al.*, 2020). Uyen *et al.* (2018) reported that the active agents propiconazole + difenoconazole, difenoconazole and axoxystrobin + difenoconazole have a strong ability to inhibit mycelial colony diameter growth. They should be applied at 7–14-day intervals where necessary (Hieu and Hoa, 2015; Mai *et al.*, 2020). Application intervals and the number of active agents should be adjusted based on disease incidence and severity. Rotation of fungicides among different groups was suggested during the growth of young plant parts for better protection of the shoots, young branches, flower buds, flowers and fruit (Mai *et al.*, 2022). Repeated use of pesticides for pest control causes environmental, ecological and human health risks, so should be combined with other measures in an integrated manner to reduce the amount of pesticides used, thus reducing the amount of residues in the environment.

11.2.4 Sooty moulds

Sooty moulds are very common in dragon fruit and can be found in almost all orchards in Vietnam (Mai *et al.*, 2020). Symptoms of sooty moulds were reported to affect about 30% of pitaya in a survey in June 2016 in Qionghai, Hainan, China (Xu *et al.*, 2019).

These fungi grow on honeydew secreted by sap-sucking insects and also by plants. They do not infect the plant tissue but negatively affect photosynthesis and plant growth by preventing sunlight from reaching the plant chloroplasts. Discoloration can occur on infected fruit underneath the fungal layer. Infected fruit have reduced fruit quality and marketability. Therefore, sooty moulds on fruit at harvest need to be washed off before sorting and packing the fruit.

Symptoms

Growth of sooty mould can occur on stems, branches, flower buds, flowers and fruit of dragon fruit plants and also on the supporting trellis, with a layer of dark grey to black mycelia on the surface (Fig. 11.9). In the initial stage of the disease, small patches of black fungal growth are found on the branches, flower buds, flowers and fruit, gradually expanding and covering the surface (Xu *et al.*, 2019; Mai *et al.*, 2020). Sooty mould growth can be a fine soot-like or powdery deposit, or a thick layer that can crack or peel away from the surface (Fig 11.10A).

Causal agent

Sooty mould growth is reported to be caused by *Cladosporium cladosporioides* (Fresen.) G.A. de Vries (unpublished data) and/or *Phaeosaccardinula javanica* (Zimm.) W. Yamam in Taiwan (Yeh *et al.*, 2016), while *Cladosporium tenuissimum* is reported as the causal agent of sooty mould on dragon fruit in Hainan, China (Xu *et al.*, 2019). In Vietnam, *Capnodium* sp. was reported to cause sooty mould disease (Hieu *et al.*, 2016). The fungi produce numerous spores on the canopy, which can be spread by wind and rain. The fungi can survive saprophytically as mycelium or spores on plant debris or on objects such as tools, equipment, containers and vehicles.

Fig. 11.9. (A) Sugary, sticky exudate on a flower bud. (B–D) Sooty mould growing on plant exudate on a young fruit (B), a branch (C) and on a fruit at harvest (D).

Fig. 11.10. (A) Sooty mould growth on the sugary, sticky exudate on a young fruit. (B) A water gun is used to wash the sooty mould, exudate and sap-feeding insects from the canopy, and also to water the plants. (C) A young fruit from which sooty mould has been removed by washing.

Epidemiology

These fungi do not directly attack the dragon fruit plants. They need a nutrient resource to grow such as honeydew, excreted by sap-sucking or phloem-feeding insects or produced as a sugary, sticky exudate from the plant (e.g. from flower buds or flowers) (Fig 11.9A).

Droplets of honeydew shed by the insects or sticky exudate produced by the plant fall on to surfaces below and the fungi grow on these sites. This frequently occurs on the surface of branches, flower buds, flowers and fruit (Fig 11.9B–D), but can also occur on any other objects that the honeydew falls on. The insects associated with sooty moulds are mealybugs, aphids, scales and whiteflies. Sooty mould growth on stems and on the trellis may be difficult to remove. The mould produces spores that are readily spread to new sites where food sources may be available. Rain splashes, rain water, irrigation water, wind and insects can carry and spread the spores. Spores that land on a suitable plant surface with honeydew or plant exudates will grow.

Sooty moulds can occur year-around but are more abundant in the dry season when the activity of sap-sucking insects increases and there is less rainwater to wash away insects and plant exudate. Sooty mould is common during flowering, from the flower bud to the end of flowering stages, especially when the flowering period occurs during the dry season.

Varieties with a higher fruit sugar content usually produce more exudate and so this disease is more common in them. The severity of sooty mould growth is greater on the red-fleshed 'LD1' variety than on the white-fleshed 'Vietnam White' variety in Vietnam.

Management

The best way to manage this disease is by washing the plant to remove the sooty mould and honeydew. This practice also helps to remove the sap-sucking insects and others, while also watering the plants (Mai *et al.*, 2020). In the dry season, irrigation using an overhead sprinkler system is also useful for reducing sooty mould and sap-feeding insects. Because the honeydew, sooty mould layer and sap-feeding insects are all sticky, a strong water stream is useful to wash the plants, such as a water gun (Fig 11.10B, C). The risk of this practice is that it can spread other diseases, including stem canker. Thus, washing should be done in the morning of a dry day to allow the plants to dry. Plant washing is usually done about 1 week after removing the petals (~9–12 days after flowing). It can be repeated after 7 days if necessary (Mai *et al.*, 2022). For hidden parts under the canopy, it may take longer (weeks or months) for the sooty mould to be washed away by water from rain or sprinklers, even if the insects involved are under control.

Chemical control for sooty moulds is not required. However, controlling the sap-sucking insects responsible for producing honeydew is important in reducing sooty mould growth on dragon fruit. These sap-sucking insects also injure and damage the plants directly. Application of proper practices for management of sap-feeding insects as well as the ants that tend and protect them is recommended (Mai *et al.*, 2020). Fruit affected by sooty moulds need to be washed in the packing house to remove the mould.

11.2.5 *Rhizopus* rot

Rhizopus spp. are well known for causing soft rots of fruit crops, especially during postharvest storage. *Rhizopus* spp. cause fruit rot in postharvest dragon fruit (Oeurn *et al.*, 2015; Ngoc *et al.*, 2017; Mai *et al.*, 2020). In tropical regions, postharvest losses caused by this pathogen can be high if hygiene, sanitary and storage conditions are poor. *Rhizopus* spp. also commonly occur on flowers and young fruit in the field. Historically, this was reported as a minor disease in many growing regions (Athipunyakom *et al.*, 2015; Hieu and Hoa, 2015; Masratul Hawa *et al.*, 2015a); however, in a recent survey of dragon fruit diseases in southern Vietnam, soft rot caused by *Rhizopus* spp. was found to be common, so may become more important for dragon fruit production in the future (Mai *et al.*, 2020).

Symptoms

The initial symptom is a blackish-brown spot followed by a soft, watery, brown lesion on petals and young fruit, and also on mature fruit. The watery lesions enlarge rapidly and become water-soaked, brown rotted areas. The expanding lesions then become covered with white mycelia with a thin, powdery layer of black sporangia (Fig 11.11A). The black mass of mycelia and sporangia completely engulfs the diseased area over time. In the field, *Rizopus* rot occurs mainly on flowers during the flowering stage. In severely infected flowers, the pathogen can cause the whole petals and flower to rot. The rot on the petals gradually spreads towards the flower stalk or young fruit after fertilization during favourable conditions (Fig. 11.11B).

Causal agent

There may be several *Rhizopus* spp. affecting dragon fruit. *Rhizopus stolonifer* has been reported in Malaysia, Thailand, Taiwan and Vietnam (Hieu and Hoa, 2015; Masratul Hawa *et al.*, 2015a; Oeurn *et al.*, 2015; Ngoc *et al.*, 2017).

R. stolonifer is a saprotroph and plays an important role in early colonization of substrata in the soil. However, it can exist in both soil and air, and can behave as a parasite of plant tissues. It can colonize a variety of natural substrates because it can tolerate broad variations in the concentration of essential nutrients (Onions *et al.*, 1981).

Epidemiology

Rhizopus spp. have a wide host range including fruits and vegetables. The spores are airborne and are spread by water, winds, insects, animals and contaminated tools. The fungus can survive

Fig. 11.11. (A, B) *Rhizopus* rot on petals (A) and young fruit (B). (C) Scars on fruit at harvest reduce the number of marketable fruit. (D) Removing the petals 2–3 days after flowering helps prevent petal and young fruit rot.

in infected parts of plants including the flowers, petals, fruit and branches, and also in plant debris. The hot and humid conditions of tropical regions favour disease development.

In the field, *Rhizopus* rot commonly occurs during the flowering stage. Off-season production with many flowering periods in a year promotes disease development. The flowers secrete water on their petals during flowering, creating a wet environment that promotes infection of this pathogen.

The diseased petals then act as a primary inoculum and infect the developing fruit (Fig 11.11B) and the mature fruit at harvest. The lesions also cause scars on the fruit at harvest (Fig 11.11C). *Rhizopus* rot is considered to be widespread in all dragon fruit-growing areas in the tropics including Malaysia, Thailand, Taiwan and Vietnam (Tsai *et al.* 2010; Athipunyakom *et al.*, 2015; Hieu and Hoa, 2015; Masratul Hawa *et al.*, 2015a; Ngoc *et al.*, 2017).

Management

Management measures used for stem canker are also useful for reducing *Rhizopus* rot. The main recommended measures include: removing and destroying the severely diseased flowers and fruit; practising orchard sanitation; increasing ventilation of the orchard by proper pruning and weed management; removing the petals 2–3 days after flowering (Fig 11.11D); reducing the canopy wetness time by increasing the use of drip irrigation; watering in the morning to allow the plants to dry; reducing wounding during care, harvest, sorting, washing and packing; and controlling insects and animals that cause wounds on the plants (Suu *et al.*, 2017; Mai *et al.*, 2020). Applying copper-based fungicides to the flower buds and after fruit set is suggested, when necessary (Mai *et al.*, 2020).

To prevent postharvest rot, treatment of the fruit with hot water at 49–55°C for 2–3 min and storage at 5–6°C is useful to inhibit rot caused by *R. stolonifer* and also several other pathogens causing postharvest fruit rot (Ngoc *et al.*, 2017).

11.3 Diseases Caused by Bacteria

11.3.1 *Erwinia* soft rot

Erwinia soft rot is a common bacterial disease of dragon fruit in Vietnam (Hieu and Hoa, 2015). The disease can occur year-around, but most of the soft-rot disease occurs during the wet periods of the rainy season. The number of infected fruit can exceed 50%, and may even reach 70–80% if control measures are not effective (Hieu and Hoa, 2015). The disease incidence on red-fleshed varieties is usually two to three times higher than that of the white-fleshed ‘Vietnam White’ variety (Hieu and Hoa, 2015; Mai *et al.*, 2020). The pathogen usually infects and causes soft rot of fruits but can also attack the stems causing stem soft rot.

Symptoms

The initial symptoms are blisters with a water-soaked appearance with tiny bubbles appearing on flower buds and young fruit skin (Fig. 11.12B, C) (Hieu and Hoa, 2015). As the infection progresses, a characteristic watery exudate can be seen oozing from infected areas. An abundant black mycelium of *Rhizopus* spp. may develop on the surface of the diseased areas. Under favourable conditions, the lesion enlarges rapidly, causing complete rot after several hours, and produces a strong putrid odour that can attract chewing insects such as *Protaetia* spp. and *Hypomeces squamosus* (Hieu and Hoa, 2015). According to Grenier *et al.* (2006), once the pathogen successfully penetrates the host tissues, the cell wall is degraded and the cellular structure collapses, and it is this cell maceration that gives the characteristic ‘water-soaked’ appearance. The symptoms may begin in wounds and at leaf edges. The disease also occurs on branches (Fig. 11.12A), usually through injuries/wounds caused by insects, animals or other pathogens.

Causal agent

Erwinia chrysanthemi was reported to cause soft-rot diseases of dragon fruit in Vietnam (Hieu and Hoa, 2015). *E. chrysanthemi* has been reclassified into the genus *Dickeya* (Samson *et al.*, 2005) and hence its name has changed to *D. dadantii* (Samson *et al.*, 2005). *D. dadantii* is one of the phytopathogenic *Enterobacteriaceae* that can cause soft-rot diseases in a wide range of economically important crops (Reverchon and Nasser, 2013). The symptoms caused by *Dickeya* spp. are characterized by rapid disorganization of

Fig. 11.12. Symptoms of *Erwinia* soft rot. (A) A water-soaked, yellow area on a branch. (B) Flower bud rot with the tepals turning brown and decaying, and then sticking together to inhibit the buds from opening. (C) Young fruit rot with water-soaked,yellow areas.

parenchymatous tissues, mainly driven by pectic enzymes (Hugouvieux-Cotte-Pattat *et al.*, 1996). *D. dadantii* strain 3937 has been reported as a pathogen of the pea aphid (Grenier *et al.*, 2006). *D. dadantii* can infect many different ornamental and horticultural plants in tropical, subtropical and temperate regions. Its host range continues to grow, with more than 50 species now recorded. This species is also found in soils, rivers and irrigation water (Cother and Gilbert, 1990; Robert-Baudouy *et al.*, 2000).

Epidemiology

The pathogen is usually transmitted in and on planting material and by infected insects (e.g. *Protaetia* and *Hypomeces* spp.). It can also be spread via pruning and harvesting tools, knives, gloves and equipment, in infected soil and farmyard manure, and by water splashes. The pathogen is also able to survive in plant debris, host plant residues, infected plants and soil (Nelson (2009). It can enter plants easily and quickly through wounds caused by chewing insects and other pests. Management of the canopy environment or irrigation techniques to reduce the duration of leaf wetness may help to reduce the incidence of this disease (Mai *et al.*, 2020). The disease can be a problem year-round in the hot and humid conditions of tropical regions. Therefore, disease management should be carried out regularly.

Management

Management of this disease relies principally on cultural practices (Nelson, 2009), as described by Mai *et al.* (2020). Disease-free planting material should be used for all new plantings, and cuttings should not be taken from infected orchards or other areas.

Cultural control is to minimize the conditions that might favour bacterial growth, reproduction, infection and spread, including reducing high humidity and crowding of plants; increasing air circulation around plants; disinfecting knifes, tools, gloves, equipment and machinery to prevent the spread of pathogens; avoiding the use of overhead irrigation if possible to reduce water splashes; proper pruning and orchard sanitation; pruning the petals 2–3 days after flowering; proper weed management; avoiding intercropping with host plants; and controlling chewing pests that cause wounds on plants.

Practices that promote soil and tree health and avoid conditions that place plants under stress should be applied, including balanced fertilization, correcting soil pH and deficiencies, applying organic manures and mulching, watering plants early in the day, avoiding high-density planting, having proper drainage and preventing waterlogging.

Biological control can be used as an alternative for disease management. Currently, there are no bioproducts registered for controlling the diseases caused by *Dickeya* spp. Applying organic manures combined with antagonistic *Trichoderma* spp. is recommended (Hieu and Hoa, 2016).

Chemical control is recommended under high disease pressure. A regular spray schedule of bactericides is recommended in these conditions. Bactericides such as kasugamycin, streptomycin sulfate and oxolinic acid in combination with copper-based fungicides are recommended in Vietnam (Hieu and Hoa, 2015). However, according to Nelson (2009), spraying bactericides on plants infected with *Dickeya* spp. is not very effective, and thus cultural measures are essential for management of this disease.

11.3.2 Other bacterial diseases

During the 1990s, stem soft-rot diseases from Mexico were reported and at least two species of *Enterobacteriaceae* were involved (Valencia-Botín *et al.*, 2003, 2013). Stem soft rot is occasionally delimited by a chlorotic halo. Using artificial inoculation, the rotting appears after 15 days. The dragon fruit species *H. undatus* and *H. purpusii* were found to be susceptible to the two bacterial species, with *H. undatus* being more susceptible, and it was found that plants deficient in calcium and nitrogen developed more severe symptoms (Valencia-Botín *et al.*, 2013).

Soft rot and stem rot caused by *Enterobacter cloacae* and *Paenibacillus polymyxa*, respectively, are the two main bacterial diseases of dragon fruit (Masyahit *et al.*, 2009b; Zhang *et al.*, 2017). The former has been reported in Malaysia (Masyahit *et al.*, 2009b) and the latter in China (Zhang *et al.*, 2017).

E. cloacae is reported to cause yellow and brown soft rot on the stems/branches of *Hylocereus* spp. in Malaysia (Masyahit *et al.*, 2009b). Symptoms start with a water-soaked lesion on the fleshy stem, which then turns from yellow to brown. *E. cloacae* reportedly infects three *Hylocereus* spp. (Masyahit *et al.*, 2009b). Soft rot develops within 24 and 48 h after inoculation of the fruit and stem, respectively (Masyahit *et al.*, 2009b).

E. cloacae is a ubiquitous, Gram-negative, facultative anaerobic, rod-shaped bacterium that causes health problems in humans (Mezzatesta *et al.*, 2012), but has increasing importance as a plant pathogen. This species has been reported as an opportunistic pathogen in many plants, causing internal yellowing of the fruit of papaya (*Carica papaya*), rhizome rot of edible ginger (*Zingiber officinale*) and grey kernel disease on macadamia (*Macadamia integrifolia*) in Hawaii (Nishijima *et al.*, 2007). It also causes internal bulb decay in onion (*Allium cepa*) in California and Colorado (Bishop and Davis, 1990), decay of chilli pepper (*Capsicum annuum*) in Mexico (García-González *et al.*, 2018) and leaf rot disease in orchids (*Odontioda* spp.) in Japan (Takahashi *et al.*, 1997). The genus *Enterobacter* includes members of the *E. cloacae* complex, which have adapted to survive and proliferate in diverse environments (Sanders and Sanders, 1997). These bacteria may become opportunistic pathogens when changes in host physiology occur related to high temperatures (Sanders and Sanders, 1997). Thus, infection may be latent until environmental conditions are favourable for onset of the disease (García-González *et al.*, 2018) and/or unfavourable for growth of the host plant.

Paenibacillus polymyxa is reported to cause stem rot symptoms on dragon fruit in southern China, and symptoms develop 48 h after injecting *P. polymyxa* into the stems (Zhang *et al.*, 2017). The incidence can reach 100% in severely infected fields. The initial symptoms on fleshy stems are water-soaked, soft, dark green, suborbicular lesions. The margins gradually turn yellow and then brown. As infection progresses, the fleshy stems become completely decomposed, leaving only the central woody stem pith.

Peng *et al.* (2022) reported that the dominant bacteria in healthy and diseased pitaya stems in Guizhou, China, belong to the phyla Proteobacteria and Firmicutes. *Pseudomonas*, *Enterobacter*, *Sphingomonas* and *Sphingobacterium* spp. were the core bacteria found during the infection stages of soft rot, and both *Enterobacter* and *Pseudomonas* spp. may play an important role in soft rot of pitaya stems.

Klebsiella mobilis, *K. oxytoca*, *Pantoea dispersa* and *Rahnella aquatilis* have been also isolated from soft-rot-infected dragon fruit stems, but these bacteria were not pathogenic to dragon fruit (Masyahit *et al.*, 2009b).

Management practices as described earlier for *Erwinia* soft rot may be followed. General measures for management of bacterial diseases include preventing the spread of pathogen, orchard hygiene, proper pruning and use of copper-based fungicides.

11.4 Diseases Caused by Viruses

At least six plant virus genera that infest members of the family Cactaceae have been recorded – *Potexvirus*, *Carlavirus*, *Carmovirus*, *Tobamovirus*, *Tospovirus* and *Badnavirus* – but before 2020, only four virus species belonging to the genus *Potexvirus* had been characterized in dragon fruit (Li *et al.*, 2015). Three of these species were first identified by Taiwanese researchers: *Cactus virus X* (CVX; Liou *et al.*, 2001), *Zygocactus virus X* (ZyVX; Mao, 2008) and *Pitaya virus X* (PiVX; Mao, 2008). Single or coinfection of these potexviruses was found in most of their samples during an investigation of several dragon fruit production areas in Taiwan (Mao *et al.*, 2018). Additionally, the potexvirus *Schlumbergera virus X* (SchVX) was found in a coinfection with ZyVX in dragon fruit plants in Brazil in 2008 (Duarte *et al.*, 2008). In 2020, Zheng *et al.* (2020) isolated and sequenced the first virus belonging to the genus *Badnavirus*, *Pitaya badnavirus 1* (PiBV1), infecting red-fleshed dragon fruit (*H. polyrhizus*) in Guizhou, China. However, the epidemiology of PiBV1 has not been characterized further.

Although the damage caused by virus infection is not a serious economic issue, the risk of a viral epidemic remains a threat for the dragon fruit industry worldwide (Lobo *et al.*, 2014; Balendres and Bengoa, 2019). When establishing a dragon fruit farm, the major propagation method is cuttings, which could spread viruses from infected mother plants. The symptoms of virus infections are easily missed. Many farmers have even mistaken the mosaic or chlorosis symptoms found on dragon fruit stems for valuable natural mutants and then multiplied these infested plants as exotic crops, which could lead to rapid spread of the virus. In addition, there is no cure for plant virus diseases in general, including pitaya virus diseases, and thus pitaya viruses remain a major concern for dragon fruit production. Therefore, rapid and accurate diagnosis of virus infections is important for efficient management and timely eradication of infested plants. Methods used to detect viral infections include the polymerase chain reaction (PCR), enzyme-linked immunosorbent assay (ELISA), double antibody sandwich-ELISA, loop-mediated isothermal amplification (LAMP) and magnetic nanoparticle-capture reverse transcription-PCR (Mao *et al.*, 2018). Virus diseases remain a minor problem for dragon fruit production in Vietnam (Suu *et al.*, 2017; Mai *et al.*, 2020). Many local growers do not know how to detect the diseases and do not have management measures for their eradication (Mai *et al.*, 2020).

11.4.1 Cactus virus X disease

CVX disease was reported on dragon fruit in Brazil in 1993 and in Taiwan in 2021 (Liou *et al.*, 2001). It then became the most common dragon fruit virus worldwide, and is now found in Vietnam, Korea, the Philippines, Malaysia, India, China, Colombia, the USA, Spain and Vietnam (Reichel *et al.*, 2002; Valencia-Botín *et al.*, 2013; Kim *et al.*, 2016b; Zhang *et al.*, 2016; Masanto *et al.*, 2018; Huy *et al.*, 2019; Janssen *et al.*, 2022a). CVX was the first declared virus disease affecting dragon fruit and is considered a major threat to the dragon fruit industry worldwide (Lobo *et al.*, 2014).

Symptoms

CVX typically causes mottling, resulting in a light and dark green mosaic on dragon fruit stems (Fig. 11.13A, B), first described by Liou *et al.* (2001). CVX is also associated with various other symptoms including necrosis, mottling, chlorosis (Fig. 11.13C, D), stunting, twisted cladodes (Fig. 11.13E), distorted aureoles, deformed spines, reddening (Fig. 11.13F) and speckling. CVX-infected dragon fruit can be asymptomatic or have local or systemic symptoms (on the buds, stems, flowers and fruit). Mechanically transmitted CVX also causes necrotic and chlorotic lesions on indicator plants such as *Chenopodium* spp. The outcome of virus infection depends on the interaction among the host plant, single or multiple viruses and the surrounding environment. Thus, external CVX symptoms are unreliable for virus diagnosis, and so molecular techniques (PCR or reverse transcription-LAMP) are commonly used to detect infected dragon fruit (Zhang *et al.*, 2016).

Causal agent

CVX disease on dragon fruit is caused by CVX in the genus *Potexvirus* (family *Alphaflexiviridae*, order *Tymovirales*). Purified viral particles are seen as flexuous rods of 480–520 nm in length

Fig. 11.13. Symptoms of pitaya virus infection: (A, B) mottling of cladodes, (C, D) chlorosis, (E) deformation and (F) reddening.

and 13 nm in diameter under an electron microscope (Zechmaan and Zellnig, 2009). Its sequence was published in 2004 (Liou *et al.*, 2004). To date, three main strains of CVX have been identified, differentiated by serology, and two of them differ in symptomatology.

Epidemiology

As with other potexviruses, CVX has a restricted host range; however, it infects various species in the family Cactaceae, both cultivated and wild cacti (*Austrocylindropuntia, Cereus, Echinocereus, Echinopsis, Epiphyllum, Eriocereus, Ferrocactus, Hylocereus, Lobivia, Mammillaria, Notacactus, Opuntia, Pereskia, Platyopuntia, Saguaro, Schlumbergera* and *Zygocactus* spp.), as reviewed by Evallo *et al.* (2021b). CVX can also be artificially transmitted to members of the families Amaranthaceae, Caryophyllaceae and Chenopodiaceae. Most of these host plants can be used as indicator plants for CVX.

To date, CVX has not been found to be transmitted by known vectors, dodder or seeds, even though CVX infectivity has been found in flowers and fruit. CVX is spread naturally by plant contact or mechanically by cultural practices such as grafting, pruning, harvesting and propagation by cuttings. Commercial dragon fruit fields are commonly planted using cuttings from vegetative propagation, which has a high risk of CVX spread, leading to adverse economic effects (Lobo *et al.*, 2014).

Management

In the USA, there are no known vectors for this virus according to Lobo *et al.* (2014). Transmission occurs primarily by mechanical means, including but not limited to cultural practices such as pruning, flower and fruit thinning, harvesting and using infected planting materials. Additionally, the virus can be transmitted naturally through plant contact and root grafting beneath the soil (Lobo *et al.*, 2014).

There is no cure for infected plants. The most effective management strategy therefore involves exclusion, prevention and eradication. Planting virus-free propagules prevents the introduction of inoculum. Growing dragon fruit from seed might reduce the virus contamination risk but is more costly and takes more time to establish a dragon fruit field up to the production stage; furthermore, the dragon fruit produced may not be identical to the parents. Proper field sanitation and decontamination of culture tools (e.g. using Vime-Protex with 1,5-pentanedial and alkyldimethylbenzylammonium chloride or Chlorox/bleach) are efficient practices to prevent virus spread. However, if a dragon fruit field has areas of local infection, it is essential to detect the infected plants early and eradicate them as soon as possible.

Resistant dragon fruit varieties are a potential future prospect in the industry. Genetic engineering has introduced silencing constructs of the CVX coat protein (CVXcp-RNAi) into *Hylocereus trigonus*, generating a CVX-resistant line (Kim *et al.*, 2012). However, this CVXcp-RNAi line is commercially restricted by genetically modified organism (GMO) regulation. Therefore, there is still a need for breeding and selection of resistant varieties that are not regulated by GMO regulation and are accepted by customers.

11.4.2 Other virus diseases

Other viruses reported to infect dragon fruit are PiVX, ZyVX and SchVX. PiVX and ZyVX were first reported in Brazil and Taiwan in 2008 and in South Korea in 2020 (Duarte *et al.*, 2008; Mao, 2008; Bae and Park, 2022). ZyVX was reported in Brazil and Taiwan in 2008, in South Korea in 2020 and in India in 2022 (Duarte *et al.*, 2008; Mao, 2008; Bae and Park, 2022; Parameswari *et al.*, 2022). SchVX has been reported to infect *H. undatus* in Brazil, China and Spain (Duarte *et al.*, 2008; Zheng *et al.*, 2017; Janssen *et al.*, 2022a). PiVX and ZyVX were recorded to cause mixed infection in *H. undatus* in Brazil (Duarte *et al.*, 2008).

Symptoms

Generally, these viruses caused chlorotic spots and mosaic symptoms on branches. PiVX and ZyVX also typically cause mottling, yellowing, and light and dark green mosaic patterns on dragon fruit stems/cladodes (Bae and Park, 2022). SchVX has slightly different symptoms, causing yellow ringspots, chlorotic spots and chlorotic local lesions on young dragon fruit cladodes (Fig. 11.13) (Janssen *et al.*, 2022a). However, it is a challenge to distinguish the typical symptoms of each virus infecting host plants in the field, as the symptoms may depend on virus titre as well as the surrounding environment, and there may be coinfections with other viruses.

Causal agent

PiVX and ZyVX have both been sequenced and re-inoculated into plants to determine the typical symptoms of infection (Mao, 2008). PiVX and ZyVX are also potexviruses and have high sequence homology to CVX, being found in the same cluster (Li *et al.*, 2015; Mao *et al.*, 2018). PiVX and ZyVX are also flexuous rod-shaped viruses, around 580 × 13 nm in size (Parameswari *et al.*, 2022). SchVX was first found infecting *Schlumbergera bridgesii* (Christmas cactus) in Croatia (Plese and Milii, 1966), but in 2008, SchVX was recorded in dragon fruit cladodes producing chlorotic symptoms (Duarte *et al.*, 2008).

Epidemiology

As PiVX, ZyVX and CVX are closely related, they may have a similar host range, as described above for CVX. SchVX has been reported not only in dragon fruit (*Hylocereus* spp.) but also in other cactus species, including Christmas cactus (*S. bridgesii*), Thanksgiving cactus (*S. truncata*) and prickly pear (*Opuntia cochenillifera*) in Croatia, Brazil, China and Mexico (Plese and Milii, 1966; Duarte *et al.*, 2008; Sanches *et al.*, 2015; Zheng *et al.*, 2017; Janssen *et al.*, 2022b).

As for CVX, there are no recorded vectors of PiVX, ZyVX and SchVX. Rather, these viruses are spread by mechanical transmission. Seed transmission of these viruses has not been reported in dragon fruit.

Management

As with CVX disease, there is no treatment for these viral infections, so prevention management is essential. Propagation from virus-free stock, field sanitation and early eradication are extremely important in dragon fruit production, particularly on large-scale dragon fruit farms. Furthermore, deformed or chlorotic dragon fruit plants should not be maintained or propagated for ornamental purposes, particularly close to dragon fruit production areas, as they may be a high-risk inoculum source of these viruses.

11.5 Sunburn

Sunburn is a disorder affecting dragon fruit production in tropical and subtropical regions during summer. Sunburn has been reported to cause severe damage on stems/branches in some planting regions with low humidity or high altitude in Florida (Crane and Balerdi, 2006). It is also a serious problem in dragon fruit in southern Vietnam (Hieu and Hoa, 2015), especially in the central coastal provinces. In Taiwan, sunburn commonly occurs in summer and may reduce the yield of dragon fruit (Chen and Lin, 2016). It also causes damage in India during summer time; complete loss can occur in an orchard if the young plant is exposed to extreme temperatures (Wakchaure *et al.*, 2020).

Symptoms

The signs of sunburn are discoloration on the wings (ribs) of the branches. The symptoms include yellowing, browning, bleaching/scorching and necrosis (Fig. 11.14). A sunburn scale was proposed by Chen and Lin (2016) and modified by Chien and Chang (2019) with five levels: level 0: normal (no damage); level 1: small-scale browning; level 2: large-scale browning; level 3: small-scale bleaching; and level 4: large-scale bleaching and necrosis. The symptoms of sunburn on branches can be similar to the infection caused by diseases. On fruit, the signs of sunburn are discoloration of the peel, followed by a change from its normal colour to yellowish or brownish areas; the flesh underneath may also become soft or mushy. In severe cases, the texture will change, making it inedible. The vigour of the plant and quality of the fruit are affected by sunburn. Therefore, sunburn can cause economic losses in dragon fruit. On young branches, the symptoms usually cause bleaching/scorching on the edge of the wings or the tips of branches (Fig. 11.15), causing rot of the tips, which is similar to bacterial rot. Damage caused by sunburn on the branches, if mild, will leave scars; if severe, it will cause tissue death, and this damage cannot be recovered (Figs 11.16 and 11.17).

Causes of sunburn

Dragon fruit are sunlight-loving plants, and dry tropical climates are their natural habitat. However, sunburn may occur with prolonged exposure to excessive sunlight. If they are exposed to direct sunlight during the hottest hours of the hottest months, they will become stressed and tend to develop signs of sunburn. The hottest hours in the summer day are usually from 12:00 to 15:00. In the year, sunburn usually occurs in the earlier months in the tropics compared with the subtropics. Severe sunburn occurs from March to April in India and Vietnam (Mai *et al.*, 2020; Wakchaure *et al.*, 2020), and from July to August in Taiwan (Chien and Chang, 2019). In tropical regions, temperatures of 35°C or above can cause sunburn.

The severity of sunburn depends on climatic conditions, cultural practices, variety and plant age. Necrosis sunburn usually occurs when the air temperatures are 38°C or higher, while yellowing and browning sunburns usually occur with lower temperatures. Young plant parts and young plants are more susceptible to severe sunburn, while fruit that grow inside the plant canopy are less affected. Severe sunburn occurs in orchards with poor care, unbalanced fertilization and incorrect irrigation, pruning and mulching.

Insufficient soil moisture due to improper watering makes sunburn more severe, and water loss may also occur rapidly from leaves due to high temperatures and dry winds. Sunburn is usually found on the horizontal branches and on the south and west side of the canopy. Mulching around the tree base typically leads to less sunburn than orchards with bare soil. Sunburn also occurs when shaded plant parts are suddenly exposed to sunlight,

Fig. 11.14. Signs of sunburn on mature branches: (A) yellowing, (B, C) browning, (D) bleaching and (E, F) necrosis.

Fig. 11.15. Signs of sunburn on young branches: (A, B) Branch scorch on the edges of the wings. (C) Tip necrosis on the end of the branch.

Fig. 11.16. (A–I) Old sunburn lesions on branches of dragon fruit showing various levels of damage. Lichens can grow on these lesions (H). Photographs courtesy of Mai Van Tri.

such as overpruned plants. Sunburn lesions are usually associated with bacterial and fungal infections. Stem canker lesions are usually found on sunburned areas.

Management

Pruning should be done carefully, and not during a hot and dry summer. Soil moisture should be

Fig. 11.17. (A) A yellow halo may indicate that a secondary pathogen is present. (B) Black fungal mycelium may grow on a sunburn lesion.

maintained through a proper irrigation schedule. Using an overhead sprinkler system for watering during the hottest hours of the day also helps to reduce sunburn. Soil mulching to reduce soil moisture loss and to cool the soil also helps. Maintaining grasses as a free ground cover is one measure for shading the soil surface from sunlight and has a cooling effect on the soil (Fig. 11.18A). Grass cover requires cutting regularly to maintain a mowing height of 8–12 cm. These practices have been recommended and are practised in dragon fruit production in Vietnam.

Diseases usually occur on the injuries/wounds caused by sunburn (Fig. 11.17B). *Erwinia* bacterial rot can be found on sunburn lesions, and stem canker can also appear on the wounds from sunburn lesions or on the yellowing/browning areas of the branch. Sunburn usually occurs on fruit growing on the higher branches on the canopy. If affected, they should be pruned immediately.

Spraying the canopy with a kaolin-based product mixed with water to create a white cover to reduce the surface temperature of branches and fruit has been recommended (Fig. 12.18B). The product instructions should be followed carefully. Growing dragon fruit under netting or shading or using a shade cloth to protect the plants from heat and sunlight during the hot and dry months have also been suggested as measures for sunburn management (Chen and Lin, 2016; Chien and Chang, 2019; Mai *et al.*, 2020).

11.6 Conclusion

In terms of commercial production, dragon fruit is not an easy plant to grow. Many diseases have been reported on this plant, causing economic loss for growers. Stem canker (*N. dimidiatum*) is considered the most common and economically important disease. Anthracnose, *Bipolaris* rot and *Erwinia* soft rot can also a problem for areas with intensive farming systems. Control of these diseases is difficult once the plants are infected. Integrated application of available and appropriate practices to reduce disease progress and keep disease development below an acceptable level to limit losses and reduce the amount of chemicals used is the correct approach for management.

Priority should be given to stem canker as the most important disease once it becomes established in an orchard. Measures for management of stem canker are also helpful for reducing other diseases. The status and levels of other diseases in the orchard should be monitored regularly throughout the year, taking additional measures for management of these diseases as needed.

Fig. 11.18. Measures for management of sunburn: (A) Grass cover. (B) Spraying of a kaolin-based product on to the canopy.

It is necessary to regularly and rigorously apply cultural practices to promote plant health, increase plant tolerance and reduce environmental conditions that lead to disease transmission and development, as well as reducing pathogens in the canopy, trees and orchards. Using tolerant/resistant varieties and disease-free planting materials, using an appropriate planting density, mulching and weed management, regular pruning in combination with removal and destruction of infected parts, suitable irrigation combined with management of leaf wetness duration, practising good sanitation, and applying organic manures and soil amendments in combination with balanced fertilization will ensure a healthy plant with less chance of disease problems. A spraying programme with protectant and curative fungicides designed according to the status of diseases, weather conditions and plant stages is crucial for disease management of dragon fruit. Pesticide registrations are constantly changing, making it difficult to compile a list of registered pesticides for different countries. A combination of cultural measures and an appropriate spraying programme using the most effective fungicides is important to reduce the amount of fungicides used, thereby reducing the risk of exceeding the maximum residue levels in the fruit.

11.7 References

Abirami, K., Sakthivel, K., Sheoran, N., Baskaran, V., Gutam, K. *et al.* (2019) Occurrence of anthracnose disease caused by *Colletotrichum siamense* on dragon fruit (*Hylocereus undatus*) in Andaman Islands, India. *Plant Disease* 103: 768.

Al Hamad, B.M., Al Raish, S.M., Ramadan, G.A., Saeed, E.E., Alameri, S.S.A. *et al.* (2021) Effectiveness of augmentative biological control of *Streptomyces griseorubens* UAE2 depends on 1-aminocyclopropane-1-carboxylic acid deaminase activity against *Neoscytalidium dimidiatum*. *Journal of Fungi* 7: 885.

Al Raish, S.M., Saeed, E.E., Sham, A., Alblooshi, K., El-Tarabily, K.A. *et al.*, (2020) Molecular characterization and disease control of stem canker on royal poinciana (*Delonix regia*) caused by *Neoscytalidium dimidiatum* in the United Arab Emirates. *International Journal of Molecular Sciences* 21: 1033.

Al Raish, S.M., Saeed, E.E., Alyafei, D.M., El-Tarabily, K.A. and AbuQamar, S.F. (2021) Evaluation of streptomycete actinobacterial isolates as biocontrol agents against royal poinciana stem canker disease caused by *Neoscytalidium dimidiatum*. *Biological Control* 164: 104783.

Al-Bedak, O.A., Mohamed, R.A. and Seddek, N.H. (2018) First detection of *Neoscytalidium dimidiatum* associated with canker disease in Egyptian *Ficus* trees. *Forest Pathology* 48: e12411.

Ali, A., Bordoh, P.K., Singh, A., Siddiqui, Y. and Droby, S. (2016) Post-harvest development of anthracnose in pepper (*Capsicum* spp): etiology and management strategies. *Crop Protection* 90, 132–141.

An, N.N., Thao, H.H.M.T., Yen, H.N.H., Hanh, N.T.D., Hoa, N.L.H. *et al.* (2020) Isolation, identification and characterization of bacterial antagonists of the dragon fruit fungal pathogen *Neoscytalidium dimidiatum*. *Journal of Science and Technology* 44B, 76–85.

André, J. (2018) Histopathology of onychomycosis. In: Rigopoulos, D., Elewski, B. and Richert, B. (eds) *Onychomycosis: Diagnosis and Effective Management*. Wiley, Hoboken, New Jersey, pp. 60–65.

Astuti, A., Muftiyatunnisa, S. and Trisnawati, D.W. (2020) Evaluation of antagonistic activity of *Trichoderma harzianum* against various pathogenic fungi infecting dragon fruit stem. *Advances in Engineering Research* 199, 218–221.

Atghia, O., Alizadeh, A., Fotouhifar, K.B., Damm, U., Stukenbrock, E.H. *et al.* (2015) First report of *Colletotrichum fructicola* as the causal agent of anthracnose on common bean and cowpea. *Mycologia Iranica* 2, 139–140.

Athipunyakom, P. and Likhitekaraj, S. (2010) Fruit rot diseases of dragon fruit (*Hylocereus* spp.) in Thailand. In: *Book of Abstracts of the 9th International Mycological Congress, 1–6 August, Edinburgh, UK*, p. 2.216.

Athipunyakom, P., Seemadua, S. and Doungsa-ard, C. (2015) Dragon fruit (pitaya) diseases in Thailand: incidence and management strategies. In: *Proceedings of an International Workshop on Improving Pitaya Production and Marketing, 7–9 September, Kaohsiung, Taiwan*, pp. 95–106.

Avner, S., Nir, N. and Henri, T. (2005) Combination of oral terbinafine and topical ciclopirox compared to oral terbinafine for the treatment of onychomycosis. *Journal of Dermatological Treatment* 16, 327–330.

Awang, Y., Ghani, M.A.A. and Sijam, K. (2010) Effects of *Colletotrichum gloeosporioides* and *Monilinia fructicola* on quality of red flesh dragon fruit (*Hylocereus polyrhizus*). *Acta Horticulturae* 880, 431–437.

Bae, M. and Park, M.R. (2022) First report of pitaya virus X and zygocactus virus X infection of dragon fruit (*Selenicereus undatus*) in Korea. *Journal of Plant Pathology* 104: 1581.

Bae, S., Kim, S.G. and Kim, Y.H. (2013) Biocontrol characteristics of *Bacillus* species in suppressing stem rot of grafted cactus caused by *Bipolaris cactivora*. *Plant Pathology Journal* 29, 42–51.

Balendres, M.A. and Bengoa, J.C. (2019) Diseases of dragon fruit (*Hylocereus* species): etiology and current management options. *Crop Protection* 126: 104920.

Ben-Ze'ev, I.S., Assouline, I., Levy, E. and Elkind, G. (2011) First report of *Bipolaris cactivora* causing fruit blotch and stem rot of dragon fruit (pitaya) in Israel. *Phytoparasitica* 39, 195–197.

Bishop, A.L. and Davis, R.M. (1990) Internal decay of onions caused by *Enterobacter cloacae*. *Plant Disease* 74, 692–694.

Bordoh, P.K., Ali, A., Dickinson, M., Siddiqui, Y. and Romanazzi, G.A. (2020) Review on the management of postharvest anthracnose in dragon fruits caused by *Colletotrichum* spp. *Crop Protection* 130: 105067.

Chang, M., Hyun, I.H. and Lee, Y.H. (1998) *Bipolaris* stem rot of cactus caused by *Bipolaris cactivora* (Petrak) Alcorn. *Korean Journal of Plant Pathology* 14, 661–663.

Chase, A.R. (1982) Stem rot and shattering of Easter cactus caused by *Drechslera cactivora*. *Plant Disease* 66, 602–603.

Chen, Y.C. and Lin, Y.Y. (2016) Research on preventing sunburn damages on the cladodes of pitaya by shading treatments. *Research Bulletin Taitung District Agriculture Improvement Station* 26, 41–58.

Chien, Y.C. and Chang, J.C. (2019) Net houses effects on microclimate, production, and plant protection of white-fleshed pitaya. *Hortscience* 54, 692–700.

Choi, M.O., Kim, S.G. and Kim, Y.H. (2010) Suppression of *Bipolaris* stem rot on cactus by heat-inactivated conidial suspension of *Bipolaris cactivora*. *Plant Pathology Journal* 26, 231–237.

Chuang, M.F., Ni, H.F., Yang, H.R., Shu, S.L., Lai, S.Y. and Jiang, Y.L. (2012) First report of stem canker disease of pitaya (*Hylocereus undatus* and *H. polyrhizus*) caused by *Neoscytalidium dimidiatum* in Taiwan. *Plant Disease* 96: 906.

Cother, E.J. and Gilbert, R.L. (1990) Presence of *Erwinia chrysanthemi* in two major river systems and their alpine sources in Australia. *Journal of Applied Bacteriology* 69, 729–738.

Crane, J.H. and Balerdi, C.F. (2006) Pitaya growing in the Florida home landscape. HS1068/HS303, 11/2005. *EDIS* 2006 (21). Available at: https://doi.org/10.32473/edis-hs303-2005.

Crous, P.W., Slippers, B., Wingfield, M.J., Rheeder, J., Marasas, W.F.O. *et al.* (2006) Phylogenetic lineages in the Botryosphaeriaceae. *Studies in Mycology* 55, 235–253.

Deng, T.C., Lin, C.P., Chou, C.M., Tsai, J.N. and Ann, P.J. (2016) Pitaya diseases in Taiwan. In: *Regional Workshop on the Control of Dragon Fruit Diseases, 4 September, Khan Kaen, Thailand*. Dragon Fruit Network, Taipei, Taiwan.

Duarte, L.M.L., Alexandre, M.A.V., Rivas, E.B., Harakava, R., Galleti, S.R. *et al.* (2008) Potexvirus diversity in *Cactaceae* from Sao Paulo state in Brazil. *Journal of Plant Pathology* 90, 545–551.

Dung, D.T.T., Binh, L.T., Truc, V.T.T. and Quang, V.D. (2018) Antagonistic activity of *Bacillus* sp. isolates against fungus *Neoscytalidium dimidiatum* causing the white spot disease in pitahaya. *Journal of Science: Natural Sciences and Technology* 15, 32–42.

Dy, K.S., Wonglom, P., Pornsuriya, C. and Sunpapao, A. (2022) Morphological, molecular identification and pathogenicity of *Neoscytalidium dimidiatum* causing stem canker of *Hylocereus polyrhizus* in Southern Thailand. *Plants* 11: 504.

Espinoza-Lozano, L., Sumba, M., Calero, A., Jiménez, M.I. and Quito-Avila, D.F. (2023) First report of *Neoscytalidium dimidiatum* causing stem canker on yellow dragon fruit (*Hylocereus megalantus*) in Ecuador. *Plant Disease* 107(6), https://doi.org/10.1094/PDIS-06-22-1403-PDN.

Evallo, E., Taguiam, J.D., Bengoa, J., Maghirang, R. and Balendres, M.A. (2021a) First report of *Colletotrichum fructicola*, causing anthracnose of *Hylocereus* plants, in the Philippines. *Czech Mycology* 73, 79–90.

Evallo, E., Taguiam, J.D. and Balendres, M.A. (2021b) A brief review of plant diseases caused by Cactus virus X. *Crop Protection* 143: 105566.

Evallo, E., Taguiam, J.D., Bengoa, J., Maghirang, R. and Balendres, M.A. (2022a) First report of *Colletotrichum tropicale* on dragon fruit and the response of three *Selenicereus* species to anthracnose. *International Journal of Pest Management*. https://doi.org/10.1080/09670874.2022.2027551.

Evallo, E., Taguiam, J.D., Tumolva, J.A., Bengoa, J., Maghirang, R. *et al.* (2022b) First report of *Cactus virus X* in *Selenicereus monacanthus* in the Philippines. *Indian Phytopathology* 75, 1237–1239.

Ezra, D., Liarzi, O., Gat, T., Hershcovich, M. and Dudai, M. (2013) First report of internal black rot caused by *Neoscytalidium dimidiatum* on *Hylocereus undatus* (Pitahaya) fruit in Israel. *Plant Disease* 97: 1513.

Farr, D.F., Bills, G.F., Chamuris, G.P. and Rossman, A.Y. (1989) *Fungi on Plants and Plant Products in the United States*. APS Press, St Paul, Minnesota.

Fuentes-Aragón, D., Juárez-Vázquez, S.B., Vargas-Hernández, M. and Silva-Rojas, H.V. (2018) *Colletotrichum fructicola*, a member of *Colletotrichum gloeosporioides sensu lato*, is the causal agent of anthracnose and soft rot in avocado fruits cv. "Hass". *Mycobiology* 46, 92–100.

Fullerton, R.A., Sutherland, P.A., Rebstock, R.S., Hieu, N.T., Thu, N.N.A. *et al.* (2018). The life cycle of dragon fruit canker caused by *Neoscytalidium dimidiatum* and implications for control. In: *Proceedings of Dragon Fruit Regional Network Initiation Workshop*. Food and Fertilizer Technology Center for the Asian and Pacific Region, Taipei, Taiwan, pp. 71–80.

García-González, T., Sáenz-Hidalgo, H.K., Silva-Rojas, H.V., Morales-Nieto, C., Vancheva, T. *et al.* (2018) *Enterobacter cloacae*, an emerging plant-pathogenic bacterium affecting chili pepper seedlings. *Plant Pathology Journal* 34, 1–10.

Grenier, A.M., Duport, G., Pagès, S., Condemine, G. and Rahbé, Y. (2006) The phytopathogen *Dickeya dadantii* (*Erwinia chrysanthemi* 3937) is a pathogen of the pea aphid. *Applied Environmental Microbiology* 72, 1956–1965.

Guo, L.W., Wu, Y.X., Ho, H.H., Su, Y.Y., Mao, Z.C. *et al.* (2014) First report of dragon fruit (*Hylocereus undatus*) anthracnose caused by *Colletotrichum truncatum* in China. *Journal of Phytopathology* 162, 272–275.

Hai, N.T. and Thuy, Q.H. (2021) Isolation and evaluation of antagonistic activities of *Bacillus* spp. against *Neoscytalldium dimidiatum* causing brown spot disease on dragon fruits. *Journal of Agriculture and Rural Development* 1, 60–64.

Hajlaoui, M.R., Nouri, M.T., Hamrouni, N., Trouillas, F.P., Ben Yahmed, N. *et al.* (2018) First record of dieback and decline of plum caused by *Neoscytalidium dimidiatum* in Tunisia. *New Disease Reports* 38: 20.

Hay, R.J. (2002) *Scytalidium* infections. *Current Opinion in Infectious Diseases* 15, 99–100.

He, P.F., Ho, H., Wu, X.X., Hou, M.S. and He, Y.Q. (2012) *Bipolaris cactivora* causing fruit rot of dragon fruit imported from Vietnam. *Plant Pathology & Quarantine* 2, 31–35.

Hien, P.T.T. and Oanh, V.T.T. (2014) [Identification, morphological and biological characterizations of *Neoscytalidium dimidiatum* causing canker on dragon fruit]. In: *Proceedings of the 16th National Conference of Phytopathological Society of Vietnam, 6–7 May, Nong Lam University, Ho Chi Minh City, Vietnam* (in Vietnamese).

Hieu, N.T. and Hoa, N.V. (2015) Management strategies of major pitaya diseases in Vietnam. In: *Proceedings of the International Workshop 'Improving Pitaya Produciton and Marketing, 7–9 September, Kaohsiung, Taiwan*, pp. 129–142,

Hieu, N.T., Thu, N.N.A. and Hoa, N.V. (2014a) [The study identified factors, morphological and biological properties of pathogenic fungi *Neoscytalidium dimidiatum* causing brown spot on the dragon fruit plants (*Hylocereus undatus*)]. In: *Proceedings of the 13th National Workshop on Plant Diseases in Vietnam, 6–7 May, Nong Lam University, Ho Chi Minh* City, Vietnam, pp. 114–120 (in Vietnamese).

Hieu, N.T., Thu, N.N.A., Cuong, N.H. and Hoa, N.V. (2014b) [The evaluation of some agrochemicals against canker (*Neoscytalidium dimidiatum*) on dragon fruit crop (*Hylocereus undatus*)]. In: *The 13th National Workshop on Plant Diseases in Vietnam, 6–7 May, Nong Lam University, Ho Chi Minh City, Vietnam*, pp. 191–199 (in Vietnamese).

Hieu, N.T., Thu, N.N.A., Linh, D.T., Uyen, D.K., Hoa, N.V. *et al.* (2016) Pitaya diseases in Vietnam. In: *Papers of a Regional Workshop on the Control of Dragon Fruit Diseases, 4–8 September, Khan Kaen, Thailand*.

Hieu, N.T., Son, N.H., Ngoc, N.T.B., Thu, N.N.A., Thanh, N.T.K. *et al.* (2018) [*The Protocol for Integrated Management of Stem Canker (*Neoscytalidium dimidiatum*) on Dragon Fruit*]. Southern Horticultural Research Institute, Ministry of Agriculture and Rural Development, Hanoi, Vietnam (in Vietnamese).

Hieu, N.T., Thu, N.N.A., Thanh, N.T.K., Hoa, N.V., Linh, D.T. *et al.* (2022) [Study on using antagonistic microorganisms to control *Neoscytalidium dimidiatum* fungus causing canker disease on dragon fruit]. *Journal of Vietnam Agricultural Science and Technology* 1, 95–101 (in Vietnamese).

Hoa, N.V., Hieu, N.T., Uoc, T., Linh, D.T., Man, V.M. *et al.* (2011) *Management of Dragon Fruit Diseases in Binh Thuan province*. Report of ACP Project (WB funding). Southern Horticultural Research Institute, Ministry of Agriculture and Rural Development, Hanoi, Vietnam (in Vietnamese).

Hong, C.F., Zhang, S., Gazis, R., Crane, J.H. and Wasielewski, J. (2019) *Stem and Fruit Canker of Dragon Fruit in South Florida*. IFAS Extension, University of Florida, Gainesville, Florida. Available at: https://edis.ifas.ufl.edu/publication/PP355 (accessed 12 November 2023).

Hong, C.F., Gazis, R., Crane, J.H. and Zhang, S. (2020) Prevalence and epidemics of *Neoscytalidium* stem and fruit canker on pitahaya (*Hylocereus* spp.) in South Florida. *Plant Disease* 104, 1433–1438.

Howard, D.H. (ed.) (2003) *Pathogenic Fungi in Humans and Animals*, 2nd edition, Marcel Dekker, New York.

Hugouvieux-Cotte-Pattat, N., Condemine, G., Nasser, W. and Reverchon, S. (1996) Regulation of pectinolysis in *Erwinia chrysanthemi*. *Annual Review of Microbiology* 50, 213–257.

Hung, T.N., Quynh, D.N., Anh, N.V.H.A. and Huong, D.Q. (2018) Studying the capability to use *Trichoderma* sp. to controls *Scytalidium* sp. causing the brown-spot disease on the dragon fruit. *Thu Dau Mot University Journal of Science* 2, 25–31.

Huy, N.D., Son, N.H., Nguyen, N.T.B.N. and Hieu, N.T. (2019) Detection and identification of *Cactus virus X* (CVX) infecting *Hylocereus undulatus* in Viet Nam. *Plant Protection Journal* 1, 37–44.

Hyun, I.H., Lee, S.D., Hwang, B.C., Ko, K.I., Chung, H.S. *et al.* (2001) Occurrence of stem rot caused by *Bipolaris cactivora* on different species of cactus and its pathogenicity. *Research in Plant Disease* 7, 56–59.

Ismail, S.I., Dahlan, K.A., Abdullah, S. and Zulperi, D. (2021) First report of *Neoscytalidium dimidiatum* causing fruit rot on guava (*Psidium guajava*) in Malaysia. *Plant Disease* 105, 220.

Janssen, D., Garcia, C. and Ruiz, L. (2022a) First report of *Cactus virus X* in Spain. *Australasian Plant Disease Notes* 17, 34.

Janssen, D., García, C. and Ruiz, L. (2022b) First report of *Schlumbergera Virus X* in dragon fruit (*Hylocereus* spp.) in Spain. *Plant Disease* 106: 2004.

Kim, B.S., Kim, S.G. and Kim, Y.H. (2016a) Biocontrol characteristics of *Bacillus* species in suppressing stem rot of grafted cactus caused by *Bipolaris cactivora*. *Plant Pathology Journal* 29, 42–51.

Kim, J.B., Roh, K.H., Kim, H.U., Lee, K.R., Kim, S.H. *et al.* (2012) Reverification of transgenic resistance of trigonal cactus (*Hylocereus trigonus*) to *Cactus virus X* (*CVX*). *Journal of the Korean Plant Biotechnology Society* 2012, 94.

Kim, J.H., Jeoung, M.I., Hyun, I.H. and Kim, Y.H. (2004) Potential biotypes in Korean isolates of *Bipolaris cactivora* associated with stem rot of cactus. *Plant Pathology Journal* 20, 165–171.

Kim, J.S., Park, C.Y., Nam, M., Lee, J.S., Kim, H.G. *et al.* (2016b) First report of *Cactus virus X* infecting *Hylocereus undatus* in Korea. *Plant Disease* 100: 2544.

Kuruppu, M., Siddiqui, Y., Ahmad, K. and Ali, A. (2021) First report of postharvest stem end rot disease on MD2 pineapple fruits caused by *Neoscytalidium dimidiatum* in Malaysia. *Plant Disease* 105, 1564.

Lai, K., Chen, S., Hu, M., Hu, Q., Geng, P. *et al.* (2012). Control of postharvest green mold of citrus fruit by application of endophytic *Paenibacillus polymyxa* strain SG-6. *Postharvest Biology and Technology* 69, 40–48.

Lan, G.B., He, Z.F., Xi, P.G. and Jiang, Z.D. (2012) First report of brown spot disease caused by *Neoscytalidium dimidiatum* on *Hylocereus undatus* in Guangdong, Chinese Mainland. *Plant Disease* 96: 1702.

Le, T.B., Truong, M.N., Ho, T.N. and Le, T.H. (2022) *Trichoderma asperellum* T10 – a potential microbial strain against the *Neoscytalidium dimidiatum* causes white spot disease on dragon fruit in Binh Thuan province. *Journal of Science and Technology* 64, 14–19.

Li, Y.S., Mao, C.H., Kuo, T.Y. and Chang, Y.C. (2015) Viral diseases of pitaya and other Cactaceae plants. In: *Workshop on Improving Pitaya Production and Marketing, 7–9 September, Fengshan Tropical Horticultural Experiment Branch, Taiwan Agricultural Research Institute, Kaohsiung, Taiwan*, pp. 119–127.

Lin, C.P., Ann, P.J., Huang, H.C., Chang, J.T., Tsai, J.N. (2017) Anthracnose of pitaya (*Hylocereus* spp.) caused by *Colletotrichum* spp., A new postharvest disease in Taiwan. *Journal of Taiwan Agriculture Research* 66, 171–183.

Liou, M.R., Hung, C.L., and Liou, R.F. (2001) First report of *Cactus virus X* on *Hylocereus undatus* (Cactaceae) in Taiwan. *Plant Disease* 85: 229.

Liou, M.R., Chen, Y.R. and Liou, R.F. (2004) Complete nucleotide sequence and genome organization of a *Cactus virus X* strain from *Hylocereus undatus* (Cactaceae). *Archieve Virology* 149, 1037–1043.

Lobo, R., Fernandez de Soto, J., Aguiar, J.L., Mathews, D.M. and Tanizaki, G. (2014). Cactus virus X (CVX) a new threat to pitahaya/dragon fruit (*Hylocereus* spp.) production in California. *In: Abstracts of the 2014 ASHS Annual Conference, 31 July, Orlando, Florida*. Available at; https://ashs.confex.com/ashs/2014/webprogram/Paper20310.html (accessed 12 November 2023).

Luong, H.T., Kieu, B.T.N., Vu, T.N., Ha, T.T., van Tong, H. *et al.* (2016) Study on the possibility of using microorganisms as biological agents to control fungal pathogens *Neoscytalidium dimidiatum* causing disease of brown spots on the dragon fruit. *Journal of Vietnamese Environment* 8, 41–44.

Ma, W.J., Yang, X., Wang, X.R., Zeng, Y.S., Liao, M.D. *et al.* (2014) First report of anthracnose disease on young stems of bawanghua (*Hylocereus undatus*) caused by *Colletotrichum gloeosporioides* in China. *Plant Disease* 98, 991–992.

Machouart, M., Menir, P., Helenon, R., Quist, D. and Desbois, N. (2013) *Scytalidium* and scytalidiosis: what's new in 2012? *Journal de Mycologie Médicale* 23, 40–46.

Mai, V.T., Vu, T.H, Nguyen, T.V., Chu, T.A., Tran, T.V. *et al.* (2020) [*Investigation of the Status of Production and Cultivation of Off-season-production Dragon Fruits in Southern Vietnam*]. Project Report. Southern Horticultural Research Institute, Ministry of Agriculture and Rural Development, Hanoi, Vietnam (in Vietnamese).

Mai, V.T., Ha, V.T., Van, N.T., An, C.T., Tran, T.V. *et al.* (2022) [*Research on Improvement of Practices of Of-season Production of Some Main Fruits in South Vietnam (Dragon fruit, Longan, Rambutan, Durian, and Mango)*]. Project Report, Southern Horticultural Research Institute, Ministry of Agriculture and Rural Development, Hanoi, Vietnam (in Vietnamese).

Mao, C.H. (2008) [Molecular characterization and detection of new *Zygocactus virus X* and *Pitaya virus X* from pitaya]. MSc Thesis, National Taiwan University, Taipei City, Taiwan (in Chinese).

Mao, C.H., Kuo, T.-Y., Lu, Y., Li, Y.-S. and Chang, Y.-C. (2018) Pitaya viral diseases and their detection methods in Taiwan. In: *Dragon Fruit Regional Network Initiation Workshop, 23–24 April, Taipei, Taiwan*.

Masanto, M., Sijam, K., Awang, Y. and Satar, M.G.M. (2018) First report of necrotic spot disease caused by *Cactus Virus X* on dragon fruit (*Hylocereus* spp.) in Peninsular Malaysia. *Jurnal Perlindungan Tanaman Indonesia* 22: 1–12.

Masratul Hawa, M., Salleh, B. and Zakaria, L. (2013) Identification and molecular characterizations of *Neoscytalidium dimidiatum* causing stem canker of red-fleshed dragon fruit (*Hylocereus polyrhizus)* in Malaysia. *Journal of Phytopathology* 161, 841–849.

Masratul Hawa, M., Salleh, B. and Latiffah, Z. (2015a) An overview of fungal diseases of pitaya in Malaysia. In: *Proceedings of an International Workshop on Improving Pitaya Production and Marketing*, 7–9 September, Kaohsiung, Taiwan, pp. 87–93.

Masratul Hawa, M., Salleh, B. and Latiffah, Z. (2015b) Identification and molecular characterizations of *Neoscytalidium dimidiatum* causing stem canker of red-fleshed dragon fruit (*Hylocereus polyrhizus)* in Malaysia. *Journal of Phytopathology* 163, 67–71.

Masyahit, M., Sijam, K., Awang, Y. and Satar, M.G.M. (2009a) The first report of the occurrence of anthracnose disease caused by *Colletotrichum gloeosporioides* (Penz.) Penz. & Sacc. on dragon fruit (*Hylocereus* spp.) in Peninsular Malaysia. *American Journal of Applied Sciences* 6, 902–912.

Masyahit, M., Sijam, K., Awang, Y., and Ghazali, M. (2009b) First report on bacterial soft rot disease on dragon fruit (*Hylocereus* spp.) caused by *Enterobacter cloacae* in Peninsular Malaysia. *International Journal of Agriculture and Biology* 1, 659–666.

Masyahit, M., Sijam, K., Awang, Y. and Ghazali, M. (2013) The occurrence of anthracnose disease caused by *Colletotrichum gloeosporioides* on dragon fruit (*Hylocereus* spp.) in Peninsular Malaysia. *Acta Horticulturae* 975, 187–195.

Meetum, P., Leksomboon, C. and Kanjanamaneesathian, M. (2015) First report of *Colletotrichum aenigma* and *C. siamense*, the causal agents of anthracnose disease of dragon fruit in Thailand. *Journal of Plant Pathology* 97, 391–403.

Mezzatesta, M.L., Gona, F. and Stefani, S. (2012) *Enterobacter cloacae* complex: clinical impact and emerging antibiotic resistance. *Future Microbiology* 7, 887–902.

Muas, I. and Jumjunidang (2015) Status of dragon fruit cultivation and marketing in Indonesia. In: *Workshop on Improving Pitaya Production and Marketing in Indonesia, 7–9 September, Kaohsiung, Taiwan*, pp. 19–29.

Nadiah, N., Mohamed Nor, N.M.I., Latiffah, Z. and Masratul Hawa, M. (2017) First report of leaf blight on white spider lily caused by *Neoscytalidium dimidiatum* in Malaysia. *New Disease Reports* 35: 16.

Nascimento, M.B., Belle, C., Azambuja, R.H., Maich, S.L., Neves, C.G. *et al.* (2019) First report of *Colletotrichum karstii* causing anthracnose spot on pitaya (*Hylocereus undatus*) in Brazil. *Plant Disease* 103: 2137.

Nelson, S. (2009) *Bacterial Leaf Blight of Aglaonema*. PD-64. Cooperative Extension Service, College of Tropical Agriculture and Human Resources, University of Hawaii at Manoa, Honolulu, Hawaii.

Nerd, A., Gutman, F. and Mizrahi, Y. (1999) Ripening and postharvest behaviour of fruits of two *Hylocereus* species (Cactaceae). *Postharvest Biology and Technology* 17, 39–45.

Ngoc, N.K., Phong, N.V., Tung, N.T., Hoa, N.V., Woolf, A.B. and Fullerton, R.A. (2017) Postharvest diseases and effect of hot water treatments on white fleshed dragon fruit [*Hylocereus undatus* (Haw.) Britton&Rose]. *International Journal of Current Research in Biosciences and Plant Biology* 4, 30–37.

Ni, H.F., Huang, C.W., Hsu, S.L., Lai, S.Y. and Yang, H.R. (2013) Pathogen characterization and fungicide screening of stem canker of pitaya. *Journal of Taiwan Agricultural Research* 62, 225–234.

Nishijima, K.A., Wall, M.M. and Siderhurst, M.S. (2007) Demonstrating pathogenicity of *Enterobacter cloacae* on macadamia and identifying associated volatiles of gray kernel of macadamia in Hawaii. *Plant Disease* 91, 1221–1228.

Noegrohati, S., Sulasmi, S., Hernadi, E. and Asviastuti, S. (2019) *Food Quality and Safety* 3, 99–106.

Nouri, M.T., Lawrence, D.P., Yaghmour, M.A., Michailides, T.J. and Trouillas, F.P. (2018) *Neoscytalidium dimidiatum* causing canker, shoot blight and fruit rot of almond in California. *Plant Disease* 102, 1638–1647.

Oanh, V.T.T., Tuyen, B.C., Thanh, N.P., Don, L.D. and Hien, P.T.T. (2014) Identification of the causative agent of brown spot disease (*Neoscytalidium dimidiatum*) on dragon fruit based on the ITS-RADN region sequence. *Vietnam Journal of Agriculture and Rural Development* 21, 17–23.

Oeurn, S., Jitjak, W. and Sanoamuang, N. (2015). Fungi on dragon fruit in Loei province, Thailand and the ability of *Bipolaris cactivora* to cause post-harvest fruit rot. *Asia-Pacific Journal of Science and Technology* 20, 405–418.

Onions, A.H.S., Allsopp, D. and Eggins, H.O.W. (1981) *Smith's Introduction to Industrial Mycology*, 7th edn. Edward Arnold, London.

Palmateer, A.J., Ploetz, R.C., van Santen, E. and Corell, J.C. (2007) First report of anthracnose caused by *Colletotrichum gloeosporioides* on pitahaya. *Plant Disease* 91: 631.

Parameswari, B., Bhaskar, B., Karthikaiselvi, L., Sivaraj, N., Mangrauthia, S.K. *et al.* (2022) First report of the association of Zygocactus virus X with dragon fruit (*Hylocereus* spp.) plants from Telangana, India. *Plant Disease* 107, 1249.

Peng, Z., Guan, J., Mou, D., Zou, X., Wang, B. *et al.* (2022) Analysis of bacterial community composition and ecological function during soft rot process in pitaya (*Hylocereus* spp.) stems. *Journal of Chemistry* 2022: 9169433.

Plese, N. and Milii, D. (1966) Vergleichende untersuchungen an isolaten des Kakteen-X-Virus mit Testpflanzen. *Journal of Phytopathology* 55, 197–210.

Prihastuti, H., Cai, L., Chen, H., McKenzie, E.H.C. and Hyde, K.D. (2009). Characterization of *Colletotrichum* species associated with coffee berries in northern Thailand. *Fungal Diversity* 39, 89–109.

Punithalingam, E. and Waterston, J.M. (1970) *Hendersonula toruloidea*. Descriptions of Pathogenic Fungi and Bacteria. No. 28. CAB International, Wallingford, UK. Available at: www.cabidigitallibrary.org/doi/10.1079/DFB/20056400274 (accessed 12 November 2023).

Qiu, F., Yang, J.S., Li, X., Xie, C.P., Li, J. and Zheng, F.Q. (2021) First report of *Bipolaris cactivora* causing flower rot of pitaya (*Hylocereus costaricensis*) in China. *Plant Disease* 105: 1205.

Ratanaprom, S., Nakkanong, K., Nualsri, C., Jiwanit, P., Rongsawat, T. and Woraathakorn, N. (2021) Overcoming encouragement of dragon fruit plant (*Hylocereus undatus*) against stem brown spot disease caused by *Neoscytalidium dimidiatum* using *Bacillus subtilis* combined with sodium bicarbonate. *Plant Pathology Journal* 37, 205–214.

Ray, J.D., Burgess, T. and Lanoiselet, V.M. (2010) First record of *Neoscytalidium dimidiatum* and *N. novaehollandiae* on *Mangifera indica* and *N. dimidiatum* on *Ficus carica* in Australia. *Australasian Plant Disease Notes* 5, 48–50.

Reckhaus, P. (1987) Hendersonula Dieback of Mango in Niger. *Plant Disease* 71: 1045.

Reichel, H., Pérez, R., Martínez, A.K., Sedano Cruz, R.E., Cuervo Ibáñez, M. *et al.* (2002) Caracterización parcial de un potexvirus aislado de pitahaya amarilla (*Selenicereus megalanthus*) afectada por mosaico en Colombia. *Fitopatología Colombiana* 26, 61–66.

Retana-Sánchez, K., Blanco-Meneses, M. and Castro-Zúñiga, O. (2019) Etiología del cáncer del tallo provocado por *Neoscytalidium dimidiatum* (Penz) en *Hylocereus costaricensis,* en Costa Rica. *Agronomía Costarricense* 43, 21–33.

Reverchon, S. and Nasser, W. (2013) *Dickeya* ecology, environment sensing and regulation of virulence programme. *Environmental Microbiology Reports* 5, 622–636.

Robert-Baudouy, J., Nasser, W., Condemine, G., Reverchon, S., Shevchik, V.E. *et al.* (2000) Pectic enzymes of *Erwinia chrysanthemi*, regulation and role in pathogenesis. In: Stacey, G. and Keen, N.T. (eds) *Plant–Microbe Interactions, Vol* 5. APS Publishing, St Paul, Minnesota, pp. 221–226.

Saini, T.J., Gupta, S.G. and Anandalakshmi, R. (2016) First report of papaya anthracnose caused by *Colletotrichum fructicola* in India. *New Disease Reports* 34: 7.

Salunkhe, V.N., Bhagat, Y.S., Chavan, S.B., Lonkar, S.G. and Kakade, V.D. (2022) First report of *Neoscytalidium dimidiatum* causing stem canker of dragon fruit (*Hylocereus spp.*) in India. *Plant Disease* 107, 1222.

Samson, R., Legendre, J.B., Christen, R., Fischer-Le Saux, M., Achouak, W. *et al.* (2005) Transfer of *Pectobacterium chrysanthemi* (Burkholder *et al.*, 1953) Brenner *et al.*, 1973 and *Brenneria paradisiaca* to the genus *Dickeya* gen. nov. as *Dickeya chrysanthemi* comb. nov. and *Dickeya paradisiaca* comb. nov. and delineation of four novel species, *Dickeya dadantii* sp. nov., *Dickeya dianthicola* sp. nov., *Dickeya dieffenbachiae* sp. nov., and *Dickeya zeae* sp. nov. *International Journal of Systematic and Evolutionary Microbiology* 55, 1415–1427.

Sanahuja, G., Lopez, P. and Palmateer, A.J. (2016) First report of *Neoscytalidium dimidiatum* causing stem and fruit canker of *Hylocereus undatus* in Florida. *Plant Diseases* 100, 1499.

Sanches, M.M., Lamas, N.S., Reis, M.B.A., Arieta-Sosa, J.G., Romano, E. *et al.* (2015) Genome assembly of *Schlumbergera virus X* infecting prickly pear (*Opuntia cochenillifera*) in Brazil. *Genome Announcements* 3: e00133-15.

Sanders, J.W. and Sanders, C.C. (1997) *Enterobacter* spp.: pathogens poised to flourish at the turn of the century. *Clinical Microbiology Reviews* 10, 220–241.

Sanoamuang, N. and Oeurn, S. (2018) The ability of *Bipolaris cactivora* to cause post-harvest fruit rot in Thailand and its prevention treatment. In: *Dragon Fruit Regional Network Initiation Workshop, 23–24 April, Taipei, Taiwan*.

Serato-Diaz, L.M. and Goenaga, R. (2021) First report of *Neoscytalidium dimidiatum* causing stem canker on dragon fruit (*Hylocereus* spp.) in Puerto Rico. *Plant Disease* 105, 2728.

Suu, T.D., Hoa, N.V., Thoai, V.H., Yen, T.T.O., Xuan, P.T.X. *et al.* (2017) [*Practices for Growing and Care of Dragon Fruit*]. Vietnam Academy of Agricultural Sciences & National Agricultural Extension Center, Hanoi, Vietnam (in Vietnamese).

Suzianti, I.V., Intan Sakinah, M.A. and Latiffah, Z. (2014) Characterization and pathogenicity of *Colletotrichum truncatum* causing stem anthracnose of red-fleshed dragon fruit (*Hylocereus polyrhizus*) in Malaysia. *Journal of Phytopathology* 163, 67–71.

Suzianti, I.V., Intan Sakinah, M.A. and Latiffah, Z. (2015) Characterization and pathogenicity of *Colletotrichum truncatum* causing stem anthracnose of red-fleshed dragon fruit (*Hylocereus polyrhizus*) in Malaysia. *Journal of Phytopathology* 163, 67–71.

Taba, S., Mikami, D., Takaesu, K., Ooshiro, A., Moromizato, Z. *et al.* (2006) Anthracnose of pitaya (*Hylocereus undatus*) by *Colletotrichum gloeosporioides*. *Japan Journal of Phytopathology* 72, 25–27.

Taba, S., Miyahira, N., Nasu, K., Takushi, T. and Moromizato, Z. (2007) Fruit rot of strawberry pear (pitaya) caused by *Bipolaris cactivora*. *Journal of General Plant Pathology* 73, 374–376.

Taguiam, J.D., Evallo, E., Bengoa, J., Maghirang, R. and Balendres, M.A. (2020a) Susceptibility of the three dragon fruit species to stem canker and growth inhibition of *Neoscytalidium dimidiatum* by chemicals. *Journal of Plant Pathology* 102, 1077–1084.

Taguiam, J., Evallo, E., Bengoa, J., Maghirang, R. and Balendres, M. (2020b) Detection of *Nigrospora sphaerica* in the Philippines and the susceptibility of three *Hylocereus* species to reddish-brown spot disease. *Journal of the Professional Association for Cactus Development* 22, 49–61.

Takahashi, L.M., Rosa, D.D., Basseto, M.A., de Souza, H.G. and Furtado, E.L. (2008) First report of *Colletotrichum gloeosporioides* on *Hylocereus megalanthus* in Brazil. *Australasian Plant Disease Notes* 3, 96–97.
Takahashi, Y., Takahashi, K., Sato, M., Watanabe, K. and Kawano, T. (1997) Bacterial leaf rot of *Odontioda* orchids caused by *Enterobacter cloacae. Annual Phytopathology Society of Japan* 63, 164–169.
Tarnowski, T.L.B., Palmateer, A.J. and Crane, J.H. (2010) First report of fruit rot on *Hylocereus undatus* caused by *Bipolaris cactivora* in South Florida. *Plant Disease* 94, 1506.
Thu, P.T.L., Thanh, L.T., Dung, L.T.M., Cuong, H.V., Thanh, N.D. *et al.* (2021) Isolation and selection of *Trichoderma* spp. antagonists against pathogens causing anthracnose in mango. *Vietnam Journal of Agricultural Sciences* 19, 1617–1627.
Truc, N.T.N. (2016) [Isolation, selection, and identification of cellulose degrading bacteria from pitaya branch]. In: *Proceedings of the 2nd National Conference on Crop Science, 11–12 August, Vietnam Academy of Agricultural Sciences, Can Tho City, Vietnam*, pp. 972–982 (in Vietnamese).
Truc, N.T.N. (2022) [*Bio-fertilisers for Sustainable Agriculture and Environment*]. Agricultural Publishing House, Hanoi, Vietnam (in Vietnamese).
Tuong, L.T., Uyen, D.T.K., Hieu, N.T. and Hoa, N.V. (2018) Study on antagonistic activity of Actinomyces isolates on anthracnose and brown spot disease on dragon fruit. *Journal of Vietnam Agricultural Sciences and Technology* 1, 78–83.
Türkölmez, Ş, Derviş, S., Çiftçi, O., Serçe, Ç.U. and Dikilitas, M. (2019) New disease caused by *Neoscytalidium dimidiatum* devastates tomatoes (*Solanum lycopersicum*) in Turkey. *Crop Protection* 118, 21–30.
Udayanga, D., Maamgoda, D., Liu, X., Chukeatirote, E. and Hyde, K. (2013) What are the common anthracnose pathogens of tropical fruits? *Fungal Diversity* 61, 165–179.
Uyen, D.K., Dung, T.N. and Hoa, N.V. (2017) Preliminary results of construction phylogenetic tree of *Colletotrichum* spp. causing anthracnose diseases on dragon fruit in Southern provinces. *Journal of Vietnam Agricultural Science and Technology* 1, 68–73.
Uyen, D.T.K., Phen, T.V. and and Hoa, N.V. (2018) Identification of *Colletotrichum trucatum* causing dragon fruit anthracnose and the efficacy of several plant extracts on mycelial growth of the fungus. *Journal of Vietnam Agricultural Sciences and Technology* 1, 83–89.
Valencia-Botín, A.J., Cruz, H.P. and Rodríguez, C.A. (2003) Avances en la etiología y manejo de la pudrición blanda de tallos de pitahaya, *Hylocereus undatus* H. (Cactaceae). *Fitosanidad* 7, 11–17.
Valencia-Botín, A.J., Kokubu, H. and Ortız-Hernandez, Y.D. (2013) A brief overview on pitahaya (*Hylocereus* spp.) diseases. *Australasian Plant Pathology* 42, 437–440.
Wakchaure, G.C., Kumar, S., Meena, K.K., Rane, J. and Pathak, H. (2020) *Dragon Fruit Cultivation in India: Scope, Marketing, Constraints and Policy Issues*. Technical Bulletin No. 46. ICAR-National Institute of Abiotic Stress Management, Baramati, India.
Wang, C.L. and Lin, C.C. (2005) Fruit rot of pitaya and stem rot of cacti in Taiwan. *Taiwan Plant Pathology Bulletin* 14, 269–274.
Weir, B.S., Johnston, P.R. and Damm, U. (2012) The *Colletotrichum gloeosporioides* species complex. *Studies in Mycology* 73, 115–180.
Wharton, P.S. and Diéguez-Uribeondo, J. (2004) The biology of *Colletotrichum acutatum*. *Anales del Jardín Botánico de Madrid* 61, 3–22.
Xie, H.H., Long, L., Huang, S., Mao, L., Huang, Q. *et al.* (2020) First report of black spot caused by *Neoscytalidium dimidiatum* on sisal in Guangxi, China. *Plant Disease* 105, 701.
Xu, G., Zheng, F., Qiu, F., Zhou, J., Zheng, F.Q. *et al.* (2019) First report of sooty mold caused by *Cladosporium tenuissimum* on pitaya in China. *Plant Disease* 104, 290.
Xu, M., Peng, Y., Qi, Z., Yan, Z., Yang, L. *et al.* (2018). Identification of *Neoscytalidium dimidiatum* causing canker disease of pitaya in Hainan, China. *Australasian Plant Pathology* 47, 547–553.
Xuyen, D.T.H., Nam, N.C., Trien, L.T. and Minh, D. (2017). Evaluation of the ability of of local *Trichoderma* spp. strains for controlling brown spot disease caused by *Neoscytalidium dimidiatum* on dragon fruit in Long An province. In: *Proceedings of the 16th National Conference of Phytopathological Society of Vietnam, 21–22 July, University of Science and Education, Da Nang, Vietnam* (in Vietnamese).
Yeh, S.T., Julius, I.P. and Shiesh, C.C. (2016) The severity and prevention of pitaya wet rot disease and pitaya sooty mould disease on pitaya fruits. In: Wu, T.C. *et al.* (eds) *Integration of Technologies for Value Chain Development of Pitaya Industry*. Taichung District Agricultural Improvement Station, Changhua, Taiwan, pp. 71–88.
Yi, R.H., Ling Lin, Q., Mo, J.J., Wu, F.F. and Chen, J. (2015) Fruit internal brown rot caused by *Neoscytalidium dimidiatum* on pitahaya in Guangdong province, China. *Australasian Plant Disease Notes* 10: 13.

Yuan, W.M. and Crawford, D.L. (1995) Characterization of *Streptomyces lydicus* WYEC108 as a potential biocontrol agent against fungal root and seed rots. *Applied and Environmental Microbiology* 61, 3119–3128.

Zakaria, L. (2021) Diversity of *Colletotrichum* species associated with anthracnose disease in tropical fruit crops – a review. *Agriculture* 11: 297.

Zechmann, B. and Zellnig, G. (2009) Rapid diagnosis of plant virus diseases by transmission electron microscopy. *Journal of Virological Methods* 162, 163–169.

Zhang, R.Y., Zhao, S.X., Tan, Z.Q. and Zhu, C.H. (2017) First report of bacterial stem rot disease caused by *Paenibacillus polymyxa* on *Hylocereus undulatus* in China. *Plant Disease* 101, 1031.

Zhang, Y., Liu, Z., Huang, Q., Luo, Y., Pennerman, K.K. *et al.* (2016) Rapid detection of *Cactus Virus X* in pitaya by efficient reverse transcription loop-mediated isothermal amplification. *Austin Journal of Biotechnology Bioengineer* 3: 1055.

Zhao, H.J., Chen, S.C., Chen, Y.F., Zou, C.C., Wang, X.L. *et al.* (2018) First report of red dragon fruit (*Hylocereus polyrhizus*) anthracnose caused by *Colletotrichum siamense* in China. *Plant Disease* 6, 862.

Zheng, Q., Wang, Z. and Peng, Z. (2017) High-throughput sequencing analysis of viruses in pitaya from Guizhou Province, China. *Acta Phytopathologica Sinica* 47, 364–369.

Zheng, Q., Wang, X., Zhou, J. and Ma, Y. (2020) Complete genome sequence of a new member of the genus *Badnavirus* from red pitaya (*Hylocereus polyrhizus*). *Archives of Virology* 165, 749–752.

12 Greenhouse Cultivation

Hamide Gübbük[1]*, Recep Balkıc[2] and Lokman Altınkaya[3]
[1]Department of Horticulture, Faculty of Agriculture, Akdeniz University, Antalya, Türkiye; [2]Department of Plant and Animal Production, Elmalı Vocational School, Akdeniz University, Antalya, Türkiye; [3]Horticulture Program, Department of Plant and Animal Production, Kumluca Vocational School, Akdeniz University, Antalya, Türkiye

12.1 Introduction

Interest in the cultivation of tropical fruit species in subtropical conditions has increased in recent years because of the effect of global climate change. Although global climate change is advantageous in terms of temperature for the growth of tropical fruit species, it has the disadvantage of causing water stress, which affects successful cultivation. Today, drought is one of the most important environmental factors affecting crop production in the Mediterranean climate (Jacobsen *et al.*, 2012). Therefore, in subtropical climate conditions, cultivation of tropical fruit species in greenhouses is an important option where the plants can grow with less water to meet production deficits and export opportunities. Among these tropical species, pitaya is considered one of the important fruit crops that can be grown in plastic greenhouses. One reason for this is that this species is suitable for soil-less cultivation in pots, and has a short timespan between planting and fruiting. Recognition is increasing worldwide of the different uses of pitaya. Le Bellec *et al.* (2006) stated that currently the demand for pitaya is increasing in Europe's exotic fruit market, and this fruit shows high potential as an ornamental agrifood commodity, as well as an industrial source of natural additives.

Pitaya is native to Vietnam and has been grown there for centuries. Currently, it is widely cultivated in tropical and warm subtropical regions, including most of the South-east Asian countries, Israel, Australia and most recently in the USA, Spain and Türkiye (Trivellini *et al.*, 2020). Cultivation of pitaya in Italy is still at the initial stages but is gaining popularity among consumers and small farmers (Trivellini *et al.*, 2020). Israel is one of the leading countries in pitaya cultivation and all plantations are cultivated under shade. The cultivation area in Israel has reached 50 ha in recent years (Y. Mizrahi, 2023, personal communication). Pitaya is also grown in Spain, with a total production area of about 75 ha, of which approximately 25 ha is in the Canary Islands (C. Méndez Hernández, 2023, personal communication). Méndez Hernández further stated that the cultivation temperature should not exceed 30°C and must not be below 13°C for successful cultivation in the Canary Islands due to their close location to the equatorial zone. Therefore, outdoor planting is promoted in the Canary Islands. However, the temperature is not the same in the Iberian Peninsula. Usually, the temperatures in summer are very high (up to 38°C), and pitaya fruit is sun-damaged within several hours. In winter, the temperature decreases to 7°C, resulting in an

* Email: gubbuk@akdeniz.edu.tr

DOI: 10.1079/9781800623156.0012

annual growth lower than that achieved in the Canary Islands. Therefore, pitaya should be grown under movable plastic greenhouses (closed in winter and open in summer). In China, pitaya is grown on an area of about 70,000 ha and plastic is used as a covering material (Y. Qin, 2023, personal communication). In Türkiye, greenhouse pitaya cultivation has increased in recent years. Initially, cultivation was started in plastic greenhouses with lower side heights, or in glass greenhouses (Fig. 12.1) where vegetables are grown. Today, it is grown in more modern plastic greenhouses (Fig. 12.2) in the Mediterranean and Aegean regions. In Türkiye, as in Israel, 40% green shade netting is used on greenhouses in summer, or shade powder is applied to the greenhouse. In Türkiye, pitaya cultivation was carried out over a total area of approximately 200 ha in 2022, with cultivation areas in Antalya, Mersin, Adana, İzmir and Mugla of 108.3, 57, 17.5, 9.6 and 8.1 ha, respectively (Agriculture and Forestry District Directorates, 2023, personal communication). In all locations in Türkiye, cultivation is generally carried out in plastic greenhouses. Total pitaya production in Türkiye in 2022 was 350 t (Agriculture and Forestry District Directorates, 2023, personal communication). Despite the high cost of construction, pitaya is grown under greenhouse conditions in subtropical conditions due to low temperatures, temperature differences between day and night, and changes in temperature in different seasons. These factors affect fruit production, quality, development and the number of flushes during the year. In addition to low temperatures, the high temperatures encountered in subtropical conditions, especially in summer (June–August), negatively affect the growth and development of plants, causing sunburn and adversely affecting pollination, fertilization and consequently yield and quality. In the Gazipasa district of Antalya, where bananas are grown in open fields, adaptation studies carried out on pitaya in open fields showed that the plants must be shaded in summer to avoid sunburn (Gübbük *et al.*, 2017). Therefore, these researchers recommended growing pitaya in plastic greenhouses in subtropical conditions and using 40% shade netting in summer. Mizrahi (2014) reported that the light intensity in Israel is over 2200 μmol photons m^{-2} min^{-1} and hence pitaya is grown completely under netting. Mizrahi also noted that the shading rate in pitaya varies according to the variety and ecology. It has been suggested that the shading rate may be lower in varieties

Fig. 12.1. An iron-framed greenhouse.

Fig. 12.2. A modern greenhouse.

with more waxy stems, and that climate plays an important role in the requirement for shading, with a shading rate along the Mediterranean coast from 20% to 60% in locations with a desert climate. Raveh *et al.* (1998) compared different shading degrees in two different pitaya species (*Selenicereus megalanthus* and *Hylocereus polyrhizus*) in the Negev Desert of Israel. The most suitable degree of shade for growth and fruit production was 30% shade for *H. polyrhizus* and 60% shade for *S. megalanthus*. High summer temperatures inhibited flower production in *Hylocereus undatus* and *H. polyrhizus* at different locations in Israel (Nerd *et al.*, 2002). Shading is also used to increase yield and quality and prevent sunburn damage in semi-tropical and tropical conditions. Chang *et al.* (2016) investigated the effects of three shading nets (25% shading with a white net, 50% and 75% shading with a black) on the growth and fruiting of red-fleshed 'Shih Huo Chuan' pitaya (*H. polyrhizus*) in Taiwan. They recommended using 50% shade for growing 'Shih Huo Chuan' pitaya during the summer. Almeida *et al.* (2018) described the effect of shading on the growth and gas exchange of red-fleshed pitaya in Köppen, Brazil. They observed that cultivation in full sunlight had a negative effect on gas exchange and growth. However, 35% shading for 180 days increased the photosynthetic rate, water-use efficiency and carboxylation with increased vegetative growth. De Oliveira *et al.* (2021) stated that red-fleshed pitaya performed better under shaded conditions, producing higher yield and fruit quality than when grown in full sunlight, especially under 35% shade.

12.2 Advantages and Disadvantage of Greenhouse Cultivation

Scientific studies have been carried out on the advantages of greenhouse cultivation of tropical fruit species under subtropical conditions, especially banana and papaya. Studies on greenhouse banana cultivation have been done in Morocco, the Canary Islands, South Africa and Türkiye (Janick and Ait-Oubahou, 1989; Galán Saúco *et al.*, 1992; Eckstein *et al.*, 1998; Pekmezci *et al.*, 1998; Gübbük and Pekmezci, 2004). Greenhouse construction and the use of new-generation plastics in Türkiye has enabled year-round production, as well as increasing the yield and quality of greenhouse banana cultivation. Studies have also been carried out on the cultivation of papaya in greenhouses in countries such as Spain, Japan, Israel and Türkiye (Pastor *et al.*, 1990; Pastor and Galán Saúco, 1995; Gunes and Gübbük, 2012; Galán Saúco and Pérez Hernández, 2020). Studies on greenhouse cultivation of papaya in Türkiye showed less physical damage to the fruit with positive effects on fruit coloration and low levels of fruit deformities compared with open-field cultivation.

In subtropical conditions, it is suggested that pitaya is grown under cover or netting to protect the plants from wind, rain and extremes of heat and cold. The temperature should not decrease below 0°C and should not exceed 38°C (Crane and Balerdi, 2005). The optimum temperature for growth and development of pitaya plants is 20–30°C, with damage to plants occurring at temperatures above 30–40°C (Mizrahi *et al.*, 2007). However, Nerd *et al.* (2002) and Jaya (2018) reported that the optimum temperature for pitaya to bear fruit is 30–32°C at a relative humidity of 60–80%. Pitaya can be grown in arid tropical and subtropical climates at an average temperature of 21–29°C, can withstand a temperature of 38–40°C, and are affected by frost at temperatures below 0°C for short periods (McMahon, 2003) The plant needs full sunlight for growth, and 600–1300 mm of annual precipitation is considered sufficient for cultivation. The plant develops rapidly when the climatic conditions are suitable and vegetative growth stops when it is not (McMahon, 2003). Therefore, pitaya orchards should be established in subtropical climatic conditions and plants should preferably be grown under greenhouse conditions or netting to avoid chilling injury.

For cultivation under nets, mesh thickness is an important issue. Chien and Chang (2019) investigated the effects of white nets of 16- and 24-mesh insect-proof screens on white-fleshed pitaya in subtropical Taiwan and recommended the use of 16-mesh net for commercial production. Mizrahi and Nerd (1999) stated that 0°C is the minimal threshold for cultivation of most species of climbing cacti, with subfreezing temperatures causing damage. However, chilling injury varies among species. Mizrahi and Nerd (1999) suggested growing pitaya in a plastic or glass greenhouse in locations exposed to cold, while Wang *et al.* (2019) stated that shading treatment reduced chilling injuries in pitaya, and also investigated the effects of shading on cold tolerance of pitaya. The results revealed that growing plants under shade prevented chilling injuries, as well as protecting from the effect of full sunlight. As well as avoiding chilling injuries, shading also protects plants from high temperatures and excessive sunburn. Merten (2003) observed that increasing the temperature above 45°C caused deformation of plants and decreased flowering. Therefore, shading must be carried out in summer to protect the plants and fruits from excessive sunburn. The above studies all demonstrate the importance of temperature, relative humidity and light in pitaya cultivation. Climatic factors that affect plant growth and fruiting can be controlled more easily by growing pitaya in greenhouses or under netting in subtropical conditions. In tropical conditions, shading is required to protect pitaya from excessive sunlight.

There are some disadvantages of greenhouse cultivation. These include the increased construction cost of the greenhouse, and the fact that the use of plastic as a covering material means that the plants do not benefit from precipitation. In addition, the most important problem in greenhouse cultivation of pitaya in Türkiye is that when the temperature in the greenhouse rises above 45–50°C in summer, new shoots occur and the main branches show burn injuries as the temperature continues to rise (Fig. 12.3). Visible and infrared light cause heating in plants and fruits, while ultraviolet light causes burning and darkening of the plant stem and fruit peel (Fischer *et al.*, 2022). Although pitaya is known to be resistant to high temperatures, it was found that high temperatures and light intensity negatively affected the flowering, yield and quality of the fruit (Mizrahi and Nerd, 1999; Nerd *et al.*, 2002; Chang *et al.*, 2016; Chien and Chang, 2019). In greenhouse cultivation of pitaya, when iron poles are used for training the plants, wrapping the iron post in a soft and sun-proof material is required to prevent it from heating up and causing damage to the plants.

12.3 Types of Greenhouse

In Türkiye, tropical fruit species are grown mainly in plastic greenhouses. This may be due to the fact that plastic covering material can cover a

Fig. 12.3. Sunburn damage in plants in a greenhouse.

larger area using one piece, requiring fewer support systems than glass greenhouses, and relatively lower initial investment costs; there have also been improvements in the properties of plastic materials with advancements in technology and it is more economical than glass (Emekli, 2014). Greenhouses in Türkiye are suitable for small family businesses, with the average size of greenhouses varying between 1000 and 3000 m^2, and heating carried out only to protect them from frost (Emekli, 2014). However, in recent years, it has been observed that, as well as traditional greenhouses, modern greenhouses with government support, climate control, and irrigation and automation units for fertilization are becoming more widely used.

Pitaya cultivation in Türkiye began in 1999 in greenhouses where vegetables were gown (gable roof and tunnel type) with pitaya varieties introduced from Vietnam. Cultivation started to increase after the 2010s and reached its highest level in the last 5 years. Today, large-scale plantations are under galvanized tunnel-type greenhouses with a side height of

4.5 m and a middle height of 7 m, with top and side ventilation (Fig. 12.4). Plastic cover materials with ultraviolet and infrared light additives are preferred because they are long lasting and reduce the temperature difference between day and night. In the Mediterranean region of Türkiye, it has been observed that tall greenhouses have more advantages than low greenhouses in terms of ventilation, disease control and climate control. Like Türkiye, plastic is used as a covering material in China (Yonghua Qin, 2023, personnel communication). Unlike Türkiye and China, pitaya cultivation in Israel is done entirely under nets (Y. Mizrahi, 2023, personnel communication). Shading is necessary for cultivation, with the degree of shade depending on region and degree of light in the area. The recommended shade percentage near to the Mediterranean is 25%, while in the Negev Desert and areas between the Negev Desert and Mediterranean, it is 60%. Plants grown in the open air show bleaching due to photoinhibition and die (Y. Mizrahi, 2023, personal communication). In the Canary Islands of Spain, pitaya cultivation is carried out

Fig. 12.4. A galvanized tunnel-type greenhouse.

mainly under nets, as in Israel (C. Méndez Hernández, 2023, personal communication). In mainland Spain, mesh is used as a cover material in cultivation (C. Méndez Hernández, 2023, personal communication). In the Canary Islands, pitaya plants grow faster than those in the Iberian Peninsula, and winter temperatures have a very limited effect on plant growth. Therefore, in the Canary Islands, only the top is covered with mesh, but in places where the winter temperatures are low, a cover is also used on all sides of the greenhouse. In general, it is suggested that mesh should be used as the covering material in warmer places and plastic in cooler places.

12.4 Varieties Grown in Greenhouses

In Türkiye, greenhouse cultivation of pitaya began after the introduction of the Vietnamese varieties 'Red Jaina' and 'White Jaina' in 1999, and continues today with many other varieties introduced from Asia and America. However, there are currently no ideal cultivars/varieties recommended. Pitaya varieties grown in Türkiye bear fruit mainly in summer and autumn. Unlike Israel, there are no varieties in Türkiye that bear fruit in winter. The varieties grown under greenhouse conditions show differences in terms of earliness, yield, size, fruit skin and flesh colour, and fruit shape. Growers generally prefer varieties that bloom and mature at different times, are highly productive, have a fruit weight of more than 250 g, have red flesh colour, and are suitable for transportation and storage, while consumers prefer red-coloured varieties with a good taste, aroma, long shelf life and large fruit size.

The species grown in greenhouses in Türkiye consist of *H. undatus*, *H. polyrhizus* and *H. guatemalensis* and their hybrids.

There are two registered varieties in Türkiye, 'Cosmic Charlie' (Fig. 12.5) and 'Bloody Mary'. 'Bloody Mary' is self-sterile and 'Cosmic Charlie' is self-fertile. Adaptation studies showed that the 'Bloody Mary' variety successfully set fruit with foreign pollination. Although 'Cosmic Charlie' gave good results in terms of fruit size, it was found to be a low yielder. In Türkiye, Vietnamese 'Jaina' is the preferred variety, especially for export, followed by varieties such as 'Red Jaina', 'Dark Star', 'Thai Red' (Fig. 12.6), 'Philippine Purple', 'Purple Haze', 'Ruby Red' and 'Costa Rica White'. Among the varieties grown in Türkiye, the varieties that develop the earliest flower buds are 'American Beauty, 'Dark Star', 'Thai Red' and 'Vietnamese Jaina'. Among the varieties evaluated under greenhouse conditions,

Fig. 12.5. 'Cosmic Charlie' variety growing under greenhouse conditions.

Fig. 12.6. 'Thai Red' variety growing in a greenhouse.

'Costa Rica White' has the highest total soluble solids content followed by 'Thai Red'. In Israel, which has important potential for greenhouse cultivation, both summer and winter clones are used (Y. Mizrahi, 2023, personal communication). Summer clones are '04-25' (= 'Yossi'), 'Golden 89-338', 'Venus' (= '34-07'), 'New Pinky 70-11-04' and 'Mouyal' (= 'S91'). Winter clones are 'Desert King', 'S-75' (= 'Dragon Eggs'), '12-14' (= 'Dragon Balls') and 'Z-11' (= 'Dudai').

In Spain, greenhouse cultivation first started with varieties introduced from Reunion Island and El Salvador (C. Méndez Hernández, 2023, personal communication). However, an adaptation study with different varieties is ongoing in the Canary Islands to determine the best varieties (C. Méndez Hernández, 2023, personal communication). They started with the famous *H. undatus*, *H. hybridum*, *H. purpusii* and *H. triangularis* from Reunion Island and 'JC01', 'JC02', 'JC03' and 'JC05' from El Salvador, introduced by Dr Victor Galan between 2000 and 2002 (C. Méndez Hernández, 2023, personal communication). Since then, 36 varieties of pitaya have been introduced, and more than 45 varieties were evaluated in 2023. Among these varieties, 'Korean White', 'Caribbean Pink', 'Royal Red' and 'Purple Haze' are the most important (C. Méndez Hernández. 2023, personal communication).

In China, approximately 90% of cultivated varieties have red peel with red pulp, and 10% have yellow peel with white pulp or red peel with white pulp (Y. Qin, 2023, personal communication).

12.5 Cultivation Techniques

In greenhouse cultivation of pitaya, cultural practices such as irrigation, fertilization, pruning, mulching, dusting, and control of diseases and pests are carried out in a similar way to outdoor cultivation. There are no public or private institutions in Türkiye that produce only seedlings, so planting materials are usually obtained from production plots that have yielded fruit (Fig. 12.7). Rooted or unrooted cuttings taken from 1-year-old branches with a length of about 25–35 cm are used as planting material in the greenhouse. In plantings with rooted cuttings, the cuttings are grown in approximately 1 l pots in a mixture of peat, perlite and soil (1.5:1:0.5, v:v:v), and are ready for planting after about 3 months. Planting of both rooted and unrooted

Fig. 12.7. New planting materials are raised inside the greenhouse.

cuttings is usually done in March. Horizontal and vertical support systems are needed in greenhouse cultivation (Le Bellec *et al.*, 2006). The training system affects the application of cultural processes, as well as plant growth and development, yield and quality of fruit. Mizrahi (2014) reported that poles with a length of 160 cm are used in cultivation under nets in Israel, with two wires on the poles and the plants wrapped around these wires; this makes pollination and harvesting easy, as well other cultural practices. In Israel, the distance between plants and rows is set at 3 × 1.5 m (Y. Mizrahi, 2023, personal communication). Row planting and column planting are used in pitaya cultivation in China (Y. Quin, 2023, personal communication). In the column-planting method, the distances between plants and rows are 2–2.5 × 2.5–3 m, and 6600–7200 plants ha^{-1} are grown. In the row-planting method, the distance between plants and rows is 0.15–0.5 × 2.2–3 m and 12,000–37,500 plants ha^{-1} are grown. In the Canary Islands, plants are planted with 3 m between rows and 60 cm apart in the row, while in the Iberian Peninsula, the spacing between plants is 40 cm and the spacing between rows is 2.5 m (C. Méndez Hernández, 2023, personal communication).

In Türkiye, pole and T systems are generally used in greenhouse cultivation of pitaya. In the pole system (Fig. 12.8), the poles are made of concrete with a length of 2 m and a diameter of 10 × 10 cm. The pillars are placed on embankments with a height of approximately 20–30 cm. Circles of 80 cm diameter are prepared around the base of the poles and four plants are planted for each main pole. The distance between the poles is generally 1.8–2.0 m and the distance between the rows is maintained at 2.0–2.5 m (7000–10,000 plants ha^{-1}). In the T system, a 40 cm-long 'T' is placed on both sides of the concrete poles and the plants are planted crosswise at 40–50 cm spacing. In the T system, the spacing between rows is generally set at 2.0–2.5 m, as in the pole system (8000–12,500 plants ha^{-1}). In both planting systems, vegetative growth of the plant starts in the first year after planting, and is 10% in the second year, 70% in the third year and 100% after the third year.

Although vegetative growth continues throughout the year in the greenhouse, flower buds form in the second week of April, and flowering (Fig. 12.9) and fruit set occur in May. The fruit season continues until December–January. In Türkiye, flowering occurs in three to five different periods in a year, depending on the climate and variety used.

In greenhouse cultivation, plantations are established with at least two different varieties. These include self-fertile, partially self-fertile and

Fig. 12.8. The pole system of training plants in the greenhouse.

Fig. 12.9. Plants in full bloom in a greenhouse. Photograph courtesy of Dr Y. Mizrahi, Israel, 2023.

self-infertile varieties. However, in order to increase yield and quality, pollination is done by hand using different pollen. The time from fruit set to harvest varies from 30 days (in the summer months) to 35 days (in the autumn months) (Gübbük *et al.*, 2017). See Fig. 12.10 for a general view of fruits at harvesting time.

Most of the roots of pitaya are located at a depth of 15–30 cm in the soil and irrigation is applied when required so that the soil does not dry out (Luders and McMahon, 2006). Irrigation in pitaya is extremely important for fruit quality, as well as for plant growth and development (Mizrahi and Nerd, 1999). The irrigation regime is also extremely important in greenhouse cultivation in terms of fruit cracking. Mizrahi *et al.* (2007) reported that pitaya and prickly pear use only 10% of the water used by citrus fruits, peach, avocado and pear. Irrigation should be done only in the summer season and in very small amounts every day. When plants are overirrigated, water has been reported to slip under the root system (Mizrahi *et al.*, 2007). A drip irrigation system is generally used in greenhouse pitaya cultivation. Drip irrigation pipes are placed on the embankments in the pole and T systems in two rows. In plantations, 16 mm drippers are placed at 20 or 40 cm spacing providing 4 l of water h^{-1}. In addition to the drip irrigation system, in some plantations, a bottom sprinkler system is used early in the morning and in the evening when the relative humidity drops excessively during the summer months. In greenhouse cultivation of pitaya in Türkiye, the amount of irrigation varies according to season, cultivation practices, training systems and the number of plants per hectare. Irrigation is done once or twice a month in the winter when the plants have a minimum water requirement, four times a month in the spring, and eight times a month during the fruiting period when the water requirement is at its maximum level (average 4 l per plant per irrigation). Irrigation practices in pitaya also vary among producers. For example, during the summer months, an average of 32 l of water (eight irrigations of 4 l each) is applied to each plant per month. If this amount of water remains constant, there may be up to 16

Fig. 12.10. General view of fruits at harvesting time. Photograph courtesy of Dr C. Méndez Hernández, Canary Islands, Spain 2023.

irrigations per month by planning 2 l per plant for each irrigation. Fertilization can also be provided via irrigation, with every alternate irrigation containing fertilizer.

In greenhouse cultivation of pitaya, the fertilization programme should be based on soil analysis. Due to the very low organic matter content in the soils in Türkiye, leonardite (1–1.5 t ha^{-1}) or burnt farm manure (25–30 t ha^{-1}) is generally applied as organic matter. In the first months after planting, humic and fulvic acids (20 l ha^{-1}) are applied in spring, autumn and summer to promote root development in plants. Nitrogen-based fertilizer is applied during the vegetative growth and development period. Nitrogen fertilization, especially during the development periods, increases the water content of the fresh shoots and can cause these shoots to be damaged by excessive heat and strong sun in the greenhouse. After the plants reach the height of the pole, the growth tip is cut off and calcium is applied, which helps in the hardening of the tissues. This is applied in addition to nitrogen until side-branch development is complete. Before the formation of flower buds, fertilization is continued with a focus on phosphorus. After the flower buds are formed, boron/zinc is applied to increase pollination and fertilization. After fruit set is achieved, fertilization is continued with a focus on potassium. In Türkiye, it is suggested that an NPK (nitrogen, phosphorus, potassium) mixture of 125:70:80 g per plant is applied over the year.

Weed control is very important in greenhouse cultivation of pitaya. In Türkiye, natural plant waste or inorganic mulch is used for this purpose. Luo *et al.* (2021) studied the effects of different mulching treatments on physicochemical properties, enzyme activity and the soil microbial community in Hainan, China. According to their results, all types of mulch treatment improved the characteristics of the soil compared with the control, and invertase enzyme activity was highest using a natural grass mulch compared with the other treatments. In addition to weed control, pruning is also important. Plants are grown with a single stem until they reach the desirable height for training. The tops are then removed and approximately four branches are left for each plant in the first year. The number of branches is increased exponentially in the second and third years. The ends of the branches left are cut so that they are approximately 20 cm from the ground. In plants, pruning is usually done during the dormant period. Additionally, diseased and undercanopy branches and unproductive branches that may prevent air movement around the plant are removed. In the period close to the formation of flower buds, those that will represent the next year are left from the new shoots emerging from the branches, and the others are removed.

12.6 Pests and Diseases in Greenhouse Cultivation

Aphids, ants, grasshoppers, mealybugs and to some extent thrips are among the most common pests in greenhouse cultivation of pitaya. Aphids, which cause damage to the fruit during the ripening period, also secrete a sugary, sticky liquid called honeydew, causing other insects such as ants to attack the plant. Ants feeding on honeydew secreted by aphids not only damage the plant but also help transfer aphids to another plant (Carrillo *et al.*, 2019).

Due to the soft texture of the new shoots in pitaya, they provide a food source for stinging and sucking insects. Mealybugs thrive in hot and humid environments. Mealybugs (Fig. 12.11) and thrips (Fig. 12.12) in particular cause damage to the fruit, significantly reducing its marketability. Thrips spread in weedy areas (Childers and Nakahara, 2006; Carrillo *et al.*, 2019). In Türkiye, pitaya is also grown in locations where greenhouse vegetables (especially pepper, tomato and aubergine) are widely grown, and damage by thrips is seen more intensely in these locations.

In pitaya cultivation, greenhouse weeds are a significant host to pests. Therefore, frequent cleaning and removal of weeds in the greenhouse and the use of yellow and blue sticky traps are suggested to reduce the pest population.

During cloudy and rainy periods, the inability to fully ventilate causes an increase in relative humidity in the greenhouse, which can lead to infestation with fungal diseases and decreases the quality of cultivation. Grey mould (Fig. 12.13) and stem rot are the leading diseases observed inside greenhouses, especially where ventilation is insufficient. For control of pests and diseases, the measures followed in open-field cultivation are used (discussed in detail in Chapters 10 and 11, this volume).

12.7 Conclusion and Future Prospects

Currently, the growth of pitaya in greenhouses or under netting in subtropical conditions is practised mainly in countries such as Israel, Türkiye, China and Spain. Greenhouse cultivation is expected to increase greatly in the coming years, because pitaya is easily propagated by cuttings, has a relatively low water requirement compared with many other fruit crops, starts to bear fruit in the second year after planting, and has relatively low labour costs, with production costs being covered within a few years of planting. In addition, the fact that pitaya is suitable for long-term storage and transportation, has a high export potential and can be used in many different areas in the food industry as well as for table consumption, along with its high and stable price, indicates that cultivation of this species under greenhouse conditions will increase further in the coming years.

Fig. 12.12. Damage caused by thrips on the fruit.

Fig. 12.11. Mealybug damage on the fruit.

Fig. 12.13. Grey mould damage on the fruit.

12.8 References

Almeida, E.I.B., de Medeiros Corrêa, M.C., Mesquita, R.O., Queiroz, R.F., Cajazeira, J.P. *et al.* (2018) Growth and gas exchanges of red pitaya under different shading conditions. *Revista Brasileira de Ciências Agrárias* 13, 1–8.

Carrillo, D., Duncan, R. and Peña, J. (2019) Pitaya (dragon fruit) (*Hylocereus undatus*) Pests and Beneficial Insects. Tropical Research and Education Center, Homestead, Florida. Available at: https://trec.ifas.ufl.edu/media/trecifasufledu/public-notices/Pitaya-EDIS.pdf (accessed 13 November 2023).

Chang, P.T., Hsieh, C.C. and Jiang, Y.L. (2016) Responses of 'Shih Huo Chuan' pitaya (*Hylocereus polyrhizus* (Weber) Britt. & Rose) to different degrees of shading nets. *Scientia Horticulturae* 198, 154–162.

Chien, Y.C. and Chang, J.C. (2019) Net houses effects on microclimate, production, and plant protection of white-fleshed pitaya. *HortScience* 54, 692–700.

Childers, C.C. and Nakahara, S. (2006) *Thysanoptera* (thrips) within citrus orchards in Florida: species distribution, relative and seasonal abundance within trees, and species on vines and ground cover plants. *Journal of Insect Science* 6, 1–19.

Crane, J.H. and Balerdi, C.F. (2005) Pitaya growing in the Florida home landscape. University of Florida/IFAS Extension, Gainsville, Florida. Available at: https://edis.ifas.ufl.edu/publication/HS303 (accessed 13 November 2023).

De Oliveira, M.M.T., Albano-Machado, F.G., Penha, D.M., Pinho, M.M., Natale, W. *et al.* (2021) Shade improves growth, photosynthetic performance, production and postharvest quality in red pitahaya (*Hylocereus costaricensis*). *Scientia Horticulturae* 286: 110217.

Eckstein, K., Fraser, C. and Joubert, W. (1998) Greenhouse cultivation of banana in South Africa. *Acta Horticulturae* 490, 135–147.

Emekli, N.Y. (2014) Antalya koşullarında sera örtü malzemesi olarak kullanılan polietilen örtülerin bazı fiziksel ve mekanik özelliklerinin zamana bağlı değişimi ile ışınım geçirgenliğinin bitki gelişimi üzerine etkisi. PhD Thesis, Akdeniz University, Antalya, Turkey.

Fischer, G., Orduz-Rodríguez, J.O. and Amarante, C.V.T. (2022) Sunburn disorder in tropical and subtropical fruits. A review. *Revista Colombiana de Ciencias Hortícolas* 16: e15703.

Galán Saúco, V. and Pérez Hernández, E. (2020) Greenhouse cultivation. In: Mitra, S. (ed.) *The Papaya: Botany, Production and Uses*. CAB International, Wallingford, UK, pp. 226–236.

Galán Saúco, V., Cabrera Cabrera, J. and Hernandez Delgado, P.M. (1992) Phenological and production differences between greenhouse and open-air banana (*Musa accuminata* Colla AAA cv. Dwarf Cavendish) in Canary Islands. *Acta Horticulturae* 296, 97–11.

Gübbük, H. and Pekmezci, M. (2004) Comparison of open-field and protected cultivation of banana (*Musa* spp. AAA) in the coastal area of Turkey. *New Zealand Journal of Crop and Horticultural Science* 32, 375–378.

Gübbük, H., Biner, Ş.B., Dal, B., Yıldırım, I., Taşgın, D. and Buhur, L. (2017) Değişik tropik meyve türlerinin Antalya koşullarına adaptasyonu üzerinde araştırmalar. Proje Sonuç Raporu, Antalya, Turkey.

Gunes, E. and Gübbük, H. (2012) Growth, yield and fruit quality of three papaya cultivars grown under protected cultivation. *Fruits* 67, 23–29.

Jacobsen, S.E., Jensen, C.R. and Liu, F. (2012) Improving crop production in the arid Mediterranean climate. *Field Crops Research* 128, 34–47.

Janick, J. and Ait-Oubahou, A. (1989) Greenhouse production of banana in Morocco. *HortScience* 24, 22–27.

Jaya, I.K D. (2018) Morphology and physiology of pitaya and its future prospects in Indonesia. *Jurnal Ilmiah Budidaya* 3, 44–50.

Le Bellec, F., Vaillant, F. and Imbert, E. (2006) Pitahaya (*Hylocereus* spp.): a new fruit crop, a market with a future. *Fruits* 61, 237–250.

Luders, L. and McMahon, G. (2006) The pitaya or dragon fruit (*Hylocereus undatus*). Crops, Forestry and Horticulture, Darwin, Australia. Available at: https://raredragonfruit.com.au/blogs/news/the-pitaya-or-dragon-fruit-hylocereus-undatus-nt-gov (accessed 13 November 2023).

Luo, J., Xu, M., Liu, C., Wei, S. and Tang, H. (2021) Effects comparation of different mulching methods on soil in pitaya orchards. *International Agrophysics* 35, 269–278.

McMahon, G. (2003) *Pitaya (dragon fruit)*. Australia Fact Sheet, No. 12. Department of Primary Industry, Fisheries and Mines, Crop Forestry and Horticulture Division, Darwin, Australia.

Merten, S. (2003) A review of *Hylocereus* production in the United States. *Journal of the Professional Association for Cactus Development* 5, 98–105.

Mizrahi, Y. (2014) Vine-cacti pitayas: the new crops of the world. *Revista Brasileira de Fruticultura* 36, 124–138.

Mizrahi, Y. and Nerd, A. (1999) Climbing and columnar cacti: new arid land fruit crops. *Perspectives on New Crops and New Uses* 1, 358–366.
Mizrahi, Y., Raveh, E., Yossov, E., Nerd, A. and Ben-Asher, J. (2007) New fruit crops with high water use efficiency. In: *Creating Markets for Economic Development of New Crops and New Uses*. ASHS Press, Alexandria, Virginia, pp. 216–222.
Nerd, A., Sitrit, Y., Kaushik, R.A. and Mizrahi, Y. (2002) High summer temperatures inhibit flowering in vine pitaya crops (*Hylocereus* spp.). *Scientia Horticulturae* 96, 343–350.
Pastor, R. and Galán Saúco, V. (1995) Papaya in the Canary Islands. A promising future. *FruiTrop* 12: 10.
Pastor, R., Galán Saúco, V. and Herrero Romero, M. (1990) Evaluation of papaya autogamy. *Fruits* 45, 387–391.
Pekmezci, M., Gübbük, H. and Erkan, M. (1998) Investigation on growing possibilities of banana in Turkey. *Acta Horticulturae* 490, 599–603.
Raveh, E., Nerd, A. and Mizrahi, Y. (1998) Responses of two hemiepiphytic fruit crop cacti to different degrees of shade. *Scientia Horticulturae* 73, 151–164.
Trivellini, A., Lucchesini, M., Ferrante, A., Massa, D., Orlando, M. *et al.* (2020) Pitaya, an attractive alternative crop for Mediterranean region. *Agronomy* 10: 1065.
Wang L., Zhang, X., Chen W., Xiao, T., Zhao, X. *et al.* (2019) Shading reduced the injury caused by winter chill on Pitaya plant. *Notulae Botanicae Horti Agrobotanici Cluj-Napoca* 47, 470–477.

13 Postharvest Management

Sisir Mitra*
Former Professor, Faculty of Horticulture, Bidhan Chandra Krishi Viswavidyalaya (State Agriculture University), Mohnapur, West Bengal, India

13.1 Introduction

Dragon fruit is an exotic fruit that exhibits a non-climacteric pattern of respiration. This indicates it needs to be harvested from the tree when it is fully ripe, as it cannot ripen after harvest. The fruit is attractive in appearance when fresh. However, in most end markets, the appearance of this fruit deteriorates, mainly due to yellowing or browning of the fruit scales. This is because dragon fruit is susceptible to mechanical injury, water loss, decay and chilling injury. Skin colour is the most frequently used criteria to judge fruit maturity. Other harvest indices also include: total soluble solids (TSS), titratable acidity (TA) and days from flowering/anthesis to maturity. Storage of dragon fruit at temperatures above 20°C leads to rapid softening and loss of both sugar and acidity; therefore, the fruit are generally stored at 10–14°C. As well as low-temperature storage, several other treatments such as coatings, 1-methylcyclopropene (1-MCP) and irradiation have been used to extend the shelf life of fruit with varying success. The availability of an approved fruit-fly disinfection treatment for USA imports in 2012 increased the possibilities of market expansion, and Vietnam has installed a vapour heat-treatment facility to meet the quarantine requirement for exporting fruits from Vietnam to Japan.

13.2 Maturity Indices

Dragon fruit are generally harvested close to the fully ripe stage as they do not continue to ripen after the skin colour changes to red or yellow. Premature fruit are susceptible to cell-wall rupture and physiological disorders caused by cellular disorganization; however, overripe fruit rapidly senesce, causing quality losses and reduced shelf life (Chitarra and Chitarra, 2005). In red-fleshed pitaya ('Zihonglong' cultivar), pigmentation appears earlier in the pulp than the peel, and the pulp turns red within a day (on day 29 postanthesis) due to the sharp increase of betalain, and the green colour in the peel fades and it turns red (Wu *et al.*, 2019). Phebe *et al.* (2009) reported a significant quadratic relationship between the days after flower anthesis (DAA) and $h°$ values of red pitaya flesh colour ($R^2 = 0.81$). As DAA progressed, the $h°$ values of peel colour increased gradually, showing the development of a red-violet colour in the fruit flesh, shifting from reddish-purple to purplish-red. The formation of pigment causing colour changes in the

* Email: sisirm55@gmail.com

DOI: 10.1079/9781800623156.0013

peel as well as in the flesh of the fruit is one of the important changes that the fruit go through in the last stage of development or maturation.

In addition to pericarp colour, the soluble solids content (SSC), TA and DDA until the fruit is fully developed are also used to determine the timing of harvest (Nerd *et al.*, 1999). Fruit development depends on seasonal temperatures, and fruit harvested at full colour maturity has the best flavour (Nerd and Mizrahi, 1998). In California, ripening of pitaya occurred at 40–45 days after flowering (DAF) (Merten, 2003). *Hylocereus undatus* grown in Brazil achieved physiological maturity at 30–32 DAA (Ortiz and Takahashi, 2015). In India, red-fleshed fruits needed comparatively fewer days (29–31 DAF) than white-fleshed fruits (31–33 DAF) for optimum maturity (Deep *et al.*, 2022). Le *et al.* (2002) suggested harvesting of yellow-fleshed dragon fruit at 28–30 DAF in Vietnam.

The increase in fruit acidity before any colour change indicates the start of the maturation processes (Arévalo-Galarza and Ortíz-Hernández, 2004). Menezes *et al.* (2015) observed a marked increase in acidity before any colour changes. Ortiz and Takahashi (2015) stated that acidity increased until 27 DAA, when it reached maximum values of 1.4% malic acid, subsequently decreasing to 0.27% malic acid by 32 days. Junior *et al.* (2021), working with white-fleshed dragon fruit, observed a quadratic behaviour in the percentage of malic acid, ranging from 0.27% to 1.36%, increasing in early development stages, with a significant increase at 28 DAA, before pigmentation of the fruit skin.

TSS was found to be related to the outer peel colour of *H. undatus* (Centurion *et al.*, 2000). Fruit with fully pigmented outer colour had a TSS content of 12.2% at 31 DAA (Ortiz and Takahashi, 2015). Castillo-Martinez *et al.* (2003) recorded a TSS content of 11.9°Brix, while Marquez-Guzman *et al.* (2005) reported values ranging from 10.9°Brix to 14.1°Brix in four pitaya genotypes. A high TSS content (14.5–17.6°Brix) was reported by Martínez Chávez (2011) in six pitaya genotypes. Fruit harvested at full or three-quarters red colour had higher TSS values than fruit harvested at a less mature stage (red colour of peel: 25–50% or 50–75%) (Osuna Enciso *et al.*, 2011). Deep *et al.* (2022) recorded higher TSS in red-fleshed fruit than in white-fleshed fruit at maturity. The TSS content increased up to optimum maturity and started decreasing during the overmature stage. The highest TSS was recorded at 31 DAF in red-fleshed (14.2°Brix) and white-fleshed fruit (13.2°Brix). The rise in TSS during maturity indicates that it is a suitable indicator of optimum maturity of dragon fruit (Nerd *et al.*, 1999). Deep *et al.* (2022) reported a significantly higher TSS:acid radio at mature (61.3 and 41.9) and overmature (76 and 55) stages in red- and white-fleshed dragon fruit, respectively. A SSC:TA ratio of 40 has been suggested as a harvest index (Paull, 2014).

13.3 Grading, Packaging and Pre-cooling

13.3.1 Grading

After harvesting, dragon fruit are classified according to the standards of different countries in order to export them. The current classification is carried out manually (Trieu and Thinh 2021). Khuriyati *et al.* (2018) stated that sorting dragon fruit is based on colour, size and physical defects using the human senses. However, measurement using human senses cannot reflect the nutritional content present in the fruit. Using a non-destructive approach to sorting of dragon fruit by colour using a chroma meter, fruit age can be predicted using a regression model, which shows a correlation between fruit age and colour. Trieu and Thinh (2021) described an automatic dragon fruit classifying system using non-destructive measurements, based on a convolutional neural network (CNN). The classifying system uses a combination of a model of machine learning and image processing using a CNN to identity the external features of dragon fruit; the fruit are then classified and evaluated by group. In Vietnamese factories, the processing speed of the system increases the sorting capacity of export packing facilities to six times higher than the manual method, with an accuracy of more than 90%.

The size grades suggested for dragon fruit in Vietnam are: extra-large >500 g, large 380–500 g, regular 300–380 g, medium 260–300 g and small <260 g (Le *et al.*, 2000). Fruit exported from Israel to Europe are graded by the number of fruit (6, 8, 10, 12, 14 or 16) per 4 kg cardboard box (Paull and Chen, 2019). Dragon fruit for the Chinese market exported from Vietnam are often packaged in cartons containing 20.5 kg of fruit; for the large-sized fruit, this is around 31–37 per carton and for the small-sized fruit it is 48–52 per carton. Dragon fruit destined for India, Thailand, Malaysia and Singapore are packaged in 9 kg cartons, which will contain 16, 18 or 20 large-sized fruit and 22 or 24 small-sized fruit.

13.3.2 Packaging

Individual fruit should be wrapped with netting to avoid mechanical damage due to abrasion during transportation. Some producers in Malaysia use low-density polyethylene plastic to wrap the fruit to reduce water loss (Sarip, 2020). The fruit are packed in cardboard boxes, corrugated fibreboard boxes or buckets (mostly for the domestic market) and sealed with adhesive tape to prevent insect infiltration. Different postharvest operations are followed in a small-scale packing house in Vietnam (Fig. 13.1). Excessive shrivelling during storage as a result of transpiration is another critical problem that adversely affects the visual quality of dragon fruit. Packing in perforated plastic bags is a common approach that is practised in Malaysia to reduce water loss and rapid shrivelling. However, when the fruit are exposed to high relative humidity environments in plastic bags, they are more susceptible to decay (Shin *et al.*, 2007). In addition, there is increasing concern about the non-renewable and/or non-biodegradable nature of plastic bags, paving the way for the development of greener alternatives (Sarip, 2020).

In Malaysia, dragon fruit are not subjected to any washing or fungicide treatment after harvest as the fruit are sensitive to mould growth under high humidity conditions, and there are no postharvest fungicides are approved for use on dragon fruit after harvest (Sarip, 2020).

13.3.3 Pre-cooling

Pre-cooling slows the ripening process, respiration rate, senescence and water loss from the harvested fruit, which helps to retain overall fruit quality and extend shelf life. Pre-cooling of dragon fruit is necessary, particularly for those destined for export. Phong *et al.* (2013) conducted a survey on the postharvest quality and management of dragon fruit exported from Vietnam to Holland. Commercially packed dragon fruit in 5 kg boxes (nine fruit per box) were pre-cooled to about 14°C, 18 h before loading, while another set was not pre-cooled. Both containers were exported from a commercial packing house in Vietnam to a packing house in Holland by sea. The storage temperature during transportation was set at 5°C and the voyage took 25 days. The temperatures for the containers showed that the boxes of dragon fruit that were pre-cooled before loading retained a better quality in terms of the green colour of the bracts and a lower incidence of rot compared with the non-pre-cooled fruit. The temperature of the fruit recorded by data loggers set at various locations in the non-pre-cooled container indicated that it did not achieve the set temperature (5°C) and only reached 7°C 5–7 days after loading, while in the pre-cooled container, the temperature of the fruit achieved 6°C only 1–2 days after loading.

A CoolBot air conditioning facility (Store It Cold, Florida, USA) took around 95 min to remove field heat and reach the lower limit of room temperature (10±1°C). Khatun *et al.* (2022) evaluated a pre-cooling system (without cooling, with hydrocooling and using CoolBot air conditioning cooling) on dragon fruit in Bangladesh. They observed reduced weight loss and extended shelf life with pre-cooling of the fruit, and the effect was more pronounced when the pre-cooling was done using the CoolBot air conditioning cooling system.

13.4 Postharvest Physiology

13.4.1 Respiration and ethylene production

Rates of respiration and ethylene production in pitaya fruit vary with species, variety/cultivar, temperature, season and maturity stage (Nerd *et al.*, 1999; Le *et al.*, 2000, 2002; Jamaludin *et al.*, 2011; de Freitas and Mitcham, 2013; Franco *et al.*, 2022).

Fig. 13.1. The postharvest operations followed in a small-scale packing house in Vietnam. (A) Collection of fruit in plastic baskets after harvesting in the field. (B) Storage of the fruit in the packing house before sorting. (C, D) Sorting of fruit for uniformity in size, shape, colour and other characters following the requirements of each market. (E) Washing of the fruit. (F) Chemical treatment (if required), drying of the fruits on a conveyer belt and packing in polypropylene bags. (G, H) Packing in cardboard boxes. (I) Boxes of fruit stacked on pallets. (J, K) Cold storage of the pallets before delivery. (L) Loading of the pallets on to a refrigerated truck before delivery. Photograph courtesy of Dr Van Tri Mai.

When *H. undatus* fruit matures, the respiration rate is about 95–144 mg CO_2 kg^{-1} h^{-1} at 20°C (52–80 ml kg^{-1} h^{-1}) (Nerd *et al.*, 1999). Jamaludin *et al.* (2011) studied red-fleshed pitaya (*Hylocereus polyrhizus*) and reported a very low respiration rate during development at 26°C decreasing from 0.14 to 0.03 ml kg^{-1} h^{-1}. Le *et al.* (2000) reported 75–100 mg CO_2 kg^{-1} h^{-1} at 23°C in red-fleshed pitaya. In yellow-fleshed pitaya (*H. undatus*), Franco *et al.* (2022) did not observe any specific pattern of respiration rate and no sudden peak was detected in CO_2 production during continuous storage at 5°C and 13°C, indicating that pitaya is a non-climacteric fruit (Le *et al.*, 2002; de Freitas and Mitcham, 2013). The highest respiration rate was noted when fruit was harvested at 31 DAA and the lowest at 35 DAA. Le *et al.* (2002) recorded a maximum respiration rate (283 mg CO_2 kg^{-1} h^{-1}) in yellow-fleshed dragon fruit at 19 DAF. However, on day 22 and thereafter, the respiration rate abruptly declined and ranged from 50 to 120 mg CO_2 kg^{-1} h^{-1} as the fruit matured.

13.4.2 Physicochemical changes during storage

Storage temperature has a direct effect on fruit metabolism in terms of respiration and an indirect effect on cellular structure that determines fruit qualities such as colour, firmness and organoleptic properties (Marsh *et al.*, 2004). In dragon fruit, the changes in peel colour are associated with changes in chlorophylls and betacyanin (Esquivel *et al.*, 2007), During ripening, degradation of chlorophyll occurs in *H. polyrhizus*, followed by the betacyanin pigment becoming visible. Fruit stored at a low temperature (6°C) shows unaffected peel colour, which is thought to be due to temperature-induced activity reduction of the enzymes responsible for the formation of betacyanin or suppression of accumulation of other pigments present within the fruit. However, fruit stored at an ambient temperature shows increased colour due to accelerated ripening or increased enzymes involved in chlorophyll degradation and betacyanin accumulation (Punitha *et al.*, 2010).

Moisture loss of the fruit can reach 4.2% after storage at 20°C for 1 week, while for fruit stored at 6°C, the loss is generally less at around 2.2% (Nerd *et al.*, 1999). The firmness of the fruit is also considerably reduced from 6.3 to 3.8 N after 14 days of storage at 20°C (Van To *et al.*, 2002). The progressive loss of firmness is the result of a gradual transformation of protopectin into pectin, which is degraded by the enzyme polygalacturonase in the cell wall. Storage under elevated concentrations of CO_2 and/or low concentrations of O_2 has been reported to retard softening by limiting the activity of cell-wall-degrading enzymes (Razali *et al.*, 2021). TSS increases initially (up to 6–8 days) and then starts decreasing at an ambient temperature (Prashanth *et al.*, 2022). A large percentage of the TSS in dragon fruit is sugars, mainly glucose and fructose, with a small percentage of sucrose. The vitamin C content gradually increased until day 18 of storage (at 10°C and relative humidity (RH) of 85–90%). The TA of *Hylocereus costaricensis* fruits reduced from 0.37 g malic acid 100 g^{-1} on day 0 to 0.9 g malic acid 100 g^{-1} on day 16 of storage (at 8±1°C and 60±4% RH). The initial acidity values are attributed to the fruit maturation stage, as greener fruits have a higher acidity whose organic content tends to decrease during storage (de Faria *et al.*, 2022). After 30 days of storage, vitamin C content was 0.16 g kg^{-1} fresh weight, noticeably lower than the initial level of 0.24 g kg^{-1} fresh weight (Nguyen *et al.*, 2020). Antioxidant activity in terms of 2,2-diphenyl-1-picrylhydrazyl (DPPH) scavenging activity in dragon fruit was reduced in the initial stages of storage (10°C, 85–95% RH) and then increased gradually (Nguyen *et al.*, 2020). The betalain content remained constant during storage (12°C and 76% RH), but total phenols decreased, suggesting that antioxidant activity is mostly provided by phenolic compounds (Rosas-Benitez *et al.*, 2016).

Obenland *et al.* (2016) identified a total of 12 volatile compounds in yellow pitaya (*Selenicereus megalanthus*). Aldehydes were the most abundant aroma volatiles (90%) detected in the juice from the flesh, with storage increasing the total aldehyde amount, particularly at 10°C. Total alcohols, however, were reduced by storage, and the amount of reduction was not dependent on storage temperature. Other volatiles identified included paraffins, esters, acids, ketones and other miscellaneous compounds.

13.5 Storage

13.5.1 Low-temperature storage

The recommended storage temperature for dragon fruit is 10°C (Nerd *et al.*, 1999). At this temperature, dragon fruit has a storage life of about 14 days, while at 5°C and 90% RH, a storage life of 17 days can be achieved (Le *et al.*, 2000). However, storage at 5°C may lead to chilling injury on return to 20°C, causing deterioration of the peel and flesh and an inferior taste (Nerd *et al.*, 1999). The severity of visual chilling injury symptoms also increases as storage duration increases. *H. undatus* in California and Vietnam stored at 5°C for 20 days maintained a good quality (Van To *et al.*, 2002; Hoa *et al.*, 2006; de Freitas and Mitcham, 2013), while Crane and Balerdi (2005) recommended a storage temperature of 7–10°C for *H. undatus* grown in Florida. Conditioning of yellow pitaya at 25°C for 24 h before storage at 2°C reduced fruit susceptibility to chilling injury, allowing storage of pitaya fruit at this low temperature (Duenas *et al.*, 2009).

Dragon fruit contains a large amount of water due to the peel thickness, scales and high level of mucilage content in the pulp. The fruit undergo accelerated softening and a decrease in sugar and TA when stored above 20°C (Punitha *et al.*, 2010). Punitha *et al.* (2010) evaluated three different temperatures, low (6°C), intermediate (16°C) and high (ambient) (23 ± 2°C) to study the storage life of *H. polyrhizus*. Attributes showing the greatest tendency to be affected by storage temperature include a reduction in fruit firmness and an increase in peel colour. TSS were maintained for 14 days of storage at 6°C, while the value decreased especially at 16°C. The percentage reduction of total sugar was less in fruit kept at a low temperature, while pH values increased gradually, regardless of storage temperature. Based on the visual appearance and organoleptic properties, fruit kept at 6°C resulted in a better storage condition compared with intermediate and ambient temperatures. As the ideal storage temperature is the lowest temperature that does not trigger chilling injury, determining this temperature for each genetic background and growing environment is necessary to effectively maintain the quality of pitaya fruit after harvest (Razali *et al.*, 2016). Nerd *et al.* (1999) found that *H. undatus* lost flavour quality after 2 weeks at 14°C and developed chilling injury to the peel following storage for 2 weeks at 6°C plus 1 week at 20°C, indicating the potential for different tolerance in different species. Hoa *et al.* (2006), using *H. undatus* in Vietnam, stored fruit for 3 weeks at 5°C with almost no loss in flavour and acceptable external quality.

Chilling injury is typified by flesh translucency; other symptoms include softening, wilting, darkening of scales, browning of the outer flesh and poor flavour (Paull, 2014). The symptoms develop rapidly on *H. undatus* and *H. polyrhizus* fruit held at 6°C for 2 weeks and then transferred to 20°C (Nerd *et al.*, 1999). Fruit harvested 25 DAF are more sensitive to chilling (6°C, 7 days), but sensitivity is significantly reduced when the fruit are harvested at 30–35 DAF (6°C, 17 days).

13.5.2 Hot-water and hot-air treatments

Heat treatments including hot-water immersion, hot air and hot vapour have been utilized as quarantine treatments to control fruit-fly infestation and pathogens of many fruits (McGuire, 1997; Irtwange, 2006; Supapvanich and Promyou, 2017). Lum and Norazira (2011) studied the effect of hot-water temperature, time of immersion and storage duration of red-fleshed dragon fruit. Freshly harvested fruit were treated with hot water at 35°C, 45°C and 60°C for 15, 30 and 60 min and evaluated at 0, 5, 10 and 15 days of storage (at 25°C). The result showed that hot-water treatment significantly ($p < 0.05$) affected the percentage weight loss, TA and firmness of the fruit. The highest percentage of weight loss (12.18%) was seen in fruit treated at 60°C for 60 min, whereas the lowest percentage of weight loss (5.05%) was in fruit treated at 35°C for 60 min. Hot-water treatment at 35°C for 60 min maintained a high TA (2.44%) and the highest fruit firmness.

Main-season, white-fleshed dragon fruit variety ‘Binh Thuan’ from Vietnam were subjected to hot-air treatment and stored for 2–4 weeks at 5°C in sealed polypropylene bags; fruit quality was examined to determine subsequent shelf life at 20°C. Hot-air treatment involved a fruit core temperature of 46.5°C for 20 and

40 min, and 48.5°C for 50, 70 and 90 min. 'Binh Thuan' dragon fruit proved to be tolerant to hot-air treatment at the lower temperature of 46.5°C for 20 min. Fruit quality was not significantly different from non-heated control fruit, either after treatment at harvest or after storage for up to 4 weeks. However, metrics quantifying the appearance of both control and hot-air treated fruit generally deteriorated with storage time at 5°C, although the internal appearance of the flesh, as well as the taste, remained acceptable throughout (Hoa *et al.*, 2006).

13.5.3 Modified-atmosphere storage

Storage under a modified atmosphere (MA) such as in polybags, or MA packaging (MAP) or coating (MAC) in polymeric films, prolongs the shelf life of a wide variety of fruits (Sivakumar and Korsten, 2006; Dhalsamant *et al.*, 2017).

MA treatments ultimately result in a build-up of CO_2 and depletion of O_2 within the internal atmosphere surrounding the fruit. In many cases, respiration as well as ethylene production and perception by the fruit are reduced, ripening is delayed and hence the shelf life of the fruit is markedly extended. The benefits of MAP can further be enhanced when combined with refrigeration. Polyethylene (PE) and polypropylene (PP) are commonly used for packaging of food products (Castro *et al.*, 2020). Both plastic packagings are permeable to gases (CO_2 and O_2) and moisture to a limited extent, which makes them suitable materials for MAP (Hasbullah *et al.*, 2000).

Red-fleshed dragon fruit harvested at 25–30 DAF were sleeved in polystyrene fruit cups and individually packed in 58.8 μm thick PE and PP non-perforated plastic bags and stored at 5°C (Castro *et al.*, 2020). The MAP-stored fruit remained in excellent condition for up to 6 weeks at 5°C without any shrivelling, with the fruit remaining firm and bracts green. The use of polystyrene fruit cups generally aided in maintaining the visual quality of the fruit by protecting the bracts from breaking down during handling, Non-packed (non-MA packed) fruits, however, lasted only 4 weeks at 5°C with a noticeable change in visual quality and the bracts starting to exhibit yellowing and tip browning. The shelf life at 20°C after 4-week storage of MAP and non-MAP fruits was 5 and 2 days, respectively. TSS, TA and total phenolic content were higher in PE-packed fruits. Van To *et al.* (2002) tried storage in PE bags with an O_2 transmission rate of $4\ l^{-1}\ m^2\ h^{-1}$ and in PE with 20 holes of 0.5 mm diameter, and succeeded in extending the shelf life up to 28 days compared with 14 days for the control. Fruit in a perforated plastic bag at 4.5°C remained fresh for 25–30 days compared with less than 10 days of storage life at room temperature (Zee *et al.*, 2004). Chandran (2010) suggested storing the fruits at 6°C after packing in a clear wrap (0.8 μm thick); the fruit remained firm and the peel colour was also maintained for 15 days of storage.

Neog *et al.* (2023) evaluated different types of packaging: low-density polyethylene films, high-density polyethylene films, polyvinyl chloride (PVC) films, polypropylene films and polyolefin films. In all the treatments, a perforation of 0.76% was maintained, and the fruit were kept under refrigerated conditions. Fruit wrapped in polyolefin film had a shelf life of 27–38 days with the lowest physiological loss in weight (4.71%), along with the lowest decay incidence (12.69%) compared with the other packaging treatments.

13.5.4 Controlled-atmosphere storage

Controlled-atmosphere (CA) storage has been tested with success in many tropical fruits; however, it is not commonly used in storage of dragon fruit. CA storage may extend the storage life of many fruits by the precise regulation of low O_2 and high CO_2 levels along with maintenance of optimum temperature (Gill, 2018). CA storage was able to extend the storage life of prickly pear, a fruit belonging to the same Cactaceae family as dragon fruit, by maintaining visual quality and sugar content (Morga *et al.*, (2006).

Ho *et al.* (2021) optimized the CA storage conditions for dragon fruit. Dragon fruit (*H. undatus*) were stored under four different CA conditions and at ambient air as a control. At harvest and on days 10, 20, 30, 40 and 50, different fruit quality, sensory and aroma analyses were performed. The results revealed that CA storage significantly reduced yellowing of the scales and maintained fruit acidity up to 50 days at 6°C. The CA condition of 2 kPa O_2 + 5 kPa CO_2 at 6°C appeared to be optimal for dragon fruit storage. Normal senescence caused the loss of many volatiles, while low O_2 and high CO_2 levels had the opposite effect on the concentration of fermentation products. However, these volatile changes did not significantly impact the sensory quality of the fruit during storage.

13.5.5 Effect of 1-methylcyclopropene

1-MCP is an ethylene antagonist that interacts with the fruit ethylene receptors, thus preventing the action of ethylene (Sisler and Serek, 1997). In many countries, 1-MCP is used commercially to extend the storage life of climacteric fruits in which ethylene plays a critical role in regulation of the ripening process. In addition to its well-known effect on climacteric fruits, several studies suggest that exposure of non-climacteric fruits to 1-MCP can affect certain ripening-related processes, including colour change, respiration, softening, and development of microbial decay or physiological disorders (Cai *et al.*, 2006; Cao and Zheng, 2010; Liguori *et al.*, 2017; Vera *et al.*, 2017).

Application 1-MCP (600 mg l^{-1}) significantly decreased the loss of firmness, TSS, loss of fresh mass and respiratory rate of yellow-fleshed pitaya stored at 18°C with 75% RH (Deaquiz *et al.*, 2014). Serna *et al.* (2011) also reported a decrease in changes in acidity, respiratory rate and colour during storage of yellow-fleshed pitaya following treatment with 1-MCP. Application of 1-MCP increased the shelf life of *Hylocereus monacanthus* fruit harvested at 30–35 days after anthesis and stored for 28 days at 10°C; under this treatment. phenolic, ascorbic acid and betacyanin content were unaffected (Novita, 2008). Fruit immersed in 1-MCP at 500 μg l^{-1} for 4 h at room temperature showed a delay in ethylene production for 14 days of storage. The treatment slowed the respiration rate and fruit senescence. 1-MCP application provides greater firmness of *Hylocereus megalanthus* fruit and represents an alternative to reduce undesirable texture changes during storage (Ayala-Aponte and Serna-Cock, 2011). Application of 1-MCP at 600 mg l^{-1} in vaporized form for 24 h treatment was found to be effective in maintaining the quality characteristics of the fruit by reducing respiration and chlorophyll degradation of yellow-fleshed pitaya for around 28 days (Álvarez-Herrera *et al.*, 2016).

Vera *et al.* (2017) treated *H. undatus* fruits with 1-MCP using microbubbles technology. The fruit was dipped for 10 min in 1500 ppb of 1-MCP microbubbles and then stored at 13°C for 20 days. The results revealed that this treatment was effective in maintaining greener bracts, as expressed by the significantly higher hue values by day 20 of storage compared with a control. However, weight loss was not influenced by the treatment. These results support microbubbles technology as a viable alternative to traditional application of 1-MCP that reduces exposure times, thus allowing a faster connection within the supply chain stages. Li *et al.* (2016) reported a delay in bract senescence following 24 h exposure to 1-MCP at 1 μl l^{-1}, which was also achieved by Vera *et al.* (2017) in 10 min using microbubbles technology at a similar concentration.

Tadeo *et al.* (2018) studied the physiological responses and storage quality of fresh-cut red and white dragon fruit treated with 1-MCP using headspace gases (ethylene and CO_2) and concentration. Gaseous 1-MCP was injected into packs of fresh-cut fruit in polyethylene terephthalate trays with PVC stretch film overwraps. The packs were then stored at 5–10°C. Treatment with 1-MCP positively affected the headspace CO_2 and ethylene in both species. Fresh-cut white dragon fruit had a better response to 1-MCP compared with fresh-cut red dragon fruit in terms of headspace CO_2 concentration. There were no incidences of water soaking throughout storage for the 1-MCP-treated fresh-cut red dragon fruit, while for the fresh-cut white dragon fruit, water soaking was observed on day 12 of storage. For both species, 1-MCP delayed visual quality deterioration and increased antioxidant activity. Lower yeast and mould count, aerobic plate count and coliform count were observed in both species after 1-MCP treatment. Post-cutting treatment with 5 ppm of 1-MCP positively affected the storage quality of fresh-cut red and white dragon fruit.

13.5.6 Coatings

Edible coatings act as a physical barrier on fruit surfaces and decrease permeability to O_2, CO_2

and water vapour, leading to a reduction in respiration rate, fruit ripening and senescence (Dhall, 2013). Chitosan-based coatings have been considered an environmentally friendly coating with antimicrobial action that can extend the shelf life of several fruits, including dragon fruit (Ali *et al.*, 2013; Li *et al.*, 2014; Castro, 2017; Nguyen *et al.*, 2020, 2021).

The chitosan coating reduces stomatal conductance, size and aperture in dragon fruit, resulting in maintenance of its fresh appearance at room temperature, but not significantly affecting maintenance of weight (Chutichudet and Chutichudet, 2011).

The use of chitosan coatings and low-temperature conditioning was studied by Castro (2017) to prolong the storage life of red-fleshed dragon fruit. Temperature conditioning for 3 days at 10°C maintained the visual quality of the fruit, and reduced the rate of bract or scale yellowing, weight loss and shrivelling, which lead to a storage life of 4 weeks and a shelf life of 6 days. The study also showed that 1% chitosan coating improved fruit quality by enhancing the red colour of the peel and maintaining fruit quality even after storage. The chitosan coating also alleviates chilling injury in fruit stored at 8°C, as exhibited by its low respiration rate and ethylene production. Total phenolic content and antioxidant activity were also found to be higher in chitosan-coated than in non-coated fruit.

Conventional chitosan and submicron chitosan dispersion (SCD) (average size of droplet is 1000 nm) were evaluated for the control of postharvest anthracnose and maintenance of quality in dragon fruit during storage at 10±2°C and 80±5% RH for 28 days (Ali *et al.*, 2013). All the chitosan treatments (0.5%, 1.0%, 1.5% and 2.0%) significantly reduced anthracnose symptoms (Fig. 13.2), resulting in a reduction of disease development and thereby maintaining the quality of fresh fruit for extended periods. SCD at 1.0% with 600 nm droplets gave the best results, delaying the onset of disease and maintaining the quality of the dragon fruit for up to 28 days of storage. It was concluded that SCD has the potential to be used as an antifungal agent to control postharvest anthracnose and maintain the quality of dragon fruit during storage (Ali *et al.*, 2013).

To improve the chitosan-based composite coating, Nguyen *et al.* (2021) introduced κ-carrageenan as a component. κ-Carrageenan is a sulfated anionic polysaccharide extracted from red algae. κ-Carrageenan has excellent film-forming properties (Li *et al.*, 2014). Nguyen *et al.* (2021) studied the effect of a composite coating based on 1% (w:v) chitosan and 0.2% (w:v) κ-carrageenan with 0.75% (w:v) glycerol as a plasticizer on yellow dragon fruit during storage at 10°C. The composite coating effectively reduced the physiological weight loss and maintained the freshness of the dragon fruit, while increasing the phenolic content and maintaining the TA in the pulp. The composite coating delayed degradation by suppressing chlorophyllase and chlorophyll-degrading peroxidase, thereby maintaining the chlorophyll content and freshness of the bracts. However, the composite coating did not have a strong effect on enhancing

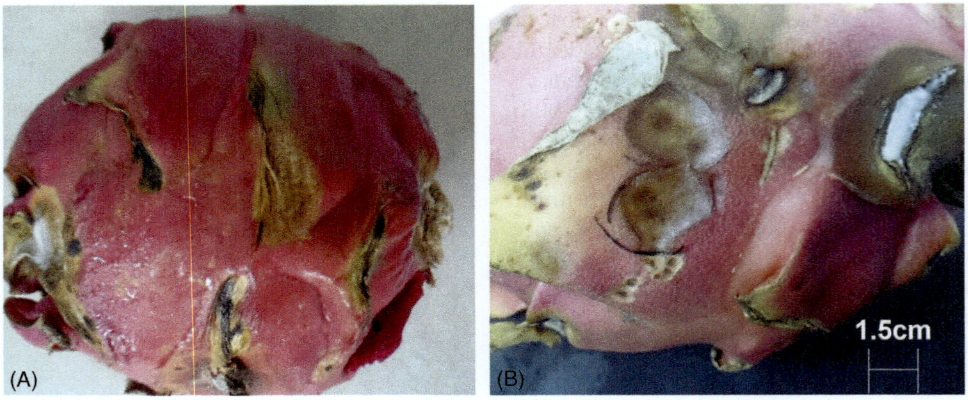

Fig. 13.2. Symptoms of anthracnose disease on fruit. Photographs courtesy of Dr Asgar Ali.

the chitinase and β-1,3-glucanase activities of the fruit during storage and controlling disease symptoms. Treatment with hot water at 50°C for 5 min followed by 1.0% chitosan and 0.2% κ-carrageenan-based composite coating enhanced antioxidant enzyme activities and reduced the accumulation of H_2O_2 in the peel, demonstrating the defence response in the tissues leading to reduced disease development. Nguyen *et al.* (2021) concluded that treatment with hot water combined with the composite coating controlled disease via the antioxidant defence response and maintained the overall quality of dragon fruit for 30 days of storage at 10°C.

Edible surface coatings such as waxes are often used to improve postharvest quality and the cosmetic features of fruits and vegetables (Arowora *et al.*, 2013). Razali *et al.* (2016) studied the effect of the vegetable oil-based coating Sta-Fresh 2981 (SF) and the carnauba wax-based coating Endra-Frese 6100 (EF) for storage of red-fleshed pitaya at 7°C for 20 days. Pitaya treated with EF coating had significantly less weight loss than those with SF coating or untreated controls over the storage period. Although the carnauba-based EF coating retained fruit weight, it did not maintain visual quality. Manioc starch (2%) coating has also been suggested as a coating for storage of pitaya (Castro *et al.*, 2017).

13.5.7 Calcium treatment

Calcium (Ca) as its chloride salt has great potential in delaying ripening, extending postharvest shelf life and improving nutritional value by increasing Ca content and maintaining fruit firmness and quality (Chuni *et al.*, 2010; Awang *et al.*, 2013a,b).

Ca plays a significant role in mechanical damage of both climacteric and non-climacteric fruit (Lamikanra and Watson, 2004). Ca application has been found to reduce the activity of cell-wall-degrading enzymes, including polygalacturonase (PG), pectin methylesterase (PME) and β-galactosidase (Awang *et al.*, 2013a). The longevity of fresh-cut dragon fruit can be extended by application of Ca, due to this reduction in cell-wall-degrading enzyme activity (Chuni *et al.*, 2010).

Awang *et al.* (2013a) examined the activity of PG and PME during storage in dragon fruit (*H. polyrhizus*) harvested at 28 days (index 3) and 34 days (index 5) after anthesis and treated postharvest with 0, 2.5, 5.0 or 7.5 g l^{-1} of $CaCl_2$. The PG activity was lower in younger fruit, while PME activity was higher in younger fruit. Increasing concentrations of $CaCl_2$ effectively reduced the activity of both enzymes. PG activity for fruit treated with 0.5 and 0.75 g l^{-1} increased linearly with storage time, while its activity in fruit treated with 2.5 g l^{-1} was lower at the start of storage. The PG activity of index 5 fruits increased almost linearly during storage, while its activity in index 3 fruit was low in the first days of storage and continued to increase up to day 7. Overall, the results indicated that $CaCl_2$ postharvest treatment reduced both PME and PG activities, thus slowing down the softening process, showing that Ca plays an important role in reducing softening of fruit, regardless of maturity index.

Application of $CaCl_2$ did not protect dragon fruit from brown rot (*Monilinia fructicola*) infection but was effective in reducing its severity. The severity of brown rot was reduced with increasing $CaCl_2$ concentrations (0–4.0 g l^{-1}), and this was supported by a strong negative correlation between fruit peel, Ca and disease severity (r = 45) (Table 13.1) (Awang *et al.*, 2013b).

Table 13.1. Correlation coefficients between disease severity, calcium content and fruit quality attributes following inoculation of *Monilinia fruticola* into red-fleshed dragon fruit. Adapted from Awang *et al.*, 2013b.

Variable	Firmness	pH	SSC	TA
Disease severity	−0.93**	0.88**	0.06^{NS}	−0.33*
Calcium (peel)	0.80**	−0.61**	0.42*	0.84*
Firmness	1	−0.80**	−0.02^{NS}	0.34*
pH	–	1	0.11^{NS}	−0.23^{NS}
Soluble solids content	–	–	1	0.15^{NS}
Titratable acidity	–	–	–	1

NS, not significant; *, $p < 0.05$; **, $p < 0.01$.

13.5.8 Irradiation treatment

Irradiation of fruit after harvest has been reported to enhance the shelf life and can also be used for postharvest disinfection (Mitra *et al.*, 2012). After harvest, the bracts of mature dragon fruit turn yellow at room temperature, which is judged a considerable detraction in appearance by the consumer. Yellowing is the loss of green colour in the plant as a result of chlorophyll catabolism (Funamoto *et al.*, 2002). Cold storage is one way of solving this problem (Razali *et al.*, 2016), as well as controlled and MA storage (Castro *et al.*, 2020; Ho *et al.*, 2021) and heat treatment (Funamoto *et al.*, 2002). Ultraviolet (UV)-C irradiation of fruit is also used to control chlorophyll degradation and maintain postharvest quality, as UV-C inhibits the chlorophyll-degrading peroxidase enzyme involved (Costa *et al.*, 2006). In addition some ripening-associated processes are also delayed by UV-C irradiation (Hallman, 2001; Razali *et al.*, 2015).

Kowitcharoen *et al.* (2010) irradiated mature dragon fruit with UV-C at 2, 4 and 6 kJ m^{-2} and stored the fruit at 10°C. The results showed that the effect on the quality of the fruit was dose dependent. At 6 kJ m^{-2}, senescence was induced, indicating a high respiration rate, water loss and low chlorophyll content, while UV-C treatment at 2 and 4 kJ m^{-2} delayed the changes in bract colour and also chlorophyll degradation. However, there was no significant difference in the ascorbic acid content of the fruit in any of the treatments. These results suggest the use of UV-C treatment at 2 kJ m^{-2} to maintain postharvest external quality of dragon fruit. Ng *et al.* (2022) showed that UV-C irradiation at 0.75 and 1.0 kJ m^{-2} was effective in reducing postharvest decay and indirectly maintained the quality of dragon fruit stored at ambient conditions (28±2°C) up to day 10. The results revealed that fruit irradiated with UV-C at 1.0 kJ m^{-2} significantly reduced the presence of rot, preserved bract appearance, slowed down the depletion of TSS and retarded fruit softening without affecting the TA content of the fruit. Thus, UV-C irradiation at 1.0 kJ m^{-2} was effective in reducing postharvest decay and maintaining the postharvest quality of red-fleshed dragon fruit under ambient conditions.

Wall and Khan (2008) determined the quality of three dragon fruit clones: clone 1 (*H. undatus* × *H. polyrhizus*) had red skin with red-purple flesh, clone 2 (*H. undatus*) had red skin with white flesh, and clone 3 (*H. undatus*) had red skin with white flesh. The fruit were harvested at 45–50 DAA and treated with irradiation doses of 0, 200, 400, 600 or 800 Gy and stored for 12 days at 10°C. Irradiation did not affect SSC, TA or fructose concentration. Glucose, sucrose and total sugar concentrations decreased linearly as dose increased. Minimal softening occurred in the outer flesh layers for fruit treated with 400 or 600 Gy irradiation. Surface colour, peel injury and bract appearance differed among the three clones with irradiation stress, but in all cases, visible changes were minor. Fruit decay was absent or minimal, and disease ratings were not affected by irradiation. Thus, irradiation treatment of dragon fruit at doses of 800 Gy or less would ensure visual and compositional quality while providing quarantine security.

Red light is an effective postharvest treatment and has been applied to the preservation of many different fruits (Yamaga *et al.*, 2016; Azuma *et al.*, 2019). Pitaya variety 'Hongshuijing' (*H. polyrhizus*) was harvested at the commercially mature stage and treated with 100 Lux red light for 24 h and stored at room temperature (25±2°C) and 75–95% RH (Wu *et al.*, 2020). Red-light irradiation significantly attenuated the variation trend of senescence traits, such as the decrease in TSS and TSS:TA ratio, the increase in TA and respiratory rate. Red-light irradiation enhanced glycolysis, the tricarboxylic acid cycle, aldehyde metabolism and antioxidant enzyme activity at the early stage of postharvest storage, leading to reductions in H_2O_2, soluble sugars, organic acids, and C-6 and C-7 aldehydes. At a later stage of postharvest storage, a large number of resistance-related metabolites and enzyme activities were induced in red-light-treated pitaya peel, such as superoxide dismutase, ascorbate peroxidase, DPPH radical scavengers, reducing power, fatty acids and volatile aroma. The volatile aroma comprising β-linalool, 1-hexanol, 2-hydroxycyclopentadecanone and palmitoleic acid increased markedly after red-light irradiation at the senescence stage (Wu *et al.*, 2020).

13.5.9 Other treatments

There is a continuing search for safer, low-cost, potent senescence inhibitors and antimicrobial technology to maintain the quality of harvested pitaya fruit (Xu *et al.*, 2021). Essential oils are now increasingly used for the preservation of a number of fruits and vegetables due to their safety and antimicrobial properties. *p*-Anisaldehyde (PAA; 4-methoxybenzaldehyde) is a main component of the essential oil derived from seeds of *Pimpinella anisum* (Shreaz *et al.*, 2011), and has shown antimicrobial activity against a number of food-borne bacteria (Shi *et al.*, 2017). The performance of PAA in preserving pitaya fruit quality was studied by Xu *et al.* (2021). The results showed that PAA treatment (the fruit was sprayed with 1 mM PAA solution until the PAA covered the fruit surface uniformly) followed by storage at 20°C for 15 days at 85–90% RH significantly reduced fruit decay, weight loss and loss of firmness, and maintained a higher content of TSS, betacyanins, betaxanthins, total phenolics and flavonoids in postharvest pitaya fruits. Compared with the control, increasing H_2O_2 content and superoxide anion (O_2^-) production was inhibited in fruit treated with PAA. Treatment with PAA also significantly improved the activity of the antioxidant enzymes superoxide dismutase, peroxidase and catalase. PAA-treated fruit maintained a higher ascorbic acid (AsA) and reduced glutathione (GSH) content but lower dehydroascorbate (DHA) and oxidized glutathione (GSSG) content, thus sustaining a higher ratio of AsA:DHA and GSH:GSSG. These findings suggest that postharvest application of PAA may be a reliable method of controlling postharvest decay and preserving the quality of harvested pitaya fruit by enhancing the antioxidant potential of the AsA–GSH cycle and activating an antioxidant defence system to alleviate reactive oxygen species accumulation.

Bordoh *et al.* (2022) reported that incorporation of turmeric (*Curcuma longa* Linn.) extract at 10 g l^{-1} with 10% gum arabic (GA) in a fruit coating delayed fruit ripening and maintained fruit quality during storage at 11± 2°C, 80% RH for 28 days and this was evident in delayed colour changes, higher fruit firmness, higher TA content, reduced respiration rate and ethylene production. Additionally, fruit coated with 10% GA + 10 g l^{-1} of turmeric extract had the best overall score in terms of external appearance, pulp colour, aroma, sweetness, texture and overall acceptability by the panel. The researchers concluded that, as an emerging hurdle technology, the incorporation of GA with a plant extract such as turmeric is a safe postharvest treatment (biofungicide) in postharvest management of disease, while also maintaining the postharvest quality of fresh produce. GA and turmeric are natural products, are generally regarded as safe and do not leave any dangerous residues on the fruit.

The Malaysian Agricultural Research and Development Institute (MARDI) developed 'EO Nature', a natural coating formulation made from plant oil with the incorporation of natural essential oils. A study on dragon fruit found that it could reduce water loss by reducing stomatal aperture, thus reducing transpiration during storage. As well as reducing water loss, EO Nature was reported to reduce anthracnose disease incidence and severity during storage. Dragon fruit coated with EO Nature can be kept for up to 21 days at 10°C compared with only 14 days without treatment (Sarip, 2020).

Propolis is a natural glue produced by honey bees with the main constituents being resins, waxes, vitamins and essential oils. Zahid *et al.* (2013) observed significant ($p \leq 0.05$) differences in dragon fruit quality when treated with different concentrations of ethanolic extract of propolis (EEP) and stored at 20±2°C and 80±5% RH for 20 days. Fruits treated with 0.50% EEP had an extended storage life without any negative effect on the quality.

13.6 Postharvest Pathology

Dragon fruit has been reported to harbour many fungal species after harvesting, such as *Bipolaris cactivora* (Petr.) Alcorn, *Botryosphaeria dothidea* (Moug. ex Fr.) Ces & De Not, *Colletotrichum capsici* (Syd. & P. Syd.) E.J. Butler & Bisby, *Colletotrichum gloeosporioides* (Penz.) Penz. and Sacc., *Colletotrichum truncatum* (Schwein.) Andrus & W.D. Moore, *Curvularia lunata* (Wakker) Boedijn, *Fusarium semitectum* (E.D. Eddy) Hesselt, *Monilinia fructicola* (Wint.) Honey and *Dothiorella* spp. (Barbeau, 1990; Oeurn *et al.*, 2015; Ngoc *et al.*, 2017; Bordoh *et al.*, 2020; Chandini and Sushma, 2021).

Colletotrichum spp. are some of the major cosmopolitan phytopathogens that cause postharvest anthracnose in dragon fruit. These pathogens attack fruit in the field, during long-distance transportation and during cold storage, leading to a shorter shelf life. Traditionally, synthetic fungicides such as prochloraz, carbendazim and benomyl have been used to control anthracnose, and these were mostly effective; however, serious health concerns raised by the scientific community and other stakeholders in the food industry have restricted the use of these chemicals. Alternatives to chemical control, which include hot-water treatment, vapour heat, forced hot air, MA storage, postharvest treatment with 1-MCP, chitosan dispersions, X-ray and UV-C irradiation, and ethanolic extracts of propolis, are being tested to reduce postharvest losses in dragon fruit due to anthracnose, while ensuring food safety for human consumption (Bordoh *et al.*, 2020).

In Vietnam, the main fungi associated with postharvest rot in dragon fruit are *Alternaria alternata*, *Bipolaris cactivora*, *Colletotrichum truncatum*, *Fusarium dimerum*, *Fusarium equiseti* and *Rhizopus stolonifer*. Except for those caused by *R. stolonifer*, all postharvest rots have their origin either as latent infections in the field or as surface-borne spores. Many of these pathogens were found to grow at low temperatures, and surface-borne *Fusarium* and *Alternaria* spp. were able to infect the fruit during storage (Ngoc *et al.*, 2017). To test the effect of hot-water treatment on postharvest rots, Ngoc *et al.* (2017) immersed freshly harvested fruit at 49°C, 51°C, 53°C and 55°C for 2–3 min and stored them at 5–6°C for 28 days followed by 3 days of shelf life at 20°C. Hot-water treatment at 49°C and 51° C for 2 or 3 min completely inhibited rots during storage, and after the shelf of 3 days, significantly fewer rots were observed than the other hot-water treatments and the untreated control. Hot-water treatment at 49°C and 51°C also led to the least weight loss, no damage to bracts, and no effect on TA, TSS, fruit firmness, or flesh or skin colour. The researchers suggested that treatments aimed at reducing postharvest rots should be considered for control in the field, such as disinfection of spores on the skin surface.

In Thailand, *B. cactivora*, *C. gloeosporioides* and *R. stolonifer* were isolated from dragon fruit in Loei Province (Oeurn *et al.*, 2015). In India, *B. cactivora*, *C. gloeosporioides*, *C. truncatum* and *Pythium acanthicum* were reported to cause fruit rot (Chandini and Sushma, 2021).

13.7 Conclusion

Dragon fruit is highly perishable and requires great attention during harvest, handling, transportation and storage. Being non-climacteric in nature, dragon fruit provides the best quality for consumption when it is harvested ripe. The recommended storage temperature of the fruit is 10°C, as 6°C can induce chilling injury. Dragon fruit has a storage life of about 14 days at 10°C, while at 5°C and 90% RH a storage life of 17 days can be achieved if the fruit is harvested 30–35 DAF. Flesh translucency is a symptom of chilling injury. Other symptoms include softening, wilting, darkening of scales, browning of the outer flesh and poor flavour. These symptoms rapidly develop on *H. undatus* and *H. polyrhizus* fruit held at 6°C for 2 weeks and then transferred to 20°C.

Current researchers are focused on shelf-life extension using appropriate packaging, low-temperature storage and coatings to delay bract senescence, and alleviate chilling injury. Establishment of shelf life at different maturity levels, postharvest disease management and quality assurance are the key areas of research that should be intensified.

13.8 References

Ali, A., Zahid, N., Manickam, S., Siddiqui, Y., Alderson, P.G. and Maqbool, M. (2013) Effectiveness of submicron chitosan dispersions in controlling anthracnose and maintaining quality of dragon fruit. *Postharvest Biology and Technology* 86, 147–153.

Álvarez-Herrera, J.G., Deaquiz, Y.A. and Herrera, A.O. (2016) Effect of different 1-methylcyclopropene doses on the postharvest period of pitahaya fruits (*Selenicereus megalanthus* Haw.). *Revista Facultad Nacional de Agronomia Medellin* 69, 7975–7983.

Arévalo-Galarza, M.L. and Ortíz-Hernández, Y.D. (2004) Comportamiento poscosecha del fruto de la pitahaya (*Hylocereus undatus*). *Cactaceas y Suculentas Mexicanas* 49, 85–90.

Arowora, K.A., Williams, J.O., Adetunji, C.O., Fawole, O.B., Afolayan, S.S. *et al.* (2013) Effects of aloe vera coatings on quality characteristics of oranges stored under cold storage. *Greener Journal of Agricultural Sciences* 3, 39–47.

Awang, Y.B., Chuni, S.H., Mohamed, M.T.M., Hafiza, Y. and Mohamad, R.B. (2013a) Polygalacturonase and pectin methylesterase activities of $CaCl_2$ treated red-fleshed dragon fruit (*Hylocereus polyrhizus*) harvested at different maturity. *American Journal of Agricultural and Biological Science* 8, 167–172.

Awang, Y.B., Ghani, M.A.A. and Hafiza, Y. (2013b) Effect of postharvest application of calcium chloride on brown rot and quality of red-flesh dragon fruit (*Hylocereus polyrhizus*). *Acta Horticulturae* 979, 763–767.

Ayala-Aponte, A.A. and Serna-Cock, L. (2011) Applicacion de 1-metilciclopropeno, Una alternative a la estacionalidad de la pitahaya amarila. *Revista de la Asociacion Colombiana de Ciencia y Technologia de Alimentos* 23, 78–97.

Azuma, A., Yakushiji, H. and Sato, A. (2019) A postharvest light irradiation and appropriate temperature treatment increase anthocyanin accumulation in grape berry skin. *Postharvest Biology and Technology* 147, 89–99.

Barbeau, G. (1990) La pitahaya rouge, un nouveau fruit exotique. *Fruits* 45, 141–147.

Bordoh, P.K., Ali, A., Dickinson, M., Siddiqui, Y. and Romanazzi, G. (2020) A review on the management of postharvest anthracnose in dragon fruits caused by *Colletotrichum* spp. *Crop Protection* 130: 105067.

Bordoh, P.K., Ali, A., Dickinson, M., Siddiqui, Y. and Ansah, F.A. (2022) Bioefficacy of composite medicinal plant extracts and gum arabic on improving postharvest quality in dragon fruit. *International Journal of Food Science* 2022: 3889563,

Cai, C., Chen, K., Xu, W., Zhang, W., Li, X. *et al.* (2006) Effect of 1-MCP on postharvest quality of loquat fruit. *Postharvest Biology and Technology* 40, 155–162.

Cao, S. and Zheng, Y. (2010) Effect of 1-methylcyclopropene on anthracnose rot caused by *Colletotrichum acutatum* and disease resistance in loquat fruit. *Journal of the Science of Food and Agriculture* 90, 2289–2294.

Castillo-Martinez, R., Livera-Munoz, M., Brechu Franco, A.E. and Marquez-Guzman, J. (2003) Sexual compatibility between two types of *Hylocereus* (Cactaceae). *Revista Biology Tropical* 51, 699–706.

Castro, A.C. (2017) Physicochemical and physiological responses of red-fleshed dragon fruit *(Hylocereus polyrhizus*) to low temperature conditioning and chitosan coating. MSc thesis, University of the Philippines, Los Baños, Laguna, Philippines.

Castro, A.C., Esguerra, E.B. and Franco, R.K.G. (2020) Modified atmosphere packaging and low temperature storage of red-fleshed dragon fruit (*Hylocereus polyrhizus* (Weber) Britton & Rose). *Philippine Journal of Crop Science* 45, 1–12.

Castro, J.C., Mardigan, L.A., Watanabe, R., Clemente, E. and de Areu Filho, B.A. (2017) [Conservation of dragon fruit (*Hylocereus undatus* (Haw.)) with use of edible coatings and refrigeration.] *Brazilian Journal of Food Research* 8, 89–103.

Centurion, A., Perez, S., Solis, S., Baez, R., Mercado, S.E. *et al.* (2000) Crecimiento, desarrollo y comercializacion de la pitahaya (*Hylocereus undatus*) durante la post cosecha. *Revisa Iberoam Technology Postcos* 2, 161–168.

Chandini, K.C. and Sushma, H.S. (2021) Isolation of pathogenic fungi causing stem and fruit rot diseases on dragon fruit grown in Kathige village, Davanagere, Karnataka. *International Journal of Creative Research Thoughts* 9, a209–a213.

Chandran, S. (2010) Effect of film packaging in extending shelf life of dragon fruit, *Hylocereus undatus* and *Hylocereus polyrhizus*. *Acta Horticulturae* 875, 172–179.

Chitarra, M.I.F. and Chitarra, A.B. (2005) *Postharvest of Fruits and Vegetables: Physiology and Handling*. Universidade Federal de Lavras, Lavras, Brazil.

Chuni, S.H., Awang, Y. and Mohamed, M.T. (2010) Cell wall enzymes activities and quality of calcium treated fresh-cut red flesh dragon fruit (*Hylocereus polyrhizus*). *International Journal of Agriculture and Biology* 12, 713–718.

Chutichudet, B. and Chutichudet, P. (2011) Effects of chitosan coating to some postharvest characteristics of *Hylocercus undatus* (Haw) Brit. and Rose fruit. *International Journal of Agricultural Research* 6, 82–92.

Costa, L., Vicente, A.R., Civello, P.M., Chaves, A.R. and Martinez, G.A. (2006) UV-C treatment delays postharvest senescence in broccoli florets. *Postharvest Biology and Technology* 39, 204–210.

Crane, J.H. and Balerdi, C.F. (2005) *The pitaya (*Hylocereus undatus *and other spp.) in Florida*. University of Florida, IFAS Extension, Florida, USA.

de Faria, R.C., Morgado, C.M.A., Vespucci, I.L. and de Campos, A.J. (2022) UV-C radiation in postharvest quality of red dragon fruit. *Comunicata Scientiae* 13: e3857.

de Freitas, S.T. and Mitcham, E.J. (2013) Quality of pitaya fruit (*Hylocereus undatus*) as influenced by storage temperature and packaging. *Scientia Agricola* 70, 257–262.

Deaquiz, Y.A., Álvarez-Herrera, J. and Fischer, G. (2014) Ethylene and 1-MCP affect the postharvest behavior of yellow pitahaya fruits (*Selenicereus megalanthus* Haw.). *Agronomía Colombiana* 32, 44–51.

Deep, L., Narayana, C.K., Karunakaran, G., Rao, D.S. and Anuradha, S. (2022) Maturity determination of red and white pulp dragon fruit. *Journal of Horticultural Sciences* 17, 157–165.

Dhall, R.K. (2013) Advances in edible coatings for fresh fruits and vegetables: a review. *Critical Reviews in Food Science and Nutrition* 53, 435–450.

Dhalsamant, K., Mangaraj, S. and Bal, L.M. (2017) Modified atmosphere packaging for mango and tomato: an appraisal to improve shelf life. *Journal of Packaging Technology and Research* 1, 127–133.

Duenas, Y.M.G., Narvaez, C.E.C. and Restrepo, L.P.S. (2009) Heat shock improves refrigeration storage performance of yellow pitaya. *Agronomia Colombiana* 27, 105–110.

Esquivel, P., Stintzing, F.C. and Carle, R. (2007) Pigment pattern and expression of colour in fruits from different *Hylocereus* spp. genotypes. *Innovative Food Science and Emerging Technologies* 8, 451–457.

Franco, R.K.G., Esguerra, E.B., Tababa, J.L. and Castro, A.C. (2022) Harvest maturity affects the quality and storage behavior of white-fleshed dragon fruit [*Hylocereus undatus* (Haworth) Britton and Rose]. *Food Research* 6, 423–433.

Funamoto, Y., Yamauchi, N., Shigenaga, T. and Shigyo, M. (2002) Effects of heat treatment on chlorophyll degrading enzymes in stored broccoli (*Brassica oleracea* L.). *Postharvest Biology and Technology* 24, 163–170.

Gill, K. (2018) Techniques for extending shelf life of guava fruits: a review. *Acta Horticulturae* 1205, 959–969.

Hallman, G.J. (2001) Irradiation as a quarantine treatment. In: Molins, R.A. (ed.) *Food Irradiation: Principles and Applications.* Wiley, New York, pp. 113–130.

Hasbullah, R., Syarief, A.M. and Akinaga, T. (2000) Gas permeability characteristics of plastic films for packaging of fresh produce. *Journal of the Society of Agricultural Structures, Japan* 31, 79–86.

Ho, P.L., Tran, D.T., Maarten, L.A., Hertog, T.M. and Nicolaï, B.M. (2021) Effect of controlled atmosphere storage on the quality attributes and volatile organic compounds profile of dragon fruit (*Hylocereus undatus*). *Postharvest Biology and Technology* 173: 111406.

Hoa, T.T., Clark, C.J., Waddell, B.C. and Woolf, A.B. (2006) Postharvest quality of dragon fruit (*Hylocereus undatus*) following disinfesting hot air treatments. *Postharvest Biology and Technology* 41, 62–69.

Irtwange, S.V. (2006) Hot water treatment: a non-chemical alternative in keeping quality during postharvest handling of citrus fruits. *Agricultural Engineering International* 5: 10.

Jamaludin, N.A., Ding, P. and Hamid, A.A. (2011) Physico-chemical and structural changes of red-fleshed dragon fruit (*Hylocereus polyrhizus*) during fruit development. *Journal of the Science of Food and Agriculture* 91, 278–285.

Junior, I.M.R., Magalhães, D.S., Rodrigues, F.A., Pasqual, M. and Pio, L.A.S. (2021) Growth and maturation of white-fleshed dragon fruit. *Research, Society and Development* 10: e119810716288

Khatun, Z., Dash, P.K. and Mannan, Md. A. (2022) Influence of precooling systems on postharvest quality and shelf life of dragon fruit (*Hylocereus polyrhizus*). *Journal of Bangladesh Agricultural University* 20, 313–322.

Khuriyati, N., Fibriato, M.B. and Nugroho, D.A. (2018) Determination of the quality of dragon fruit (*Hylocereus undatus*) with non-destructive methods. *Journal of Technology and Industry of Agricultural Products* 23, 67–74.

Kowitcharoen, L., Kammapana, L. and Srilaong, V. (2010) UV-C treatment delays chlorophyll degradation in the bract of dragon fruit cultivar 'Vietnam'. *Acta Horticulturae* 875, 105–110.

Lamikanra, O. and Watson, M.A. (2004) Effect of calcium treatment temperature on fresh-cut cantaloupe melon during storage. *Journal of Food Science* 69, C468–C472.

Le, V.T., Nguyen, N., Nguyen, D.D., Dang, K.T., Nguyen, M.V.H. *et al.* (2000) Quality assurance system for dragon fruit. *ACIAR Proceedings* 100, 101–114.

Le, V.T., Ngu, N., Duc, N. and Huong, H. (2002) Dragon fruit quality and storage life: effect of harvesting time, use of plant growth regulators and modified atmosphere packaging. *Acta Horticulturae* 575, 611–621.

Li, L., Ni, R., Shao, Y. and Mao, S. (2014) Carrageenan and its applications in drug delivery. *Carbohydrate Polymers* 103, 1–11.

Li, L., Lichter, D., Chalupowicz, D., Gamrasni, T., Goldberg, Y. *et al.* (2016) Effect of the ethylene-action inhibitor 1-methylcyclopropene on postharvest quality of non-climacteric fruit crops. *Postharvest Biology and Technology* 111, 322–329.

Liguori, G., Farina, V., Corona, O., Mazzaglia, A., Barone, E. and Inglese, P. (2017) Effects of 1-MCP on postharvest quality and internal browning of white-flesh loquat fruit during cold storage. *Fruits* 72, 67–73.

Lum, M.S. and Norazira, M.A. (2011) Effects of hot water, submergence time and storage duration on quality of dragon fruit (*Hylocereus polyrhizus*). *Journal of Agricultural Science* 3, 146–152.

Marquez-Guzman, G.J., Castillo-Martinez, R. and Livera-Munoz, M. (2005) Characterizacion morfologica y compatibilidad sexual de cinco genotipos de pithaya (*Hylocereus undatus*). *Agrociencia* 39, 183–194.

Marsh, K., Attanayake, S., Walker, S., Gunson, A., Boldingh, H. and MacRae, E. (2004) Acidity and taste in kiwifruit. *Postharvest Biology and Technology* 32, 159–168.

Martínez Chávez, R. (2011) Relaciones entre genotipo, productividad y calidad de fruto en pitahaya (*Hylocereus* spp.). MSc thesis, Colegio de Postgraduados, Montecillo, Mexico.

McGuire, R.G. (1997) Market quality of guavas after hot-water quarantine treatment and application of carnauba wax coating. *HortScience* 32, 271–274.

Menezes, T.P.D., Ramos, J.D., Lima, L.C.D.O., Costa, A.C., Nassur, R.D.C.M.R. and Rufini, J.C.M. (2015) Características físicas e físico-químicas de pitaia vermelha durante a maturação. *Semina Ciencia Agraria* 36, 631–644.

Merten, S. (2003) A review of *Hylocereus* production in the United States. *Journal of the Professional Association for Cactus Development* 1, 98–105.

Mitra, S.K., Devi, H.L., Chakraborty, I. and Pathak, P.K. (2012) Recent development in postharvest physiology and storage of guava. *Acta Horticulturae* 959, 89–95.

Morga, J.A., Bolaños, E.A. and Silva, E.M. (2006) Effect of controlled atmosphere on the preservation of minimally processed cactus pears. *Acta Horticulturae* 728, 211–216.

Neog, T., Barua, P.C., Sharma, S., Hazarika, B., Saikia, B. and Das, K. (2023) Role of modified atmospheric packaging in enhancing the shelf life and physical qualities of pitaya fruit (*Hylocereus* spp.). *Pharma Innovation Journal* 12, 146–151.

Nerd, A. and Mizrahi, Y. (1998) Fruit development and ripening in yellow pitaya. *Journal of the American Society for Horticultural Science* 123, 560–562.

Nerd, A., Gutman, F. and Mizrahi, Y. (1999) Ripening and postharvest behaviour of fruits of two *Hylocereus* species (Cactaceae). *Postharvest Biology and Technology* 17, 39–45.

Ng, L.C., Tan, J.S. and Fauziah, T.A. (2022) Efficacy of ultraviolet-C irradiation to suppress fruit decay and retain the postharvest quality of dragon fruit (*Hylocereus polyrhizus*). *Malaysian Applied Biology* 51, 119–128.

Ngoc, N.K., van Phong, N., Tung, N.T., van Hoa, N., Woolf, A.B. and Fullerton, R.A. (2017) Postharvest diseases and effect of hot water treatments on white fleshed dragon fruit (*Hylocereus undatus* (Haw.) Britton & Rose). *International Journal Current Research in Bioscience and Plant Biology* 4, 30–37.

Nguyen, H.T., Boonyaritthongchai, P., Buanong, M., Supapvanich, S. and Wongs-Aree, C. (2020) Postharvest hot water treatment followed by chitosan- and κ-carrageenan-based composite coating induces the disease resistance and preserves the quality in dragon fruit (*Hylocereus undatus*). *International Journal of Fruit Science* 20, S2030–S2044.

Nguyen, H.T., Boonyaritthongchai, P., Buanong, M., Supapvanich, S. and Wongs-Aree, C. (2021) Edible coating of chitosan ionically combined with κ-carrageenan maintains the bract and postharvest attributes of dragon fruit (*Hylocereus undatus*). *International Food Research Journal* 28, 682–694.

Novita, M. (2008) Postharvest quality of red pitaya (*Hylocereus polyrhizus*) as affected by harvest date, storage duration and 1-methylcyclopropene. PhD thesis, Universiti Putra Malaysia, Selangor, Malaysia.

Obenland, D., Cantwell, M., Lobo, R., Collin, S., Sievert, J. and Arpaia, M.L. (2016) Impact of storage conditions and variety on quality attributes and aroma volatiles of pitahaya (*Hylocereus* spp.). *Scientia Horticulturae* 199, 15–22.

Oeurn, S., Jitjak, W. and Sanoamuang, N. (2015) Fungi on dragon fruit in Loei Province, Thailand and the ability of *Bipolaris cactivora* to cause post-harvest fruit rot. *Asia-Pacific Journal of Science and Technology* 20, 405–418.

Ortiz, T.A. and Takahashi, L.S.A. (2015) Physical and chemical characteristics of pitaya fruits at physiological maturity. *Genetics and Molecular Research* 14, 14422–14439.

Osuna Enciso, T., Ibarra Zazueta, M.E., Muy Rangel, M.D., Valdez Torres, J.B., Villarreal Romero, M. and Hernández Verdugo, S. (2011) Calidad postcosecha de frutos de pitahaya (*Hylocereus undatus* Haw.) cosechados en tres estados de madurez [Postharvest quality of pitahaya (*Hylocereus undatus* Haw.) fruits harvested in three maturity stages]. *Revista Fitotecnia Mexicana* 34, 63–72.

Paull, R.E. (2014) *Dragon fruit: Postharvest Quality – Maintenance Guidelines.* College of Tropical Agriculture and Human Resources, University of Hawaii at Manoa, Honolulu, Hawaii.

Paull, R.E. and Chen, N.J. (2019) Overall dragon fruit production and global marketing. In: *Y2019 FFTC and VAAS-SOFRI Joint Workshop on 'Dragon Fruit Network: Marketing and the Whole Value Chain'*, 9–11 September, My Tho City, Vietnam. Food and Fertilizer Technology Center for the Asian and Pacific Region, Taipei, Taiwan. Available at: https://ap.fftc.org.tw/article/1596 (accessed 6 December 2023).

Phebe, D., Chew, M.K., Suraini, A.A., Lai, O.M. and Janna, O.A. (2009) Red-fleshed pitaya (*Hylocereus polyrhizus*) fruit colour and betacyanin content depend on maturity. *International Food Research Journal* 16, 233–242.

Phong, N.V., Chau, N.M. and Woolf, A. (2013) Survey on the postharvest quality and management of dragon fruits exported from Vietnam to Holland. In: *Proceedings of the International Symposium on Super Fruits: Myth or Truth, 1–3 July, Ho Chi Minh City, Vietnam*, pp. 99–104.

Prashanth, R., Kiran Kumar, A., Rajkumar, M. and Aparna, K. (2022) Studies on postharvest quality and shelf life of pink- fleshed dragon fruit (*Hylocereus* spp.) coated with chitosan and stored at ambient temperature. *Biological Forum* 14, 340–347.

Punitha, V., Boyce, A.N. and Chandran, S. (2010) Effect of storage temperature on the physiological and biochemical properties of *Hylocereus polyrhizus*. *Acta Horticulturae* 875, 137–144.

Razali, N.A., Anfunes, A.C.N., Berry, A.D. and Sargent, S.A. (2016) Postharvest storage temperature and coating effects on fruit quality of red-fleshed pitaya (*Hylocereus costaricensis*). *Proceedings of the Florida State Horticulture Society* 129, 190–194.

Razali, Z., George, D.S. and Ibrahim, N.A.N. (2015) Improvements in quality and antioxidant capacity of guava (*Psidium guajava* L.) exposed to UV-C irradiation. In: *Proceedings of the 3rd International Conference on Chemical, Agricultural and Medical Sciences (CAMS-2015), 10–11 December, Singapore*, pp. 50–53. Available at: www.iicbe.org/proceedingspdf.php?id=27 (accessed 6 December 2023).

Razali, Z., Sargent, S.A., Sins, C.A., Brecht, J.K., Berry, A.D. and Cheng, C. (2021) Potential of postharvest coatings to maintain freshness of red-fleshed pitaya (*Hylocereus costaricensis*). *Agriculture* 11: 892.

Rosas-Benitez, A., Trujillo-Cardenas, L., Valle-Guadarrama, S., Salinos-Moreno, Y. and Garcia-Cruz, L. (2016) Quality attributes of pitaya (*Stenocereus pruinosus*) fruit handled in postharvest with and without thorns under refrigerated storage. *Revista Chapingo Serie Horticultura* 22, 191–207.

Sarip, J. (2020) Dragon fruit exportation practices, challenges and country experiences: Malaysia. In: *Proceedings of the 3rd DFNet Workshop on Dragon Fruit Value Chain for Global Markets, 22–23 September, Taichung, Taiwan*, pp. 13–26. Available at: www.scribd.com/document/507647927/2020FFTC-Dragon-Fruit-Value-Chain-for-Global-Markets (accessed 6 December 2023).

Serna, L., Torres, L.S. and Ayala, A.A. (2011) Efecto del empaque y del 1-MCP sobre características físicas, químicas y fisiológicas de pitahaya amarilla. *Biotecnología en el Sector Agropecuario y Agroindustrial* 9, 139–149.

Shi, C., Zhao, X., Meng, R., Liu, Z., Zhang, G. and Guo, N. (2017) Synergistic antimicrobial effects of nisin and *p*-anisaldehyde on *Staphylococcus aureus* in pasteurized milk. *LWT – Food Science and Technology* 84, 222–230.

Shin, Y., Liu, R.H., Nock, J.F., Holliday, D. and Watkins, C.B. (2007) Temperature and relative humidity effects on quality, total ascorbic acid, phenolics and flavonoid concentrations, and antioxidant activity of strawberry. *Postharvest Biology and Technology* 45, 349–357.

Shreaz, S., Bhatia, R., Khan, N., Muralidhar, S., Basir, S.F., Manzoor, N. and Khan, L.A. (2011) Exposure of *Candida* to *p*-anisaldehyde inhibits its growth and ergosterol biosynthesis. *Journal of General and Applied Microbiology* 57, 129–136.

Sisler, E.C. and Serek, M. (1997) Inhibitors of ethylene responses in plants at the receptor level: recent developments. *Physiologia Plantarum* 100, 577–582.

Sivakumar, D. and Korsten, L. (2006) Evaluation of the integrated application of two types of modified atmosphere packaging and hot water treatments on quality retention in the litchi cultivar 'McLean's Red'. *Journal of Horticultural Science and Biotechnology* 81, 639–644.

Supapvanich, S. and Promyou, S. (2017) Hot water incorporated with salicylic acid dips maintaining physicochemical quality of 'Holland' papaya fruit stored at room temperature. *Emirates Journal of Food and Agriculture* 31, 395–404.

Tadeo, P.R.M., Castillo-Israel, K.A.T., Serrano, E.P., Gandia, J.B.L. and Absulio, W.L. (2018) Physiological responses and storage quality of fresh-cut red and white dragon fruit (*Hylocereus* spp.) treated with 1-methylcyclopropene (1-MCP). *International Food Research Journal* 25, 2090–2098.

Trieu, N.M. and Thinh, N.T (2021) Quality classification of dragon fruits based on external performance using a convolutional neural network. *Applied Science* 11: 10558.

Van To, L., Ngu, N., Duc, N.D. and Huong, H.T.T. (2002) Dragon fruit quality and storage life: effect of harvest time, use of plant growth regulators and modified atmosphere packaging. *Acta Horticulturae* 575, 611–621.

Vera, J., Pongprasert, N., Jitareerat, P., Boonyaritthongchai, P., and Srilaong, V. (2017) 1-MCP microbubbles influence in fruit color and quality of dragon fruit. *Agricultural Science Journal* 48, 237–240.

Wall, M.M. and Khan, S.A. (2008) Postharvest quality of dragon fruit (*Hylocereus* spp.) after X-ray irradiation quarantine treatment. *HortScience* 43, 2115–2119.

Wu, Q., Gao, H., Zhang, Z., Li, T., Qu, H. *et al.* (2020) Deciphering the metabolic pathways of pitaya peel after postharvest red light irradiation. *Metabolites* 10: 108.

Wu, Y., Xu, J., He, Y., Shi, M., Han, X. *et al.* (2019) Metabolic profiling of pitaya (*Hylocereus polyrhizus*) during fruit development and maturation. *Molecules* 24: 1114.

Xu, Y., Cai, Z., Ba, L., Qin, Y., Su, X. *et al.* (2021) Maintenance of postharvest quality and reactive oxygen species homeostasis of pitaya fruit by essential oil *p*-anisaldehyde treatment. *Foods* 10: 2434.

Yamaga, I., Shirai, Y., Nakajima, T. and Kobayashi, Y. (2016) Rind color development in satsuma mandarin fruits treated by low-intensity red light-emitting diode (LED) irradiation. *Food Science and Technology Research* 22, 59–64.

Zahid, N., Ali, A., Siddiqui, Y. and Maqbool, M. (2013) Efficacy of ethanolic extract of propolis in maintaining postharvest quality of dragon fruit during storage. *Postharvest Biology and Technology* 79, 69–72.

Zee, F., Yen, C.R. and Nishing, M. (2004) *Dragon Fruit, Strawberry and Pear.* Co-operative Extension Service, College of Tropical Agriculture and Human Resources, University of Hawaii at Manoa, Honolulu, Hawaii.

Index

Note: Page numbers in **bold** type refer to figures.
Page numbers in *italic* type refer to tables.

Printed and bound by CPI Group (UK) Ltd, Croydon, CR0 4YY
01/03/2026
14835366-0002